GREATEST
AIR ACES
OF ALL TIME

GREATEST AIR ACES OF ALL TIME

Air Marshal Anil Chopra (Retd)
PVSM, AVSM, VM, VSM

PENTAGON PRESS LLP

Greatest Air Aces of All Time

By Air Marshal Anil Chopra (Retd), PVSM, AVSM, VM, VSM

ISBN 978-93-90095-32-2

First Published in 2021

Copyright © RESERVED

All rights reserved. No part of this publication may be reproduced, stored in a retrieval system, or transmitted in any form or by any means, electronic, mechanical, photocopying, recording or otherwise, without the prior written permission of the Publisher.

Disclaimer: The views and opinions expressed in the book are the individual assertion of the Author. The Publisher does not take any responsibility for the same in any manner whatsoever. The same shall solely be the responsibility of the Author.

Published by
PENTAGON PRESS LLP
206, Peacock Lane, Shahpur Jat,
New Delhi-110049
Phones: 011-64706243, 26491568
Telefax: 011-26490600
email: rajan@pentagonpress.in
website: www.pentagonpress.in

Printed at Aegean Offset Printers, Greater Noida, U.P.

"Be fearless and pure; never waiver in your determination ... be self-controlled, sincere, truthful, and full of the desire to serve. Cultivate vigour, patience, will, purity ... fulfill your duty as a warrior ... then, Arjuna, you will achieve your divine destiny."

—**Bhagavad Gita**

Exceptionally skilled pilot could win the battle in the skies. The few dare-devil aces among combat aviators have historically accounted for the majority of air-to-air victories in military history. These pilots had great situational awareness, aggressive spirit and aerial shooting skills. I was always enamoured by Manfred von Richthofen "Red Baron" who is considered to be the ace-of-aces, credited with 80 air combat victories. Air Marshal Anil Chopra has chosen some of the greatest Air Aces from across the globe. The book is fascinating and well researched with historic facts and pictures. It is a must read for all aviators and those with adrenaline running in their veins.

Air Marshal Anil Khosla, PVSM, AVSM, VM
Former 41st Vice Chief of the Air Staff of the Indian Air Force

* * *

It's around 120 years since Wright brothers invented a machine which just about got of the ground for few moments. Since then to "Touch the sky with glory" is saga of bravery and innovations. Over a century now, these beautiful airborne machines have become safest and fastest mode of transportation across the world. In military aviation aircraft have contributed towards decisive outcomes but not without sacrifices. Fighter aviation is agog with heroic deeds of pilots who exploited the extreme envelopes of their machines in order to impinge severe blow on their adversaries occasionally falling during call of their duty. Air Marshal Chopra's incredible effort is a must read for, both students and practitioners, aviators. A scripted document of great value for understanding the psyche of those brave fighter pilots who mattered in political outcomes of World Wars.

Vice Admiral Shekhar Sinha, PVSM, AVSM, NM & Bar
Former Flag Officer Commanding-in-Chief Western Naval Command and Chief of Integrated Defence Staff

* * *

As the aircraft started getting employed in support of military operations, aerial combat to gain freedom of operations in the air, became one of the most important missions of the Military aviation Forces. Special fighters were developed and employed to achieve aerial superiority. While the opposing fighters almost matched in performance, it was the skill of the men in the fighter cockpits that made the difference in the outcome of aerial engagements. Aerial combat involved attacking the adversary by achieving surprise and skillfully manoeuvring the aircraft to its limits to achieve a 'kill'. 'Air Ace' was a military pilot credited with shooting down five or more enemy aircraft in the air. While the exciting stories of the Air Aces can be read in parts while studying different aerial campaigns, Air Marshal Anil Chopra has done a tremendous research to select 25 greatest air aces from all over the world and compiled them into a single book titled 'Greatest Air Aces of All Time'. The book provides a gripping account of exciting aerial engagements from different countries but with common attributes of addiction to flying, determination and mastery over the flying machine. I particularly liked Israel's Giora "Hawkeye" Epstein – ace of aces of supersonic fighters with 17 aerial victories. I compliment Air Marshal Chopra for publishing an excellent collection of the Aces exploits, a must read for young aviation enthusiasts, historians and pilots all over the world.

Air Marshal Daljit Singh PVSM AVSM VM
Former AOC-in-C South Western Air Command

* * *

"Skillful pilots gain their reputation from storms and tempest." Only the strong willed become fighter pilots. Air Aces were the intrepid men and women in their flying machines, and audacity, grit, and ingenuity were their hallmark. Air Marshal Anil Chopra, an avid writer has done extensive researched to put down the stories of these great combat aviators. Since my young days, I was particularly excited to read about the Royal Air Force "Fighter Ace Douglas Bader – Determined Dogmatic and Fearless". The book is strongly recommended for aviation aficionados and cognoscenti.

Air Marshal Tejbir Singh Randhawa PVSM VM
Former Director General (Flight Inspection & Safety) and Commandant National Defence Academy

* * *

Contents

	Acknowledgements	ix
	Forward by Chief of the Air Staff	xi
	Message by Chairman Hero Group	xiii
	Famous Fighter Aviation Quotes	xv
	Preface	xvii
1.	Erich Hartmann: Highest Scoring Air Ace	1
2.	Douglas Bader: Determined Dogmatic and Fearless	11
3.	Manfred von Richthofen "Red Baron": Ace of Aces	23
4.	"Johnnie" Johnson : Highest Scoring Western Fighter Ace in WW II against Germany	33
5.	German Ace General Adolf Galland : 104 Aerial Victories – all against the Western Allies	44
6.	Fighter Ace Lydia Litvyak "White Lily": Highest Aerial Victories by a Female Fighter Pilot	64
7.	"Triple Ace" American Fighter Pilot "Robin Olds": World War II and Vietnam War	74
8.	Hans-Joachim Marseille "Triple Ace": 17 Victories in a Day	88
9.	Charles B. DeBellevue: Top American Ace of the Vietnam War – A Non-Pilot	105
10.	Lieutenant Indra Lal Roy, DFC: The First and the Only Indian Air Ace	113
11.	Gabby Gabreski: America's Two-War Ace: WW II and Korea	118
12.	Hans-Ulrich Rudel: German Ace: Eagle of the Eastern Front	134
13.	Vietnamese Fighter Ace Nguyễn Văn Cốc: The Highest Scoring Pilot in Vietnam War	145
14.	Ivan Kozhedub: The Highest Scoring Allied and Soviet Air Ace of World War II	157
15.	"Zero" Fighter Ace Tetsuzō Iwamoto "Tiger Tetsu": Highest Scoring Japanese	166
16.	German Major Heinz-Wolfgang Schnaufer: Highest Scoring Night Ace	178

17.	Israel's Giora "Hawkeye" Epstein: Ace of Aces of Supersonic Fighter jets	194
18.	Iranian Air Ace Jalil Zandi : Highest-Scoring F-14 Tomcat Pilot	204
19.	Finnish Ilmari Juutilainen: Highest Scoring Non-German Ace	211
20.	German Gerhard Barkhorn: Second Highest Scoring Ace of All Time	220
21.	Günther Rall: Third Highest scoring Air Ace of All Time	228
22.	French Colonel René Paul Fonck: The Highest Scoring All-time Allied Ace of Aces	243
23.	Billy Bishop: The Top Canadian and British Empire Ace of all Time	255
24.	Richard Ira Bong: The most Decorated American Fighter Pilot and Top Air Ace	268
25.	Ernst Udet: The Highest Scoring Surviving German Air Ace of WW I	278
	Index	292

Acknowledgements

It was indeed a privilege to be a fighter pilot in one of the top air forces of the world, the Indian Air Force. Fighter aircraft are always the cutting edge of all aviation technologies. Fighter flying, though often associated with glamour, is a highly professional activity requiring skills, precision, guts and an aggressive spirit. Aviation had enamoured me since my childhood, and stories of valour and courage of great fighter pilots enthused me to become one. And that is how this book. Having an idea and turning it into a book is as hard as it sounds, but more rewarding than I could have ever imagined. I must thank the Indian Air Force in shaping my thoughts, mind and skills, not only for flying but also for being analytical and taking on a huge task like writing a book.

The world is a better place thanks to people who want to develop and lead others by building trust, honour, and respect. What makes it even better are people who share the gift of their time to mentor future leaders. Thank you to everyone who strives to grow and help others grow. To all the individuals I have had the opportunity to lead, be led by, or watch their leadership from afar, I want to say thank you for being the inspiration and foundation for this book. This book is in a way a tribute to my seniors and flying colleagues.

The Indian Air Force (IAF) is the primary custodian of air power in India, and its motto "Touch the Sky with Glory", says it all. I feel very privileged that the Chief of the Air Staff of the Indian Air Force, Air Chief Marshal RKS Bhadauria, PVSM, AVSM, VM, ADC has written the foreword of the book. His appreciation and visionary words mean a lot.

Writing a book with historic content requires a significant amount of research. The Heroes of the ground, Hero MotoCorp, the World's leading two-wheeler manufacturer chose to kindly support this project of recording stories of the Heroes of the air. I am grateful to Mr. Pawan Munjal, Chairman & CEO, Hero MotoCorp for supporting the research.

My sincere thanks are due to Mr Rajan Arya, and the entire team at Pentagon Press LLP, the publishers of this book, for supporting my raw script to shape into a book.

The hours one spends in research and punching keys on the laptop is a prime time taken away from the family. I thank my parents for the great upbringing and seeds of analytical thought. I am indeed grateful to my wife, Suman, and children for tolerating my incessant disappearances into my home office, and for persuasion and forbearance. The loving smiles of the grandchildren are a priceless source of encouragement. The family makes both the journey and destination worthwhile. It greatly helped bring these stories to life. The arduous experience was both internally challenging and rewarding. I hope that through this effort, I pass on the book writing legacy to our future generations.

Air Marshal Anil Chopra (Retd)
PVSM, AVSM, VM, VSM

एयर चीफ मार्शल आर के एस भदौरिया
प विसे मे अ विसे मे व मे एडी सी
Air Chief Marshal R K S Bhadauria
PVSM AVSM VM ADC

Tel : (011) Off : 23012517
       Res : 23017300
       Fax : 23018853

वायु सेना मुख्यालय
नई दिल्ली-110106
Air Headquarters
New Delhi - 110 106

FOREWORD

1. A dogfight between two or more aircraft is probably the most fascinating type of combat mission. Among the many ways that air power was used for dominance of airspace, engaging the enemy in the air, in combat, was one of the early uses of flying machines in warfare. The pilots who flew in the World Wars were gritty men, strapped in uncomfortable cockpits with their eyes peeled for predator or prey. In the days of steam gauges, weak airframes and unreliable engines, the airplane was as ready to kill the pilot as the enemy.

2. Despite being limited by their primitive machines and weapons, a few of these legends went on to achieve some extraordinary feats of heroism. Some of the tactics invented by these pilots still find a place in modern day combat. These fighter pilots of the yore took enormous risks, operated their machines at far limits, went beyond the realms of operating envelopes and scripted tales of bravery and courage which have inspired generations of combat aviators. While a number of these 'Air Aces' became national heroes, many of them did not live to tell the tale. Contrary to the glamorous portrayal of aviators, becoming an Air Ace entails rigorous training and tremendous hard work to acquire and hone special skill set for combat flying.

3. As aircraft evolved with greater speed and agility, air combat became more intense, technical and demanding. Modern day close combat engagements require a precise balance of spatial perception and technical acumen combined with the ability to bear the mental and physiological strain associated with highly manoeuvrable platforms. Even though the cockpits have become more comfortable, the technological advances have only increased the complexities of visual combat. Although the nature of air combat will continue to change with evolving technologies, one common characteristic of all 'Air Aces' that would not change is an extremely high level of Situational Awareness.

4. While the lives of famous Air Aces like 'Red Baron', Manfred von Richthofen have been converted into books, movies and documentaries, there are a large number of similar war heroes who have not been covered adequately. Very few of us may know that there is an Indian Air Ace, Lieutenant Indra Lal Roy DFC, who flew for the British Royal Flying Corps in World War

I. Also very little is spoken about the Air Ace who scored all victories by night and also about the highest scoring women Air Ace.

5. I am very happy that Air Marshal Anil Chopra has chosen to document the lives and achievements of twenty five Air Aces across wars and regions. Covering their personal accounts, anecdotes and rare historical pictures, this work of his makes an interesting and an inspirational read for all combat aviators. I strongly recommend this book to all aviation enthusiasts and combat pilots.

12 Mar 2021

Air Chief Marshal
Chief of the Air Staff

A Message from the Ace of Mobility

There is a popular saying in aviation - "You fly the plane from your head, not your hands and legs". This saying is equally apt in life too, which can be related to flying. You have to keep multiple things in control to ensure a smooth flight. Yet, despite all our efforts, we often face turbulence due to extraneous factors. Again, as in real life!

Aviation teaches us a lot. Not too many people are aware that as a child, I dreamt of becoming a fighter pilot for the Indian Air Force. Therefore, when Air Marshal Anil Chopra approached me with the idea of the book on "The Greatest Air Aces of All Time", I was immediately hooked and offered to support his research.

Our lives are sprinkled with pioneers from different walks of life, who leave an indelible mark on generations to come - just like our own Hero MotoCorp, an organization stacked with rich legacy and heritage. From revolutionizing the Indian motorbike and scooter industry nearly 40 years ago, to providing wheels to the aspirations of millions across the globe today, Hero has truly become the Ace of Mobility.

Reaching the unique milestone of 100 million cumulative production and sales recently, Hero MotoCorp continues to be the world's largest motorbike and scooters manufacturer for 20 consecutive years. Yet, not resting on our laurels, we are working relentlessly towards our new Vision – Be the Future of Mobility.

I am confident that this book will inspire, motivate and energize people and provide them with context on the evolution of aviation. I wish Air Marshal Chopra the very best for his new book. The pioneers of military aviation have been the real heroes of their nations, and it is a fitting tribute that the heroes on the ground extend support to the literary works highlighting the heroes of the air.

Air Marshal Chopra has produced an outstanding book that is very well researched and packed with interesting anecdotes and historic images. It makes a great read for anyone with a passion for aviation and history. On behalf of all of us, my gratitude to you for having the vision for doing the research and creating a book which will be treasured by millions.

Dr Pawan Munjal
Chairman & CEO, Hero MotoCorp

Famous Fighter Aviation Quotes

"Fight on and fly on to the last drop of blood and the last drop of benzene - to the last beat of the heart and the last kick of the motor."

— Manfred von Richthofen

"Everything I had ever learned about air fighting taught me that the man who is aggressive, who pushes a fight, is the pilot who is successful in combat and who has the best opportunity for surviving battle and coming home."

—Robert S. Johnson

"The first lesson is that you can't lose a war if you have command of the air, and you can't win a war if you haven't."

—Jimmy Doolittle

"It was my view that no kill was worth the life of a wingman, many of whom were young and inexperienced boys. Pilots in my unit who lost wingmen on this basis were prohibited from leading a Rotte. They were made to fly as wingman, instead."

—Erich 'Bubi' Hartmann

"There is a peculiar gratification on receiving congratulations from one's squadron for a victory in the air. It is worth more to a pilot than the applause of the whole outside world. It means that one has won the confidence of men who share the misgivings, the aspirations, the trials and the dangers of aeroplane fighting".

—Edward V. 'Eddie' Rickenbacker

"Go in close, and when you think you are too close, go in closer."

—Thomas B. 'Tommy' McGuire

Preface

"Everything I had ever learned about air fighting taught me that the man who is aggressive, who pushes a fight, is the pilot who is successful in combat and who has the best opportunity for surviving battle and coming home".

—Robert S. Johnson

The first military uses of aviation involved lighter-than-air balloons during the Battle of Fleurus in 1794, when the French observation balloon *l'Entreprenant* was used to monitor Austrian troop movements. On December 17, 1903, at 10:30 am at Kitty Hawk, North Carolina, an airplane arose for a few seconds to make the first powered, heavier-than-air controlled flight in history. The first flight lasted 12 seconds and flew a distance of 120 feet. Orville Wright piloted the historic flight while his brother, Wilbur, observed. The U.S. Army Signal Corps purchased a Wright Model-A in August 1909 which became the first military aircraft in history. In 1911, the Italians used a variety of aircraft types in reconnaissance, photo-reconnaissance, and bombing roles during the Italo-Turkish War. On October 23, 1911, an Italian pilot, Captain Carlo Piazza, flew over Turkish lines on the world's first aerial reconnaissance mission, and on November 1, the first ever aerial bomb was dropped by *Sottotenente* (Second Lieutenant) Giulio Gavotti, on Turkish troops in Libya, from an early model of Etrich Taube aircraft. The Turks, lacking anti-aircraft weapons, were the first to shoot down an airplane by rifle fire.

The first instance of plane-on-plane combat and the first instance of one plane intercepting another during an aerial conflict occurred during the Mexican Revolution on November 30, 1913, between two American soldiers of fortune fighting for opposing sides, Dean Ivan Lamb and Phil Rader. Both men had orders to kill, but neither pilot wanted to harm the other, so they exchanged multiple volleys of pistol fire, intentionally missing before exhausting their supply of ammunition.

Dogfighting became widespread in World War I. Adversary pilots at first simply exchanged waves, or shook their fists at each other. Due to weight restrictions, only small weapons could be carried on board. Intrepid pilots decided to interfere with enemy reconnaissance by improvised means, including throwing bricks, grenades and sometimes rope, which they hoped would entangle the enemy plane's propeller. Pilots quickly began firing hand-held guns at enemy planes, such as pistols and carbines. The first aerial dogfight of the war occurred during the Battle of Cer (August 15–24, 1914), when Serbian aviator Miodrag Tomiæ encountered an Austro-Hungarian plane while performing a reconnaissance mission over Austro-Hungarian positions. The Austro-Hungarian pilot initially waved, and Tomiæ reciprocated. The Austro-Hungarian pilot then fired at Tomiæ with his revolver. Tomiæ managed to escape, and within several weeks, all Serbian and Austro-Hungarian planes were fitted with machine-guns. In August 1914, Staff-Captain Pyotr Nesterov,

from Russia, became the first pilot to ram his plane into an enemy spotter aircraft. In October 1914, an airplane was shot down by a handgun from another plane for the first time over Reims, France. Once machine guns were mounted to the airplane, either on a flexible mounting or higher on the wings of early biplanes, the era of air combat began. By middle of the World War I, military aviation had rapidly embraced many specialised roles, such as artillery spotting, air superiority, bombing, ground attack, and anti-submarine patrols.

The Air Ace

A Flying Ace, or Air Ace is a military aviator credited with shooting down five or more enemy aircraft during aerial combat. The concept of the "Ace" emerged in 1915 during World War I, at the same time as aerial dogfighting. It was a term intended to provide the home population with a cult of the hero in what was otherwise a war of attrition. The combat duels of Aces were widely reported and an image created of a chivalrous knight reminiscent of the bygone era. For a brief early period in the initial years, the exceptionally skilled pilot could shape the battle in the skies. Later, the fighters fought in formation and air superiority depended heavily on the relative availability of resources. Use of the term began in World War I, when French newspapers described Adolphe Pégoud, the Ace after he became the first pilot to down five German aircraft.

The successes of such German Ace pilots as Max Immelmann and Oswald Boelcke was publicised, for the benefit of civilian morale. Prussia's highest award for gallantry, *Pour le Mérite,* was awarded to most leading German Aces. It was later nicknamed "The Blue Max", after Max Immelmann, who was the first pilot to receive this award. The few Aces among combat aviators have historically accounted for the majority of air-to-air victories in military history.

World War I

During World War I the single-seat fighter aircraft evolved with enough speed and agility to catch and maintain contact with targets in the air, and coupled with armament to destroy the targets. Thus began the real air-combat. It was also the beginning of a long-standing trend in warfare, showing statistically that approximately five percent of combat pilots account for the majority of air-to-air victories. Germany established a practice to maintain very strict guidelines for the official recognition of victory claims by its pilots. Shared victories were either credited to one of the pilots concerned or to the unit as a whole. The destruction of the aircraft had to be physically confirmed by locating its wreckage, or an independent witness to the destruction had to be found. Victories were also counted for aircraft forced down within German lines, as this usually resulted in the death or capture of the enemy aircrew.

Allied fighter pilots fought mostly in German-held airspace and were often not in a position to confirm that an apparently destroyed enemy aircraft had in fact crashed, so these victories were frequently claimed as "driven down", "forced to land", or "out of control" (called "probables" in later wars). These victories were usually included in a pilot's totals and in citations for decorations.

The British high command considered praise of fighter pilots to be detrimental to equally brave bomber and reconnaissance aircrew, so they did not publish official statistics on the successes of individuals. Nonetheless some pilots did become famous through press coverage, making the British system for the recognition of successful fighter pilots much more informal and somewhat inconsistent. Other Allied countries, such as France and Italy, fell somewhere in between the very strict German approach and the relatively casual British one. The United States Army Air Service adopted French standards for evaluating victories. Americans also believed that five victories were the minimum needed to become an Ace. While "Ace" status was generally won only by fighter pilots, bomber and reconnaissance crews on both sides also destroyed some enemy aircraft, typically in defending themselves from attack. There were thus non-fighter Aces.

Between the World Wars

Between the two world wars, there were two theaters that produced flying Aces, the Spanish Civil War and the Second Sino-Japanese War. The Spanish Ace Joaquín García Morato scored 40 victories for the Nationalists during the Spanish Civil War. Part of the outside intervention in the war was the supply of "volunteer" foreign pilots to both sides. Russian and American Aces joined the Republican Air Force, while the Nationalists included Germans and Italians. The Soviet Volunteer Group began operations in the Second Sino-Japanese War as early as December 2, 1937, resulting in 28 Soviet Aces. The Flying Tigers were American military pilots to aid the Chinese Nationalists.

World War II

In World War II many air forces adopted the British practice of crediting fractional shares of aerial victories, resulting in fractions or decimal scores, such as 11 $^1D_2$ or 26.83. Some U.S. Commands also credited aircraft destroyed on the ground as equal to aerial victories. The Soviets distinguished between solo and group kills, as did the Japanese, though the Imperial Japanese Navy stopped crediting individual victories in favour of squadron tallies in 1943. The Soviet Union had women all fighter pilot air regiments. There were two female Aces of the war, both Soviet. Fighting on different sides, the French pilot Pierre Le Gloan had the unusual distinction of shooting down four German, seven Italian and seven British aircraft, the latter while he was flying for Vichy France in Syria.

The Luftwaffe had a tradition of "one pilot, one kill", and referred to top scorers as *Experten*. There are 107 German pilots with more than 100 kills. A number of factors probably contributed to the very high totals of the top German A Battle of Cer ces. For some period, especially during Operation Barbarossa, against Soviet Union, many Axis victories were over obsolescent aircraft and either poorly trained or inexperienced Allied pilots. In addition, Luftwaffe pilots generally flew many more individual sorties (sometimes well over 1000) than their Allied counterparts. Moreover, they often kept flying combat missions until they were captured, incapacitated, or killed, while successful Allied pilots were usually either promoted to positions involving less combat flying or routinely rotated back to training bases to pass their valuable combat knowledge to younger pilots. An imbalance in the number of targets available also contributed to the apparently lower numbers on the Allied side, since the number of operational Luftwaffe fighters was normally well below 1,500, with the total aircraft number never exceeding 5,000, and the total aircraft production of the Allies being nearly triple that of the other side. A difference in tactics might have been a factor as well. World's highest scoring Air ace, Erich Hartmann, for example, stated "See if there is a straggler or an uncertain pilot among the enemy... Shoot him down." This would have been an efficient and relatively low-risk way of increasing the number of kills. At the same time, the Soviet 1943 "Instruction For Air Combat" stated that the first priority must be the enemy Commander, which was a much riskier task, but one giving the highest return in case of a success. Similarly, in the Pacific theater, one of the factors leading to the superiority of Japanese Aces could be the early technical dominance of the Mitsubishi A6M "Zero" fighter.

Korean War: Jet Aces

The Korean War of 1950–53 marked the transition from piston-engined propeller driven aircraft to more modern jet aircraft. As such, it saw the world's first jet-vs.-jet Aces.

Vietnam War

The outbreak of the largest sustained bombardment campaign in history prompted rapid deployment of the nascent Vietnamese People's Air Force (VPAF) with subsonic MiG-17 against technically superior

American fighters with sophisticated radars and missiles. Vietnamese relied on dog-fighting and manoeuvrability to score kills. VPAF later got MiG-21s, and adopted the "guerrilla warfare in the sky" utilising quick hit-and-run attacks against US targets. Quite often air-to-air losses of US fighter jets were re-attributed to surface-to-air missiles, and anti-aircraft artillery, as it was considered "less embarrassing". By the war's end, the U.S. had nevertheless confirmed 245 air-to-air US aircraft losses while the figures for North Vietnam are relatively lower. The long-running conflict produced 22 Aces: 17 North Vietnamese pilots, two American pilots, and three American weapon systems officers (WSO) sitting in the second cockpit.

Arab-Israeli Wars

The series of wars and conflicts between Israel and its neighbours began with Israeli independence in 1948 and continued for over three decades. During these wars, there were 5 Egyptian, 48 Israeli, and 7 Syrian Air Aces.

Iran-Iraq War

Iranian Brig. General Jalil Zandi was a fighter pilot who served for the full duration of the eight year Iran–Iraq War. He was the most successful Ace with eight confirmed and three probable victories against Iraqi combat aircraft. He was also the most successful Grumman F-14 Tomcat pilot worldwide. There were other Air Aces on both sides. Iranian pilot Asadullah Adeli with support of his Radar Intercept Officer Mohammed Masbough engaged a formation of three MiG-23s. Target assigned was the one in the middle. The Phoenix missile's explosive delivery was so powerful, that it downed all three aircraft. The wreckage of all three MiGs was found on Kharg Island the next day. The only time in history that a single missile shot three aircraft.

Indo-Pakistan War

Air Commodore Muhammad Mahmood Alam of Pakistan Air Force, claimed to have downed five aircraft in a single sortie on 7 September 1965 with four in less than a minute, and reportedly establishing a world record. Experts around the world, including from within Pakistan Air Force have questioned the accuracy and authenticity of the claim. It is physically not possible in modern jet fighter close air combat to achieve such a feat.

Realistic Assessment of Kills

Realistic assessment of enemy casualties is important for intelligence assessment purposes also. Therefore, most air forces expend considerable effort to ensure accuracy in victory claims. In World War II, the aircraft gun camera came into general usage, partly in hope of alleviating inaccurate victory claims. And yet, in the Korean War, both the U.S. and Communist air arms made mutually incompatible claims of 10-to-1 victory/loss ratio. Many Air Aces overstated victories. Nations backed these claims for propaganda. There also was inherent confusion of three-dimensional, high speed combat between large numbers of aircraft. The on-board cameras did help resolve this. High air victories also brought decorations and promotions. The most accurate figures usually belong to the air arm fighting over its own territory, where aircraft wreckage could be located, and even identified, and where the shot down enemy was either killed or captured. Claims can be tallied with wreckage reported or through Prisoners of War (POW). Loss of records due to enemy action, fire, wartime confusion, and/or purposeful destruction, also made it difficult to reconcile victories.

Non-pilot Aces

While Aces are generally thought of exclusively as fighter pilots, some have accorded this status to gunners on bombers or reconnaissance aircraft, observers in two-seater fighters, and navigators. Because

pilots often teamed with different air crew members, an observer or gunner might be an Ace while his pilot is not, or vice versa. Observer Aces constitute a sizable minority in many lists. Charles George Gass, who tallied 39 victories, was the highest scoring observer Ace in World War I. With the advent of more advanced technology, a third category of Ace appeared. Charles B. DeBellevue became not only the first U.S. Air Force weapon systems officer (WSO) to become an Ace but also the top American Ace of the Vietnam War, with six victories.

Ace in a day

The first military aviators to score five or more victories on the same date, thus each becoming an "Ace in a day", were pilot Julius Arigi and observer/gunner Johann Lasi of the Austro-Hungarian Air Force, on August 22, 1916, when they downed five Italian aircraft. The feat was repeated five more times during World War I. Becoming an Ace in a day became relatively common during World War II. There were many German and Allied pilots who had been Ace in a day, and some had been multiple times.

The Greatest Air Aces: The Book

Over the years, many books have been written on the lives and exploits of Air Aces. There are authorised biographies written with the permission, cooperation, and at times, participation of a subject or a subject's heirs. There are autobiographies written by the person himself or herself, sometimes with the assistance of a collaborator or ghostwriter. There are many organisations and history websites that extensively cover the stories of wars and especially of Air combat and Air Aces. Many movies, TV serials and documentaries have been made on some of these great military aviators. Individuals and organisations have tried to short list the greatest Air Aces using different criteria.

This book is aimed at documenting the lives and achievements of twenty-five Air Aces. The chosen Air Aces have operated across the globe in different wars, which includes the two World Wars, Korea, Vietnam, Arab-Israeli, and Iran-Iraq wars. There are Air Aces who have operated in different sectors, such as Western Europe, the Eastern front, North Africa, South East Asia, Pacific Ocean, West Asia, China among others. There are Air Aces from Germany, USA, UK, Russia, France, Japan, Vietnam, Israel, Iran, Canada and Finland. There is an Indian Air Ace who flew for the British Royal Flying Corps in World War I and achieved 10 air victories in just a few days. There is an Air Ace who had all victories by Night. There is the highest scoring women Air Ace. There is an Air Ace who is considered the greatest air-to-ground strike pilot, and also had aerial victories. Many Air Aces were shot themselves and became Prisoners of War. Many Air Aces died very young. A few famous Air Aces continued to fly and score aerial victories after losing limbs. Some of the dogfight tactics and firing solutions evolved by them continue to be followed even today. All the Air Aces were dare-devil pilots, and they were highly decorated and were national Heroes. This book is meant to act as an inspiration for future combat aviators.

Air Marshal Anil Chopra (Retd)
PVSM, AVSM, VM, VSM

1

ERICH HARTMANN

Highest Scoring Air Ace

Major Erich Hartmann.
Image Source: YouTube

"Of all my accomplishments I may have achieved during the war, I am proudest of the fact that I never lost a wingman."

— Erich Hartmann

Erich Alfred Hartmann was a German fighter pilot, who served with Luftwaffe, during World War II and was the most successful fighter ace in the history of aerial warfare. He flew 1,404 combat missions and faced 825 air combat engagements. He was credited with shooting down 352 Allied aircraft, which included 350 Soviet and 2 American. In his flying career, Hartmann had to crash-land his fighter 16 times, either due enemy action or technical failure. Hartmann, was a pre-war glider pilot who joined the Luftwaffe in 1940 and completed his fighter pilot training in 1942. He was posted to the veteran Jagdgeschwader[1] 52 (Fighter Wing JG 52) on the eastern front. He was fortunate to be placed under the supervision of some of the Luftwaffe's most experienced fighter pilots. Under their guidance, Hartmann steadily developed his tactics.

By 29 October 1943 he had already destroyed 148 enemy aircraft and was awarded the Knight's Cross of the Iron Cross. Later he got Oak Leaves to the Knight's Cross for destroying 202 enemy aircraft on 2 March 1944, and the Swords to the Knight's Cross with Oak Leaves, four months later for 268 enemy aircraft shot down. Ultimately, Hartmann earned the coveted Knight's Cross of the Iron Cross with Oak Leaves, Swords and Diamonds on 25 August 1944 for claiming 301 aerial victories. At the time of its presentation, this was Germany's highest military decoration.

Hartmann achieved his 352nd and last aerial victory at midday on 8 May 1945, hours before the German surrender. Along with the remainder of JG 52, he surrendered to the United States Army, and was later turned over to the Red Army. The Soviets wanted to use his expertise and initially induced him to join the East German National People's Army, which he refused. He was tried for war crimes charges and convicted. He was initially sentenced to 20 years of imprisonment, later increased to 25 years, and spent 10 years in various Soviet prison camps and gulags until he was finally released in 1955. In 1956, Hartmann joined the newly established West German Air Force. He was retired in 1970, due to his strong opposition to the German's procurement of the F-104 Starfighter. In his later years, after his military career had ended, he became a civilian flight instructor. Erich Hartmann died on 20 September 1993 aged 71.

Early Life and Career

Erich Hartmann was born on 19 April 1922. His father was a doctor, and wanted him to become one. Hartmann flying career began when he joined the glider training program of the fledgling Luftwaffe and was taught to fly by his mother, Elisabeth Hartmann, one of the first female glider pilots in Germany. The Hartmanns also owned a light aircraft but were forced to sell it in 1932 as the German economy collapsed. The mother being a licensed pilot, used to take the children up and teach things. His father was not pleased that he wanted to be a pilot, he wanted the children to follow him in medicine. The rise of Nazi party in 1933 resulted in government support for gliding, and, in 1936, Elisabeth Hartmann established the glider club for locals and served as instructress. The 14-year-old Erich Hartmann became a gliding instructor in the Hitler Youth. In 1937, he gained his pilot's license, allowing him to fly powered aircraft. During World War II, Hartmann's younger brother, Alfred, also joined the Luftwaffe, serving as a gunner on Junkers Ju 87 in North Africa. Alfred Hartmann was captured by the British and spent four years as a prisoner of war.

Erich as Hitler Youth.
Image Source: ww2gravestone.com

Hartmann during advanced flying training.
Image Source: ww2gravestone.com

Hartmann began his military training in October 1940. His advanced pilot training was completed in January 1942. By August 1942, he had learned to fly the Messerschmitt Bf 109. As a trainee in March 1942, during a gunnery training flight, he ignored regulations and performed some aerobatics in his Bf 109. His punishment was a week of confinement to quarters, and loss of two-thirds of his pay in fines. He was scheduled to go up on a gunnery flight that was now allotted to his roommate. Shortly after he took off, while on his way to the gunnery range, aircraft developed engine trouble and crashed killing the batch mate.

Hartmann evolved his own credo and practices. "Fly with your head, not with your muscles" he said. During a gunnery meet in June 1942, he hit the target drogue with 24 out of the 50 rounds of machine-gun fire, which was considered outstanding. His training had qualified him to fly 17 different types of powered aircraft.

Initial Tactics and Grooming

In October 1942, Hartmann was assigned to fighter wing JG 52, at Maykop on Eastern Front facing Soviet Union. Hartmann and other pilots were initially given the task of ferrying Junkers Ju 87 Stukas down to Mariupol. His first flight ended with brake failure, causing the Stuka to crash into and destroy the controller's hut. He was also assigned to experienced pilots to hone his combat skills. After a few days of intensive mock combat and practice flights, his senior Alfred Grislawski conceded that, although Hartmann had much to learn regarding combat tactics, he was quite a talented pilot. Hartmann was placed as wingman to Paule Roßmann (a flying ace), who acted as his teacher and mentor. Grislawski also taught Hartmann how to aim in combat. Hartmann eventually adopted the tactic "See – Decide – Attack – Break". Roßmann taught him to "stand off", evaluate the situation, then select a target that was not taking evasive action and destroy it at close range.

Messerschmitt Bf 109G JG 52.
Image Source: Wiki Commons

Early Aerial Combat

Hartmann flew his first combat mission on 14 October 1942 as Roßmann's wingman. When they encountered 10 enemy aircraft below. An impatient Hartmann opened full throttle and separated from Roßmann. He engaged an enemy fighter, but failed to score any hits and nearly collided with it. He then ran for cover in low clouds, and his mission subsequently ended with a crash landing after his aircraft ran out of fuel. Hartmann had violated almost every rule of air-to-air combat, and he was sentenced to three days of working with the ground crew. Twenty-two days later, Hartmann claimed his first victory, an Ilyushin IL-2. By the end of 1942, he had added only one more victory to his tally. As with many high-claiming aces, it took him some time to establish himself as a consistently successful fighter pilot. On 5 November 1942, an IL-2 shot up his Bf 109 G-2 engine resulting in a forced landing.

Hartmann's youthful looks got him the nickname "Bubi" (young boy in German language). Hartmann steadily improved. On 5 July Hartmann claimed four victories during one of the large dogfights that took place during

Hartmann's Leader, Edmund "Paule" Roßmann German fighter ace (93 aerial victories).
Image Source: Pinterest

the Battle of Kursk. Hartmann began to score successes regularly in a target rich environment. On 7 July he claimed four, including two IL-2s. On 8 and 9 July 1943 he claimed four on each day. The Soviet post-battle analysis acknowledged the engagement claims. From the third week of May 1943 to the first week of August, Hartmann's number of claims rose from 17 to 60. On 1 August 1943 Hartmann became an ace-in-a-day by claiming five victories. Another four followed on 3 August and five on the 4 August. Another five were claimed destroyed on the 5 August, a single on the 6 August, and a further five on 7 August. On 8 and 9 August he claimed another four Soviet fighters. Hartmann's last claim of the month came on the 20th, when he accounted for an IL-2 for his 90th victory.

Fighting Techniques

Unlike the German Ace Hans-Joachim Marseille, who was a marksman and expert in the art of deflection shooting, Hartmann was a master of stalk-and-ambush tactics, and fire at close range rather than dogfight. He held fire until extremely close (20 m or less), then unleashed a short burst at point-blank range. This technique concealed his position till the last possible moment, and compensated for the low muzzle velocity of the slower-firing 30 mm cannon on some aircraft. It also resulted in accuracy and minimum waste of ammunition. Also the adversary had no time for evasive action. His approach was summarised in "See–Decide–Attack–Reverse", wherein observe the enemy, decide how to proceed with the attack, make the attack, and then disengage to re-evaluate the situation. Once the attack was over, the rule was to vacate the area; survival was paramount. Another attack could be executed if the pilot could re-enter the combat zone with the advantage.

Knight's Cross of the Iron Cross

On 20 September 1943, Hartmann was credited with his 100th aerial victory, having claimed four that day to end it on 101. He was the 54th Luftwaffe pilot to achieve the century mark. Nine days later, Hartmann downed the Soviet ace Major Vladimir Semenishin for his 112th victory. In October 1943, Hartmann claimed another 33 aerial victories. On 29 October, he was awarded the Knight's Cross of the Iron Cross, at which point his tally stood at 148. By the end of the year, this had risen to 159.

100 Victories.
Image Source: allthatsinteresting.com

Counter Check of His Claims

In the first two months of 1944, Hartmann claimed over 50 Soviet aircraft. His spectacular rate of success raised a few eyebrows even in the Luftwaffe High Command; so his claims were double and triple-checked, and his performance closely monitored by an independent observer flying in his formation. By this time, the Soviet pilots were familiar with Hartmann's radio call sign of "Karaya 1", and the Soviet Command had put a price of 10,000 Roubles on the German pilot's head. Hartmann was nicknamed the *Cherniy Chort* (Black Devil) because of his skill and paint scheme of his aircraft. This scheme was in the shape of a black tulip on the engine cowling; though this became synonymous with Hartmann in reality he flew with the insignia on only five or six occasions. Hartmann's opponents were often reluctant to stay and fight if they noticed his personal design. As a result, this aircraft was often allocated to novices, who could fly it in relative safety. On 21 March, it was Hartmann who claimed his unit JG 52's 3,500th victory of the war. Adversely, the supposed reluctance of the Soviet airmen to engage

Erich (Centre), with Hermann Göring (right) and Adolf Galland. Image Source: ww2gravestone.com

him caused Hartmann's kill rate to drop. Hartmann then had the tulip design removed, and his aircraft painted just like the rest of his unit.

Drunk at Awards Ceremony

In March 1944, Hartmann and three other German Aces were summoned to Adolf Hitler to be honoured with their awards. According to Hartmann, all four of them got drunk on cognac and champagne. On arrival, Hartmann was reprimanded by Hitler's adjutant for intoxication and for handling Hitler's hat.

Aces meeting Hitler. Image.
Source: ww2gravestone.com

Top Scoring Ace

By the end of May 1944, Hartmann claims went up to 231. On 24 May 1944, Hartmann engaged the United States Army Air Force P-51 Mustang for the first time over Romania. He made the only other claim against P-51 in 1945. On 17 August, Hartmann became the top scoring fighter ace, surpassing fellow JG 52 pilot Gerhard Barkhorn, with his 274th victory. On 23 August, Hartmann claimed eight victories in three combat missions, an ace-in-a-day achievement, bringing his score to 290 victories. He passed the 300-mark on 24 August 1944, a day on which he shot down 11 aircraft in two combat missions, representing his greatest ever victories-per-day ratio (a double-ace-in-a-day) and bringing the number of aerial victories to an unprecedented 301.

Hartmann meeting Hitler.
Image Source: ww2gravestone.com

Diamonds to the Knight's Cross

Hartmann became one of only 27 German soldiers in World War II to receive the Diamonds to his Knight's Cross. He was the youngest recipient of the Diamonds, at twenty-two. Hartmann was summoned to Adolf Hitler's military headquarters to receive the coveted award from Hitler personally. Hartmann was asked to surrender his side arm, a security measure heightened by the aftermath of the failed assassination attempt on Hitler on 20 July 1944. Hartmann reportedly refused and threatened to decline the Diamonds if he were not trusted to carry his pistol. During Hartmann's meeting with Hitler, Hartmann discussed at length the shortcomings of fighter pilot training. Allegedly, Hitler admitted to Hartmann that he believed that, "militarily, the war is lost," and that he wished the Luftwaffe had "more like him and Rudel." After the ceremony he was allowed leave for marriage. Gerhard Barkhorn, the second highest scoring Air Ace was the best man at his wedding.

Gerhard Barkhorn, the second highest scoring Air Ace was the best man at his wedding.
Image Source: Pinterest (Zombie Cat)

Knights Cross of the Iron Cross with Golden Oak leaves Swords and Diamonds.
Image Source: Wikipedia

The Diamonds to the Knight's Cross also earned Hartmann a 10-day leave. On his way to his vacation, he was ordered by the Luftwaffe Commander Adolf Galland, who was an ace pilot himself, for a meeting. Galland wanted to transfer Hartmann to the Messerschmitt Me 262 flight test program. Hartmann declined and wanted to continue in JG 52. Hartmann also argued to Göring that he best served the war effort on the Eastern Front. On 10 September, Hartmann married his long-time teenage love, Ursula "Usch" Paetsch.

Last Combat Missions

In March 1945, Hartmann's score stood at 336 aerial victories, and was asked a second time by General Adolf Galland to join the newly forming Me 262 jet fighter units. Hartmann did attend the jet conversion program, but declined a permanent move. Hartmann claimed his 350th aerial victory on 17 April. The last wartime photograph of Hartmann known was taken in connection with this victory. Hartmann's last aerial victory occurred on 8 May, the last day of the war in Europe. Hartmann saw a Yak-9, ambushed it from his vantage point at 12,000 ft (3,700 m) and shot it down. When he landed, Hartmann learned that the Soviet forces were within artillery range of the airfield. So the unit destroyed the 24 other Bf 109s, and large quantities of ammunition. Hartmann was ordered to fly to the British sector to avoid capture by Soviet forces. Hartmann chose to surrender his unit to members of the US 90th Infantry Division.

Prisoner of War

After his capture, the U.S. Army handed Hartmann, his pilots, and ground crew over to the Soviet Union on 14 May. According to his account, Soviets attempted to convince him to cooperate with them. He was asked to spy on fellow officers, but refused, and was given ten days' solitary confinement. The Soviets threatened to kidnap and murder his wife. During interrogations about his knowledge of the Me. 262, Hartmann was struck by a Soviet officer using a cane. More subtle efforts by the Soviet authorities to convert Hartmann to communism also failed. He was offered a post in the East German Air Force, which he refused.

War Crimes Charges

During his captivity, in December 1949, he was sentenced to 20 years in prison. He was condemned for atrocities against Soviet citizens, the attack on military objects and the destruction of Soviet aircraft and thus having significantly damaged the Soviet economy. Hartmann protested multiple times against this judgment. In June 1951, he was charged for a second time, specifically the "deliberate shooting of 780 Soviet civilians" in the village of Briansk, attacking a "bread factory" on 23 May 1943, and destroying 345 "expensive" Soviet aircraft. Sentenced to 25 years of hard work, he refused to work, and was put into solitary confinement. In late 1955 Hartmann was released as a part of the last group of German prisoners sent back to Germany. In January 1997, more than three years after his death, Hartmann's case was reviewed by the Chief Military Prosecutor in Moscow of the now Russian Federation, after the dissolution of the Soviet Union, and he was acquitted of all historical charges against him in Russian Law.

Erich on return as POW from Soviet Union.
Image Source: ww2gravestone.com

The autographed photograph shows Hartmann with an American officer.
Image Source: ebay.com.au

Post War Years

During his long imprisonment, Hartmann's son, Erich-Peter, was born in 1945 and died as a three-year-old in 1948, without his father ever having seen him. Hartmann later had a daughter, Ursula Isabel, born on 23 February 1957. When Hartmann returned to West Germany, he re-entered military service and became an officer in the West German Air force, where he commanded West Germany's first all-jet unit from 6 June 1959 to 29 May 1962. This unit was equipped initially with Canadair Sabres, and later Lockheed F-104

Hartmann with his wife and daughter.
Image Source: Pinterest (Luftwaffe pilots)

Erich Hartmann.
Image Source: migflug.com

Starfighters. Hartmann had to make several trips to the United States, for flying training on Sabres and F-104. Hartmann considered the F-104 a fundamentally flawed and unsafe aircraft and strongly opposed its adoption by the air force. Events subsequently validated his low opinion of the aircraft that had 269 crashes and 116 German pilots killed on the F-104 in non-combat missions. Hartmann's outspoken criticism proved unpopular with his superiors, and he was given early retirement in 1970.

From 1971–74, Hartmann worked as a flight instructor in Hangelar, near Bonn and also flew in fly-ins with other wartime pilots. Hartmann died on 20 September 1993, at the age of 71. In 2016, Hartmann's former unit, JG 71, honoured him by applying his tulip colour scheme to their current aircraft. Hartmann's biography "The Blond Knight of Germany" was written by American authors Trevor J. Constable and Raymond F. Toliver in 1970.

Assessment of the Enemy in the Air

Hartmann felt that if an enemy pilot started firing early, well outside the maximum effective range of his guns, then he was an easy kill. But, if a pilot closed in and held his fire, and seemed to be watching the situation, then you knew that an experienced pilot was on you. Hartmann developed different tactics for various conditions, such as always turning head-on into the approaching enemy, or rolling into a negative G dive forcing him to follow or break off, then rolling out and sometimes reducing airspeed to allow him to over commit. That was when he took advantage of his failing.

Favourite Method of Attack

His favourite method of attack was coming out of the sun and getting close. Dog fight he thought was a waste of time. The hit and run with the element of surprise served best, as was the case with most of the high scoring pilots. Once a leader was shot down they became disorganized and easy to attack. This was not always the case, especially later in the war, and there were special units of highly skilled and disciplined pilots, such as the Red Banner units who would make life difficult.

Aerial Victories Authenticated

Göring, himself an Air ace from WW I, could not believe the staggering kills being recorded by pilots from 1941 on. He was doing double checks on all the kills and could find no wrong. There were people in Hartmann's airbase, like fighter Ace Fritz Oblesser, who questioned his kills. Hartmann requested such people to be transferred to fly as his wingman for a while. Oblesser soon became a believer and signed off on some kills as a witness, and he later started having regard and became friend.

German Aces Gerhard "Gerd" Barkhorn (301), Erich Hartmann (352), Johannes "Macky" Steinhoff (176), Günther Rall (275).
Image Source: ww2gravestone.com

Matthews and Foreman, authors of "Luftwaffe Aces – Biographies and Victory Claims", researched the German Federal Archives and found records for 352 aerial victory claims, plus two further unconfirmed claims. All victory claims were logged to map-references. The Luftwaffe grid map covered all of Europe, western Russia and North Africa and was composed of rectangles. All victories for all pilots were plotted in time and place.

Impressions about Hitler

When he met Hitler for the first time, he found him a little disappointing, although very interested in the war at the front and extremely well informed on events. He felt Hitler had a tendency to drone on about minor things. He was interesting yet not that imposing. He lacked sufficient knowledge about the air war in the east. He was more concerned with the Western Front's air war and the bombing of cities. Of course, the Eastern Front ground war was his area of most interest. This was evident. Hitler listened to the men from the Western Front and assured them that weapons and fighter production were increasing, and history proved this to be correct. Then he went into the U-boat war, how we were going to decidedly destroy maritime commerce and all of that. Hartmann found him an isolated and disturbed man.

Image Source: 9gag.com

Securing Release from Soviet Prison

German Chancellor Konrad Adenauer was very crucial in this. Hartmann's mother had written to Stalin and is Foreign Minister Molotov, but without any response. She wrote to Adenauer and he replied personally that he was working on the problem. The Soviets wanted a trade agreement with the West, especially West Germany, and part of this deal was the release of all the POWs.

Roll of Honour

Hartmann was never shot by an enemy plane, but he had to crash land fourteen times due to damage from victories or mechanical failure, but never took to the parachute. He never became another pilot's victory. Of all the fighter aces in history, the first names that usually come to mind are the Baron von Richthofen, the "Red Baron" of World War I and Erich "Bubi" Hartmann. They are considered by many as the top two best fighter pilots in history. His 352 confirmed kills will probably stand forever as the benchmark of success in aerial warfare. However, Erich's war in the air became a footnote to his life following the decade of Soviet imprisonment he experienced following the war. In his

Hartmann's Grave. Photo by Bob Hopmans.
Image Source: ww2gravestone.com

own words, Hartmann discussed his life, career, passions and survival in a world few have seen and even fewer survived. He gave a final interview just a little while before he died in 1993.

END NOTES

1. **Jagdgeschwader** were the series of fighter wings of initially, the German Empire's *Luftstreitkräfte* air arm of the *Deutsches Heer*, then the successor fighter wings of the Third Reich's original *Luftwaffe* air arm of its combined Wehrmacht armed forces (1935-45), and after 1949, the fighter wings of the air arm of the current Federal German Republic's *Bundeswehr* armed forces, the Luftwaffe.

REFERENCES

1. William DeLong, Erich Hartmann: The German World War II Pilot Who Was The Deadliest Flying Ace Of All Time, December 3, 2018, https://allthatsinteresting.com/erich-hartmann
2. Diane Tedeschi, Erich Hartmann, the Most Successful Fighter Pilot of All Time, AIR & SPACE MAGAZINE, October 2020, HTTPS://WWW.AIRSPACEMAG.COM/MILITARY-AVIATION/WHO-WAS-ERICH-HARTMANN-180975845/
3. *Erik Schmidt,* 352 Kills — How Germany's Erich Hartmann Became History's Deadliest Flying Ace, MilitaryHistoryNow.com, April 02, 2020 https://militaryhistorynow.com/2020/04/02/352-kills-how-germanys-erich-hartmann-became-historys-deadliest-flying-ace/
4. Erich Hartmann, Major, Aces of Luftwaffe, https://www.luftwaffe.cz/hartmann.html
5. Erich Hartmann Quotes https://quotefancy.com/erich-hartmann-quotes
6. Erich Hartmann, Wikipedia, https://en.wikipedia.org/wiki/Erich_Hartmann
7. Final Interview with Erich Hartmann, MiG Flug, https://migflug.com/jetflights/final-interview-with-erich-hartmann/
8. Larry Dwyer, Erich Hartmann, The Aviation History on Line Museum http://www.aviation-history.com/airmen/Erich_Hartmann.htm
9. Robert Jackson. *Fighter Pilots of World War II*. St. Martin's Press; New York, 1976.
10. Hartmann, Erich 'Bibi", World War II Graves, https://ww2gravestone.com/people/hartmann-erich-bubi
11. Raymond Toliver & Trevor Constable. *The Blond Knight of Germany*. TAB/AERO Books; Blue Ridge Summit, PA, 1970.
12. Mike Spick. *LuftWaffe Fighter Aces*. Mechanicsburg, Pennsylvania: Stackpole Books, 1996.
13. Edward H. Simms. *The Fighter Pilots*. Corgi Books, Transworld Publishers Ltd.; Great Britian, 1967.
14. Robert Jackson. *Fighter Pilots of World War II*. St. Martin's Press; New York, 1976.

2

Douglas Bader

Determined Dogmatic and Fearless

Picture Source: OwlCation

Group Captain Sir Douglas Robert Bader, CBE, DSO & Bar, DFC & Bar, DL, FRAeS was a Royal Air Force (RAF) flying ace of World War II. He was credited with 22 aerial victories, four shared victories, six probable, one shared probable and 11 enemy aircraft damaged. Bader joined the RAF in 1928, and was commissioned in 1930. In December 1931, while flying aerobatics, he crashed and lost both his legs. He recovered from the brink of death, got artificial legs, retook flight training, passed his check flights and then requested reactivation as a pilot. There were no regulations to allow this and he was retired against his will on medical grounds.

After the outbreak of WW II in 1939, however, Douglas Bader returned to the RAF and was accepted as a pilot. He scored his first victories over Dunkirk during the Battle of France in 1940. During the Battle of Britain he became a friend and supporter of Air Vice Marshal Trafford Leigh Mallory and his 'Big Wing' experiments. In August 1941, Bader bailed out over German-occupied France and was captured. Soon he met and became friends with Adolf Galland, the already famous German fighter ace. Despite his disability, Bader made a number of escape attempts and was eventually sent to the prisoner of war (POW) camp at Colditz Castle, near Leipzig, Dresden, where he remained until April 1945 when the camp was liberated by the US Army.

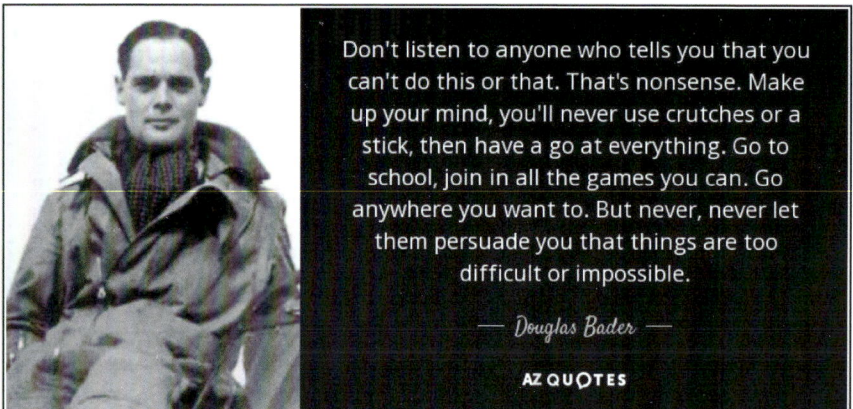

AZ Quotes.
Image Source: Pinterest

Early Years

Bader was born on 21 February 1910 in London, the second son of a civil engineer father. His first two years were spent with relatives, as his father with the rest of the family was in India. At the age of two, Bader joined his parents in India for a year. When his father resigned from his job in 1913 the family moved back to London. Bader's father saw action in WW I in the Royal Engineers, and was wounded in action in 1917. He remained in France after the war, where, having attained the rank of Major, he died in 1922 of complications from the war wounds near Saint-Omer, the same area where Bader would bail out and be captured in 1941. Meanwhile, Bader's mother remarried.

Bader was subsequently brought up in the rectory of the village. Bader's mild-mannered stepfather did not become the father figure he needed. His mother showed little interest in Bader and sent him to his grandparents on occasion. Without guidance, Bader became unruly. Bader played rugby and often enjoyed physical battles with bigger and older opponents. Fellow RAF night fighter and bomber pilots Guy Gibson and Adrian Warburton also attended the same school. Bader's sporting interests would later continue into his military service. He was selected for the RAF Cricket team. He played cricket in a German POW camp after his capture in 1941, despite his disability.

Douglas Bader with the 1st Cricket XI, Royal Air Force College Cranwell, 1929. Standing Far Right.
Image Source: rafmuseum.org.uk

In mid-1923, Bader, at the age of 13, was introduced to an Avro 504 during a trip to visit his aunt, Hazel, who was marrying RAF Flight Lieutenant Cyril Burge at RAF Cranwell. Due to his new connection with Cyril Burge, Bader learned of the six annual prize cadetships offered by RAF Cranwell each year. Out of hundreds of applicants, he finished fifth.

Joining the RAF

In 1928, Bader joined the RAF as an officer cadet at RAF Cranwell. He continued to excel at sports, and added hockey and boxing to his repertoire. Bader was involved in banned high speed motorcycling, and was close to expulsion after being caught too often. He was also coming 19th out of 21 in his class academic examinations. On 13 September 1928, Bader took his first flight in an Avro 504. After just 11 hours and 15 minutes of flight time, he flew his first solo, on 19 February 1929. Bader competed for the "Sword of Honour" award at the end of his two-year course, but finally came second. On 26 July 1930, Bader was commissioned as a Pilot Officer into No. 23 Squadron RAF. Bader became a daredevil while training, often flying illegal and dangerous stunts. While very fast for its time, the Bulldog aircraft had directional stability problems at low speeds, which made such stunts exceptionally dangerous. Strict orders were issued forbidding unauthorised aerobatics below 2,000 feet. Douglas took this as an unnecessary safety rule rather than an order to be obeyed.

Air Crash and Amputation of Legs

Bader continued to perform unauthorised low-level aerobatics to show-off his skill. The unit CO gave his pilots more latitude. Under another CO, he would have been court-martialed. No. 23 Squadron had won the Hendon Air Show "pairs" event in 1929 and 1930. In 1931 Bader, teamed with Harry Day, successfully defended the squadron's title. In late 1931, Bader undertook training for the 1932 Hendon Air Show. Two pilots had been killed attempting aerobatics. The pilots were warned not to practice these manoeuvres under 2,000 feet and to

Avro 504 Image Credit: Flickr
Air Crash and Amputation of Legs

keep above 500 feet at all times. Nevertheless, on 14 December 1931, while visiting Reading Aero Club, Bader attempted some low-flying aerobatics in a Bulldog Mk. IIA. His aircraft crashed when the tip of the left wing touched the ground. Bader was rushed to the hospital, where both his legs were amputated, one above and one below the knee. Bader made the following entry in his logbook after the crash: "Crashed slow-rolling near ground. Bad show".

Bader, Flt.Lt. Harry Day and Fg.Offr. Geoffrey Stephenson during training for the 1932 Hendon airshow, with a Gloster Gamecock.
Image Source: Wikipedia

Invalided From Service

In 1932, he was given a new pair of artificial legs. His determination paid off, and he was able to drive a specially modified car, play golf, and even dance with his artificial legs. He met and fell in love with Thelma Edwards, a waitress at a tea room. Bader got his chance to prove that he could still fly when, in June 1932, Air Under-Secretary Philip Sassoon arranged for him to take up an Avro 504, which he piloted competently. A subsequent medical examination proved him fit for active service. But they decided that his case was not covered by regulations, and in May, Bader was invalided out of the RAF. He took a job with the Asiatic Petroleum Company (now Shell) and, on 5 October 1933, married Thelma Edwards.

Return to RAF

In view of the increasing tensions in Europe in 1937–39, Bader repeatedly requested the Air Ministry to accept him back into the RAF. He was initially offered a ground job. But later Air Vice Marshal Halahan,

Commandant of RAF Cranwell in Bader's earlier days, personally endorsed him and asked the Central Flying School (CFS) to assess his capabilities. In October 1939 Bader undertook refresher courses. Despite reluctance of the establishment to allow him to apply for full flying category status, his persistent efforts paid off. Bader regained a medical categorisation for operational flying at the end of November 1939 and was sent for conversion on modern types of aircraft. On 27 November, eight years after his accident, Bader flew solo again. Once airborne, he could not resist the temptation to turn the biplane upside down at 600 feet inside the circuit area.

The inspirational, legless, Douglas Bader (centre) pictured with pilots of 92 and 222 Squadrons whilst a flight commander with 222 and flying Spitfires in support of the Dunkirk evacuation.
Image Source: ourfinesthour.net

Getting into Action

In January 1940, Bader was posted to No. 19 Squadron, where, at 29, he was older than most of his fellow pilots. Here he got a first glimpse of a Spitfire. It was thought that Bader's success as a fighter pilot was partly because of his having no legs. Pilots pulling high g-forces in combat turns often blacked out as the flow of blood from the brain drained to other parts of the body, usually the legs. As Bader had no legs he could remain conscious longer, and thus had an advantage over more able-bodied opponents. Initial months, Bader practiced formation flying and air tactics, as well as undertaking patrols over convoys out at sea. Bader found opposition to his ideas about aerial combat. He favoured using the sun and altitude to ambush the enemy, but the RAF did not share his opinions. Official orders/doctrine dictated that pilots should fly line-astern and attack singly. Despite this being at odds with his preferred tactics, Bader obeyed orders, and his skill saw him rapidly promoted to section leader. Bader was subsequently promoted from Flying Officer to Flight Lieutenant, and appointed as a Flight Commander of No. 222 Squadron RAF.

Battle of France

On 10 May German Army invaded Luxembourg, Netherlands, Belgium and France. RAF squadrons were ordered to provide air supremacy to the Royal Navy during Operation Dynamo (evacuation from Dunkirk). While patrolling the coast near Dunkirk, on 1 June 1940, at around 3,000 ft, Bader saw a Messerschmitt Bf 109 in front of him, flying in the same direction and at approximately the same speed. He believed that the German must have been a novice, taking no evasive action even though

Squadron Leader Douglas Bader with pilots of No. 242 Squadron in front of his Hawker Hurricane at Duxford, September 1940.
Image Source: commons.wikimedia.org

it took more than one burst of gunfire to shoot him down. Bader claimed five victories in that particular dogfight, and damaged a Bf 110. In the next patrol Bader was credited with a Heinkel He 111 damaged.

After flying operations over Dunkirk, on 28 June 1940 Bader was posted to command the No. 242 Squadron, flying Hurricanes, as acting Squadron Commander. The Squadron was mainly made up of Canadians who had suffered high losses in the Battle of France and was suffering from low morale. Bader's strong personality and perseverance, especially in cutting through red tape, made the squadron operational again.

Battle of Britain

After the French campaign, Luftwaffe intended to achieve air supremacy and then launch Operation Sea Lion, codename for the invasion of Britain. The Battle of Britain officially began on 10 July 1940. On 11 July, Bader scored his first victory

Pilot Officer William McKnight (left) Canada's highest scoring ace during the Battle of Britain, who flew with the Royal Air Force's 242 Squadron, The unit's commander during the Battle of Britain Squadron Leader Douglas Bader (centre). Saved by Aviators on Pinterest

with his new squadron. It was overcast and drizzling with cloud base was down to just 600 ft. Bader was alone on patrol, and was soon directed toward an enemy aircraft flying north up the Norfolk coast. He spotted a Dornier Do 17 at 600 yards. When he closed to 250 yards its rear gunner opened fire. Bader continued his

Bader.
Image Source: Wiki Commons

attack and fired two bursts into the bomber before it vanished into the cloud. The Dornier, crashed into the sea, and was later confirmed by a member of the Royal Observation Corps. On 21 August, a similar engagement took place. Later in the month, Bader scored a further two victories over Messerschmitt Bf 110s. On 30 August 1940, No. 242 Squadron was moved to Duxford again. On this date, the squadron claimed 10 enemy aircraft, Bader scoring two victories against Bf 110s. On 7 September, two more Bf 110s were shot down, but in the same engagement Bader was badly hit by a Messerschmitt Bf 109. Bader almost bailed out, but recovered the Hurricane. Other pilots witnessed one of Bader's victims crash. On 7 September, Bader claimed two Bf 109s shot down, followed by a Junkers Ju 88. On 9 September, Bader claimed another Dornier. During the same mission, he attacked a He 111 only to discover he was out of ammunition. On 14 September, Bader was awarded the Distinguished Service Order (DSO) for his combat leadership.

On 15 September, also known as the Battle of Britain day, Bader damaged a Do 17 and a Ju 88, while destroying another Do 17 in the afternoon. Bader flew several missions that day, which involved heavy air combat. Another Do 17 and a Ju 88 were claimed on 18 September. On 24 September, Bader was promoted to the war substantive rank of Flight Lieutenant. A Bf 109 was claimed on 27 September. Bader was gazetted on 01 October 1940.

The "Big Wing" Tactic

Bader was a supporter of his 12 Group commander, Air Vice Marshal Mallory's controversial "Big Wing" theory. Bader was an outspoken critic of the careful "husbanding" tactics being used by

German Fighter Bf 109.
Image Source: Wikipedia

Air Vice Marshal Keith Park, the Commander of 11 Group. Bader vociferously campaigned for an aggressive policy of assembling large formations of defensive fighters north of London ready to inflict maximum damage on the massed German bomber formations. Achievements of the Big Wing were hard to quantify, as the large formations often took too long to form up, over claimed victories, and too often did not provide timely support. The claims of the RAF and Big Wings were often exaggerated. RAF ace Johnnie Johnson felt that there was room for both tactics – the Big Wings and the small Squadrons. For not only it took longer to gather and get large numbers to their height, but sixty or seventy packed climbing fighters could have been seen for miles and would have been sitting ducks for higher Bf 109s. Also nothing would have pleased Göring more than for his Bf 109s to pounce on large numbers of RAF fighters. Keith Park's brilliance on the other hand was that by refusing to concentrate his force he preserved it throughout the battle.

Bader's Hawker Hurricanes

During the Battle of Britain, Bader used three Hawker Hurricanes. The first was P3061, in which he scored six air victories. The second aircraft (number unknown), Bader score one victory and two aircraft damaged on 9 September. The third was V7467, in which he destroyed four more and added one probable and two damaged by the end of September.

Distinguished Flying Cross (DFC)

On 12 December 1940, Bader was awarded the DFC for his services during the Battle of Britain. His unit, No. 242 Squadron, had claimed 62 aerial victories. Bader became an acting Squadron Leader by 7 January 1941.

Distinguished Flying Cross (DFC)

Douglas Bader's Supermarine Spitfire W3185 "D-B" written on fuselage.
Image Source: airshowsamerica.biz

Wing Leader

On 18 March 1941, Bader was promoted to acting Wing Commander and became one of the first "Wing Leaders" at Tangmere with three squadrons under his command. Bader led his Wing of Spitfires on sweeps over north-western Europe throughout the summer campaign. These were missions combining bombers and fighters designed to lure out and tie down German Luftwaffe fighter units that might otherwise serve on the Russian front. One of the wing leader's "perks" was permission to have his initials marked on his aircraft as personal identification, thus "D-B" was painted on the side of Bader's Spitfire.

During 1941 his wing was re-equipped with Spitfire VBs, which had two Hispano 20 mm cannons and four .303 machine guns. Bader preferred to fly a Mk VA equipped with eight .303 machine guns, as he insisted that these guns were more effective against fighter opposition. Since he believed in a close-in approach, the lower calibre weapons had a more devastating effect. Bader's combat missions were mainly fought against Bf 109s over France and the Channel.

Douglas Bader with Adolf Galland and Other Pilots after Bail-Out on 09 August 1941.
Image Source: Pinterest. Christophe Le Guenic Saved to Luftwaffe Pilots.

On 7 May 1941 he shot down one Bf 109 and claimed another as a probable victory. The German formation belonged to the JG 26 (Fighter Wing 26), which on that date was led in action by German Ace Adolf Galland, and was also when Galland claimed his 68th victory. Bader and Galland met again 94 days later. On 21 June 1941, Bader shot down a Bf 109E off the coast near Desvres. His victory was witnessed by two other pilots who saw a Bf 109 crash and the German pilot bailout. On 25 June 1941 Bader shot down two more Bf 109Fs.

Bar to DSO

On 2 July 1941 he was awarded the bar to his DSO. Later that day he claimed one Bf 109 destroyed and another damaged. On 4 July, Bader fired on a Bf 109E which slowed down so much that he nearly collided with it. On 6 July another Bf 109 was shot down and the pilot bailed out. This victory was witnessed by Pilot Officers Johnnie Johnson and Alan Smith (Bader's usual wingman). On 9 July, Bader claimed one probable and one damaged, both trailing coolant and oil. Between 10 and 23 July, Bader claimed another 6 Bf 109 and four probable. Bader had been pushing for more sorties to fly in late 1941 but his Wing was tired, and in near-mutinous state. Mallory, Bader's immediate superior as OC No. 11 Group, Fighter Command, relented and allowed Bader to continue even though his score of 20 and the accompanying strain was evident.

Last Combat: Who Shot Bader Controversy

Between 24 March and 9 August 1941, Bader flew 62 fighter sweeps over France. On 9 August 1941, Bader was flying a Spitfire Mk VA serial W3185 "D-B" on an offensive patrol over the French coast, without his trusted wingman Alan Smith. Smith, who was described by fellow pilot Johnnie Johnson as "leechlike" and the "perfect number two", was unable to fly on that day due to a cold. Just after Bader's section of four aircraft crossed the coast, 12 Bf 109s were spotted flying in formation approximately 2,000 to 3,000 feet below them and travelling in the same direction. Bader dived on them too fast and too steeply to be able to aim and fire his guns, and barely avoided colliding with one of them. He levelled out at 24,000 feet to find that he was now alone, separated from his section, and was considering whether to return home when he spotted three pairs of Bf 109s a couple of miles in front of him. He dropped down below them and closed up before destroying one of them with a short burst of fire from close range. Bader was just opening fire on a second Bf 109, which trailed white smoke and dropped down, when he noticed the two on his left turning

Illustration from Channel 4 History Episode, Who Downed Douglas Bader.
Image Source: distribution.channel4.com

towards him. At this point he decided it would be better to return home; however, made a mistake of banking away from them. Bader believed he had a mid-air collision with the second of the two Bf 109s on his right. Bader's fuselage, tail and fin were gone from behind him, and he lost height rapidly at what he estimated to be 400 mph in a slow spin. He jettisoned the cockpit canopy, released his harness pin, and the air rushing past the open cockpit started to suck him out, but his prosthetic leg was trapped. Part way out of the cockpit and still attached to his aircraft, Bader fell for some time before he released his parachute, at which point the leg's retaining strap snapped under the strain and he was pulled free. Subsequent research showed no Bf 109 was lost to a collision that day. Max Meyer of JG 26 flying a Bf 109 had claimed him shot down. Furthermore, Meyer mentioned that he had followed the downed Spitfire and watched the pilot bail out. Bader met Max Meyer in Sydney in 1981 during the Schofields Air Show. Adolf Galland went through every report, even those of German pilots killed in the action, to determine Bader's victor. Each case was dismissed. In 2003 air historian Andy Saunders wrote a book "Bader's Last Flight". Saunders' research suggests that Bader may have been a victim of friendly fire, shot down by one of his fellow RAF pilots after becoming detached from his own squadron. RAF combat records indicate Bader may have been shot down by Flight Lieutenant "Buck" Casson who had claimed a Bf 109 that day. In a letter to Bader on 28 May 1945, Casson explained the action. Saunders stated that this was not absolute proof, and that it would be helpful to find the "Bader Spitfire".

Combat Credo

Bader attributed his success to the belief in the three basic rules, shared by the German Ace Erich Hartmann. "If you had the height, you controlled the battle." "If you came out of the sun, the enemy could not see you." "If you held your fire until you were very close, you seldom missed."

Prisoner of War

The Germans treated Bader with great respect. When Bader was taken prisoner, he was sent to a hospital near Saint-Omer, near the place where Bader's father's grave is located. On leaving the hospital, Colonel Adolf Galland and his pilots invited him on to their airfield and they received him as a friend. Bader was cordially invited to sit in the cockpit of Galland's personal Me109. Bader asked Galland if it was possible to test the 109 by "a flight around the airfield". Galland refused him – with laughter!

Bader had lost a prosthetic leg when escaping his disabled aircraft. General Adolf Galland notified the British, and offered them safe passage to drop off a replacement. Hermann Goring himself gave the green light for the operation. The British responded on 19 August 1941 with the "Leg Operation". An RAF bomber was allowed to drop a new prosthetic leg by parachute to St Omer, a Luftwaffe base in occupied France.

Escape From the Hospital

Bader escaped from the hospital where he was recovering by tying together sheets. A French maid at the St. Omer hospital attempted to get in touch with British agents to enable Bader to escape to Britain. Eventually, he escaped out of a window. The plan worked initially. Bader completed the long walk to the pre-fixed safe house despite wearing a British uniform. Unfortunately for him, the plan was betrayed by another woman at the hospital. He hid in the garden when a German staff car arrived at the house, but was found later. Bader denied that the couple had known he was there. They, along with the French woman at the hospital, were sent for forced labour in Germany. The couple survived. After the war, French authorities sentenced the woman informer to 20 years in prison.

Bader sitting middle in Colditz.
Image Source: Wikipedia

Bader made so many escape attempts that the Germans threatened to take away his legs. In August 1942, Bader escaped with Johnny Palmer and three others from the camp in Sagan, but was recaptured. He was finally dispatched to the "escape-proof" Colditz Castle near Leipzig, Dresden on 18 August 1942, where he remained until 15 April 1945 when it was liberated by the US Army.

Last years in the RAF

After his return to Britain, Bader was given the honour of leading a victory flypast of 300 aircraft over London in June 1945. On 1 July, he was promoted to temporary Wing Commander. Soon after, Bader was looking for a post in the RAF. Bader was made

Douglas Bader leaves Buckingham Palace after receiving new bars to his DSO and DFC, 27th November 1945. Accompanying him are his wife Thelma (right) and Thelma's sister Jill Addison. Photo by Keystone/Hulton Archive/Getty Images.
Image Source: doncasterfreepress.co.uk

Commanding Officer of the Fighter Leader's School. He received a promotion to war substantive Wing Commander on 01 December, and soon after was promoted to temporary Group Captain. Unfortunately for Bader, the fighter aircraft's roles had now expanded significantly and he spent most of his time instructing on ground attack and co-operation with ground forces. Also, Bader did not get on with the newer generation of Squadron Commanders who considered him to be "out of date". Bader's enthusiasm for continued service in the RAF waned. On 21 July 1946, Gp Capt Bader retired from the RAF.

Bader entering the Cockpit with Artificial Leg.
Image Source: cinconoticias.com

Post RAF Career

Bader considered politics, and standing as a Member of Parliament for his home constituency in the House of Commons. He despised how the three main political parties used war veterans for their own

Douglas with second wife Joan after being knighted in 1976.
Image Source: sundaypost.com

political ends. Instead, he resolved to join Shell Company, who had taken him at age 23, after his accident, though some others offered more money. Joining Shell would allow him to continue flying a company-owned Percival Proctor and later a Miles Gemini. Bader became Managing Director of Shell Aircraft until he retired in 1969. That same year, he also served as a technical advisor to the film, Battle of Britain. Bader travelled to every major country outside the Communist world becoming internationally famous and a popular after-dinner speaker on aviation matters.

Personal Charm and Traits

When the film "Reach for the Sky" was released, people associated Bader with the quiet and amiable personality of actor Kenneth More. The producers had deleted all those habits he displayed when on operations, particularly his prolific use of bad language. In reality many thought that he was a somewhat 'difficult' person. Nevertheless, Bader was viewed as a legendary figure by the wider public. He had a force of his personality. It slightly unsettled him that people indignantly questioned his overbearing personality and then applied normal standards on a man who had lost both his legs and yet came back to fly in the cockpit of wartime aircraft. Never a person to hide his opinions, Bader also became controversial for his political interventions. Bader was known, at times, to be head-strong, blunt and unsophisticated when he made his opinion known.

Sir Douglas Bader pictured in 1982 with a remote controlled spitfire.
Image Source: http://veteransinasia.weebly.com/

Personal Life

Douglas Bader House in Fairford is now the headquarters for the RAF Charitable Trust.
Image Source: Wikipedia

Bader's first wife, Thelma, developed throat cancer in 1967. Thelma was a smoker, and although she stopped smoking, it did not save her. After a long battle, she died on 24 January 1971, aged 64. Bader married Joan Murray on 3 January 1973. Joan was the daughter of a steel tycoon. She had an interest in riding and was a member of the British Limbless Ex-Servicemen's Association. She also helped associations involved in riding for the disabled. Bader campaigned vigorously for people with disabilities and set an example of how to overcome a disability.

Honours and Awards

01 October, 1940, Acting Squadron Leader Bader (26151) appointed a Champion of the Distinguished Service Order, for displaying gallantry and leadership of the highest order. Led the squadron with such skills that 33 enemy aircraft were destroyed, six by himself. On 7 January 1941, Acting Squadron Leader Bader, DSO, No.242 Squadron was awarded the Distinguished Flying Cross for continuing to lead his squadron and wing with utmost gallantry, and by now destroying 10 and damaging many hostile aircraft. On 15 July 1941, acting Wing Commander Bader, DSO, and DFC was awarded a bar to the DSO, for consistently

successful sorties over enemy territory, and by now destroying 15 hostile aircraft. 9 September 1941, acting Wing Commander Bader, DSO & Bar, and DFC was awarded a bar to the DFC in recognition of gallantry displayed in flying operations against the enemy. On 02 January 1956, Group Captain Bader, DSO & Bar, DFC & Bar was appointed a Commander of the Most Excellent Order of the British Empire for services to the disabled. On 12 June 1976—Group Captain Bader, CBE, DSO, DFC was made a Knight Bachelor for services to disabled people. In 1977 he was made a fellow of the Royal Aeronautical Society. He also received a Doctorate of Science from Queen's University Belfast.

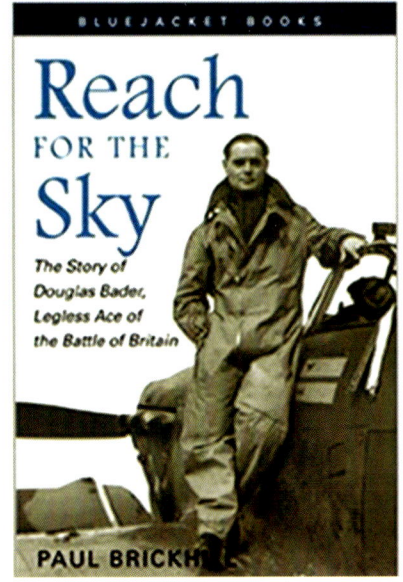

Reach for the Sky: The Story of Douglas Bader.
Image Source: Amazon.com

Last Flight

Bader's health was in decline in the 1970s, and he soon gave up flying altogether. On 4 June 1979, Bader flew his Beech 95 Travel Air for the last time, the aircraft having been gifted to him on his retirement from Shell. He had recorded 5,744 hours and 25 minutes flying time.

With Adolf Galland. Screenshots of Douglas Bader "This Is Your Life". An event to honour Bader.
Image Source: bigredbook.info

On 5 September 1982, after a dinner honouring Marshal of the Royal Air Force Sir Arthur "Bomber" Harris at the Guildhall, at which he spoke, Bader died of a heart attack while being driven on his way home. Among the many dignitaries and personalities at his funeral was Adolf Galland. Galland and Douglas Bader had shared a friendship that spanned more than 42 years since their first meeting in France. On the 60th anniversary of Bader's last combat sortie, his widow Joan unveiled a 6 ft bronze sculpture statue at Goodwood, the aerodrome from which he took off. The Douglas Bader Foundation was formed in honour of Bader in 1982 by family and friends, and many former RAF pilots who had flown with Bader. One of Bader's artificial legs is kept by the RAF Museum at their warehouse, and is not on public display. Airfields, roads, institutions, and pubs are named after Bader.

Group Captain Douglas Bader playing golf, with his wife Joan Murphy (on the Right) and to the left is Mr Peter Cadbury and Mrs John Beck.
Image Source: dailymail.co.uk

REFERENCES

1. Douglas Bader https://www.cinconoticias.com/douglas-bader-piloto-sin-piernas/
2. Prisoner of War, Royal Air Force Museum, https://www.rafmuseum.org.uk/research/online-exhibitions/douglas-bader-fighter-pilot/prisoner-of-war.aspx
3. Douglas Bader: Fighter Pilot, Royal Air Force Museum, https://www.rafmuseum.org.uk/research/online-exhibitions/douglas-bader-fighter-pilot.aspx

4. Biographies Douglas Bader, Sky History, https://www.history.co.uk/biographies/douglas-bader
5. Battle of Britain: how the British press found a hero in Douglas Bader – the amputee fighter ace, The Conversations, July 6, 2020 https://theconversation.com/battle-of-britain-how-the-british-press-found-a-hero-in-douglas-bader-the-amputee-fighter-ace-142069
6. Douglas Bader, Wikipedia https://en.wikipedia.org/wiki/Douglas_Bader
7. Bader, Sir Douglas (1910-1982), English Heritage Blue Plaques, https://www.english-heritage.org.uk/visit/blue-plaques/douglas-bader/
8. Douglas Bader Story, Royal Air Force Benevolent Foundation, https://www.rafbf.org/leavealegacy/douglas-bader
9. Brad Lendon, CNN, August 30, 2020 https://edition.cnn.com/2020/08/29/europe/british-world-war-ii-pilot-douglas-bader-intl-hnk-dst/index.html
10. Sir Douglas Bader – December 1981, YouTube, https://www.youtube.com/watch?v=6UQgm39HNrQ
11. Group Captain Douglas Bader http://www.littleshelfordhistory.co.uk/little-shelford-people/sir-douglas-bader
12. Greg Syers IMDb Mini Biography, Douglas Robert Steuart Bader, https://www.imdb.com/name/nm0046035/bio
13. Douglas Bader Foundation https://www.douglasbaderfoundation.com/about-us/sir-douglas-bader/
14. This Day in Aviation, Group Captain Sir Douglas R.S. Bader, February 21, 2020. https://www.thisdayinaviation.com/tag/douglas-bader/

3

Manfred von Richthofen "Red Baron"

Ace of Aces

Portrait photo of the romantic and daring young 'Baron' Manfred von Richthofen.
Image Source: flickr.com

"Of course, with the increasing number of aeroplanes one gains increased opportunities for shooting down one's enemies, but at the same time, the possibility of being shot down one's self increases".

– **Manfred von Richthofen, the Red Baron**

Manfred Albrecht Freiherr von Richthofen, more commonly known as Baron von Richthofen, and famous as the "Red Baron", was a fighter pilot with the Deutsche Luftstreitkräfte (German Air Force, air arm of the Imperial German Army) during World War I. He is considered the ace-of-aces for his 80 air combat victories. Originally a cavalryman, Richthofen transferred to the Air Service in 1915, and became one of the first members of fighter squadron Jagdstaffel 2 in 1916. He quickly distinguished himself as a fighter pilot, and during 1917 became leader of the larger fighter wing better known as "The Flying Circus" or "Richthofen's Circus" because of the bright colours of its aircraft, and perhaps also because

Fokker Tri Plane.
Image Source: History.com

of the way the unit was transferred from one area of allied air activity to another, moving like a travelling circus, and frequently setting up in tents on improvised airfields. By 1918, Richthofen was regarded as a national hero in Germany, and respected by his enemies. Richthofen was shot down and killed near Vaux-sur-Somme on 21 April 1918. He remains one of the most widely known fighter pilots of all time, and has been the subject of many books and films.

Richthofen was a Freiherr (literally "Free Lord"), a title of the nobility often translated as "baron". All male members of the family were entitled to it. Richthofen painted his aircraft red, and this combined with his title led to him being called "The Red Baron". Some also called him the "The Red Battle Flyer" or "The Red Fighter Pilot".

Early Years and Initial War Service

Richthofen was born in Kleinburg, now part of the city of Wroclaw Poland, on 2 May 1892 into a prominent Prussian aristocratic family. His father was a Major. As a child, he enjoyed riding horses, hunting and gymnastics at school. After initial school, he began military training when he was 11. After cadet training in 1911, he joined a cavalry unit. When WW I began, Richthofen served as a cavalry reconnaissance officer, and saw action on both fronts, in Russia, France, and Belgium. With the advent of trench warfare, traditional cavalry operations became inefficient, and Richthofen's regiment was dismounted, serving as dispatch runners and field telephone operators.

Manfred von Richthofen with his father. Image Source: Pinterest Anna Leas saved to Aerodrome

Richthofen as Cadet in Cavalry Regiment Nr. 1, 1912.
Image Source: spiegel.de

Disappointed and bored at not being able to directly participate in combat, his interest in the Air Service got aroused when he saw German military aircraft. He applied for a transfer to Imperial German Army Air Service, later called Luftstreitkräfte. He reportedly wrote in his application for transfer, "I have not gone to war in order to collect cheese and eggs, but for another purpose." In spite of this unmilitary attitude, his request was granted, and he joined the flying service at the end of May 1915 as an observer. From June to August 1915, Richthofen served as an observer on reconnaissance missions over the Eastern front. Later, on being transferred to the Champagne front, he is believed to have shot down an attacking French Farman aircraft with his observer's machine gun in a tense battle. He was not credited with the kill, since it fell behind Allied lines and therefore could not be confirmed.

Flying Pilot Career

Manfred had a chance meeting with German ace fighter pilot Oswald Boelcke which led him to enter

Albatros D.III.
Image Source: Wikipedia

training as a pilot in October 1915. The following month, Manfred joined the No. 2 Bomber Squadron flying a two-seater Albatros D.III. Initially, he appeared to be a below-average pilot. He struggled to control his aircraft, and he crashed during his first flight at the controls. Despite this poor start, he rapidly became attuned, and on 26 April 1916 he shot down a French Nieuport aircraft, although he received no official credit. Richthofen met Oswald Boelcke again in August 1916, after another spell flying two-seaters on the Eastern Front. Boelcke was in search of candidates for his newly formed Jasta 2 (Squadron 2), the most well-known squadron of WW I, and he selected Richthofen to join this unit. Jasta 2 was also known as "Jasta Boelcke" and it was the incubator of several notable aviation careers. Boelcke was unfortunately killed during a mid-air collision with a friendly aircraft on 28 October 1916, and Richthofen was an eyewitness.

First Confirmed Aerial Victory and Initial Tactics

Richthofen scored his first confirmed aerial victory over Cambrai, France, on 17 September 1916. His autobiography states, "I honoured the fallen enemy by placing a stone on his beautiful grave." He contacted a jeweller in Berlin and ordered a silver cup engraved with the date and the type of enemy aircraft. He continued to celebrate each of his victories in the same manner until he had 60 cups, by which time the dwindling supply of silver in blockaded Germany meant that silver cups could no longer be supplied. His brother Lothar who had 40 victories flew risky aggressive tactics. Manfred was not a spectacular or aerobatic

The brothers Manfred and Lothar von Richthofen
Image Credit: The Imperial War Museum in London. Image Source: kumc.edu

pilot like his brother. He was a noted tactician and formation leader and a fine marksman. Typically, he would dive from above to attack with the advantage of the sun behind him, with other pilots of his squadron covering his rear and flanks.

Shooting British Ace Major Hawker

On 23 November 1916, Richthofen shot down his most famous adversary, British Air Ace Major Lanoe Hawker, Victoria Cross, whom he described as "the British Boelcke". Richthofen was flying an Albatros D.II and Hawker was flying the older Airco DH.2. After a long dogfight,

British Air Ace Major Lanoe Hawker.
Image Source: Wikipedia

Hawker was shot in the back of the head as he attempted to escape back to his own lines. After this combat, Richthofen was convinced that he needed a fighter aircraft with more agility, even if with lesser speed. He switched to the Albatros D.III in January 1917, scoring two victories before suffering an in-flight crack in the spar of the aircraft's lower wing on 24 January, and he had to revert back to the Albatros D.II.

The Aircraft of His Choice

On 6 March, flying his Halberstadt aircraft in combat with British F.E.8, Richthofen's aircraft was shot through the fuel tank. He was able to force land without his aircraft catching fire. He switched back to Albatros D.III on 2 April 1917 and scored 22 victories by June. Richthofen flew the celebrated Fokker Dr.I triplane from late July 1917, the distinctive three-winged aircraft with

Actual photo of 'The Red Baron' landing his Fokker DL1 tri-plane
Image Credit: The Imperial War Museum in London. Image Source: kumc.edu

which he is most commonly associated. He asked for the strengthening of wings in November. Only 19 of his 80 kills were made in this type of aircraft, despite the popular link between Richthofen and the Fokker Dr.I. It was his Albatros D.III Serial No. 789/16 that was first painted bright red, in late January 1917, and in which he first earned his name and reputation. Meanwhile Richthofen championed the development of the Fokker D.VII with suggestions to overcome the deficiencies of the current German fighter aircraft. However, he never had an opportunity to fly the new type in combat, as he was killed before it entered service.

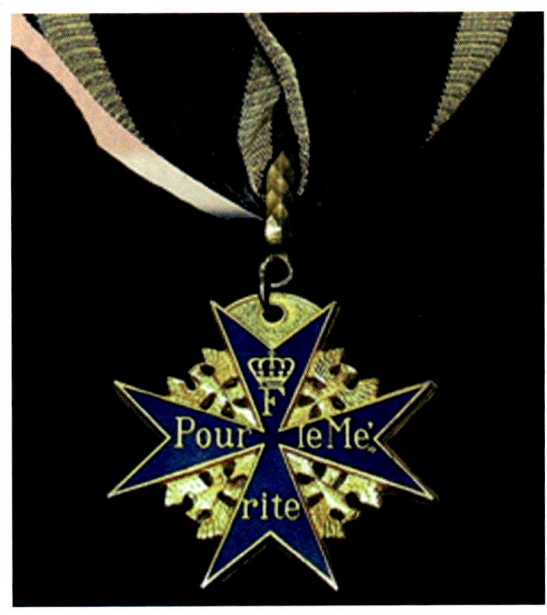

The Blue Max

Richthofen received the coveted Pour le Merite in January 1917 after his 16th confirmed kill, the highest military honour in Germany at the time and informally known as "The Blue Max". The medal was called "Blue Max" in honour of the first aviator to win the medal 'Max Immelmann' (15 Aerial victories), on whose name is also the common flying tactic, the Immelmann turn. That same month, he assumed command of Jasta 11 which ultimately included some of the elite German pilots, many of whom he trained himself, and several of whom later became leaders of their own squadrons. When Lothar joined the unit, the German high command appreciated the propaganda value of two Richthofens fighting together to defeat the enemy in the air.

Pour Le Merite "Blue Max".
Image Source: Wikipedia

Aircraft Painted Red

Richthofen took the flamboyant step of having his Albatros painted red when he became a Squadron Commander. His autobiography states, "For whatever reasons, one fine day I came upon the idea of having my craft painted glaring red. The result was that absolutely everyone could not help but notice my red bird. In fact, my opponents also seemed to be not entirely unaware of it". Thereafter he usually flew in red-painted aircraft, although not all of them were entirely red, nor was the "red" necessarily the brilliant scarlet beloved of model- and replica-builders. Other members of the unit soon took to painting parts of their aircraft red. Their official reason was to make their leader less conspicuous, and to avoid having him singled out in a fight. In practice, red colour became a unit identification. Other units soon adopted their own squadron colours. The German high command permitted this practice (in spite of obvious drawbacks from the point of view of intelligence), and German propaganda made much of it by referring to Richthofen as "the Red Fighter Pilot."

Jasta 11 Pilots, 1917. From left to right around Sebastian Festner, Karl Emil Schäfer, Manfred von Richthofen, Lothar von Richthofen, and Kurt Wolff.
Image Source: New York Public Library, sciencesource.com

The Bloody April and the Flying Circus

Richthofen led his unit from the front, and with unparalleled success. In the "Bloody April" 1917, alone, he shot down 22 British aircraft, including four in a single day, raising his official tally to 52. By June, he had become the Commander of the first of the new larger "fighter wing" formations; these were highly mobile, combined tactical units that could move at short notice to different parts of the front as required. Richthofen's new command, J.G.1, was composed of fighter squadrons No. 4, 6, 10, and 11. J.G. 1 became widely known as "The Flying Circus" due to the unit's brightly coloured aircraft and the mobility, including the use of tents, trains, and caravans.

The brightly painted Albatros fighters of Jagdgeschwader 1, Richthofen's "Flying Circus."
Image Source: historynet.com

Brilliant Tactician

Richthofen was a brilliant tactician, who built on Boelcke's tactics. Unlike Boelcke, however, he led by example and force of will rather than by inspiration. He was often described as distant, unemotional, and rather humourless, though some colleagues contended otherwise. He taught his pilots the basic rule which he wanted them to fight by: "Aim for the man and don't miss him. If you are fighting a two-seater, get the observer first; until you have silenced the gun, don't bother about the pilot."

Red Baron.
Credit: ullstein bild/Getty Image Source: HistoryNet

Richthofen in the cockpit, with other members of Jasta 11, including his brother Lothar (sitting, front).
Image Source: Wikipedia

Why He Remained a Captain Only

Although Richthofen was now performing the duties of a Lieutenant Colonel (Wing Commander), he was never promoted past the relatively junior rank of Rittmeister, equivalent to Captain in the British army. In the German army, it was not unusual for a wartime officer to hold a lower rank than his duties. It was also the custom for a son not to hold a higher rank than his father, and Richthofen's father was a reserve Major.

Wounded in Combat

Richthofen sustained a serious head wound on 6 July 1917, during combat against a formation of F.E.2d British fighters, causing instant disorientation and temporary partial blindness. He regained his vision in time to ease the aircraft out of a spin and execute a forced landing in a field in friendly territory. The injury required multiple operations to remove bone splinters from the impact area. The Red Baron returned to active service against doctor's orders on 25 July, but went on convalescent leave from 5 September to 23 October. His wound had caused lasting damage, and he later often suffered from post-flight nausea and headaches.

The Baron prepares for a flight over British lines in his Fokker Dr.I Triplane.
Image Source: historynet.com

The Legend and Hero

By 1918, Richthofen had become such a legend that it was feared that his death would be a blow to the morale of the German people. He refused to accept a ground job after his wound, stating that "every poor fellow in the trenches must do his duty" and that he would therefore continue to fly in combat. Certainly he had become part of a cult of officially supported hero-worship. German propaganda circulated various rumours, including that the British had raised squadrons specially to hunt Richthofen and had offered large rewards and an automatic Victoria Cross to any Allied pilot who shot him down.

"Von Richthofen's notoriety grew with each new victory. Eventually, he became the most famous (and feared) pilot of the war."
Image Source: militaryhistorynow.com

Final Combat and Fatal Wound

Richthofen received a fatal wound just after 11:00 am on 21 April 1918 while flying. At the time, he had been pursuing, at very low altitude, a Sopwith Camel piloted by novice Canadian pilot Wilfrid May. May had just fired on the Red Baron's cousin Lt. Wolfram von Richthofen. On seeing his cousin being attacked, Manfred flew to his rescue and fired on May, causing him to pull away. Richthofen pursued May. The Baron was spotted and briefly attacked by other aircraft. Richthofen disengaged and then resumed his pursuit of May. It was during this pursuit, a single .303 bullet hit Richthofen, damaging his heart and lungs so severely that it may have caused a quick death. In the last seconds of his life, he managed to retain sufficient control to make a rough landing in a field on a hill near in a sector defended by the Australian Imperial Force. There were several witnesses, and many claimed to have been the first to reach the triplane, and Richthofen's last words, generally including the word "kaputt". His Fokker Dr.I (425/17) was not badly damaged by the landing, but it

Australian soldiers pose with remains of von Richthofen's famous tri-plane.
Image Source: militaryhistorynow.com

was soon taken apart by souvenir hunters. In 2009, Richthofen's death certificate was found in the archives in Ostrów Wielkopolski, Poland. He had briefly been stationed in Ostrów before going to war, as it was part of Germany until the end of World War I. The document is a one-page, handwritten form in a 1918 registry book of deaths. It simply states that he had "died 21 April 1918, from wounds sustained in combat".

Dramatisation of the final moments of the Red Baron's Life.
Image Source: Flickr (militaryhistorynow.com)

Who Shot Red Baron?

Controversy continues to surround the identity of the person who fired the shot that actually killed Richthofen. The RAF credited Lt. Arthur Roy Brown with shooting down the Red Baron, but it is now generally agreed that the bullet which hit Richthofen was fired from the ground. Richthofen died following an extremely serious and inevitably fatal chest wound from a single bullet, penetrating from the right armpit and resurfacing next to the left nipple. Brown's attack was from behind and above, and from Richthofen's left. Even more conclusively, Richthofen could not have continued his pursuit of May for as long as he did (up to two minutes) had this wound come from Brown's guns. Brown himself never spoke much about what happened that day, claiming, "There is no point in me commenting, as the evidence is already out there." Following an autopsy most believed that some AA machine gunner had killed Richthofen, but "Who" has not been established.

Theories about Last Combat

Richthofen was a highly experienced and skilled fighter pilot, and fully aware of the risk from ground fire. Further, he concurred with the rules of air fighting created by his late mentor Boelcke, who specifically advised pilots not to take unnecessary risks. In this context, Richthofen's judgment during his last combat was clearly unsound in several respects. Some contend that Richthofen's earlier brain injury would have affected

his lack of judgement on his final flight, and thus flying too low over enemy territory and suffering target fixation. Richthofen may have been suffering from cumulative combat fatigue, which made him fail to observe some of his usual precautions.

Ceremonial Burial and Final Cemetery

No. 3 Squadron RAF officers were pallbearers during the Red Baron's funeral.
Image Source: Wikipedia

Major Blake, who was responsible for Richthofen's body, regarded the Red Baron with great respect, and he organised a full military funeral. The body was buried in the cemetery at the village of Bertangles, near Amiens, on 22 April 1918. RFC became the Royal Air Force (RAF) on 01 April 1918. RAF Squadron officers served as pallbearers. There was a guard of honour and other ranks fired a salute. Allied squadrons stationed nearby presented memorial wreaths, one of which was inscribed with the words, "To Our Gallant and Worthy Foe". In the early 1920s the French authorities created a military cemetery at Fricourt, in which a large number of German war dead, including Richthofen, were reinterred. In 1925 von Richthofen's youngest brother, Bolko, recovered the body from Fricourt and took it to Germany. The German Government requested that the body should be interred at the Invalidenfriedhof Cemetery in Berlin, where many German military heroes and past leaders were buried, and the family agreed. Richthofen's body received a state funeral. Later the Third Reich held a further grandiose memorial ceremony at the site of the grave, erecting a massive new tombstone engraved with the single word: Richthofen. During the Cold War the Invalidenfriedhof was on the boundary of the Soviet zone in Berlin, and the tombstone became damaged by bullets fired at attempted escapees from East Germany. In 1975 the body was moved to a Richthofen family grave plot at the Sudfriedhof in Wiesbaden.

Richthofen family grave at the Südfriedhof in Wiesbaden.
Image Source: Wikipedia

Red Barons Painting.
Credit: Deutsches Historisches Museum, Berlin

Richthofen's Victories Authenticated

Some authors initially questioned Richthofen's 80 victories, insisting that his record was exaggerated for propaganda purposes. Some said that he took credit for aircraft downed by his Squadron or Wing. Truth is that Richthofen's victories are unusually well documented. A full list was published as early as 1958, with documented RFC/RAF squadron details, aircraft serial numbers, and the identities of Allied airmen killed or captured. 73 of the 80 listed matched records with British

losses. There were also unconfirmed victories that would put his actual total as high as 100 or more. For comparison, the highest-scoring Allied ace, the Frenchman Rene Fonck, achieved 75 confirmed victories. The highest-scoring British Empire fighter pilot was Canadian Billy Bishop, who was officially credited with 72 victories. Richthofen's early victories and the establishment of his reputation coincided with a period of German air superiority, but he achieved many of his successes against a numerically superior enemy, who flew fighter aircraft that were, on the whole, better than his own.

Honours and Tributes

Captain Roy Brown donated the seat of the Fokker tri-plane in which the German flying ace made his final flight to the Royal Canadian Military Institute (RCMI) in 1920. The engine of Richthofen's Dr. I was donated to the Imperial War Museum in London, where it is still on display. The museum also holds the Baron's machine guns. The control column (joystick) of Richthofen's aircraft can be seen at the Australian War Memorial in Canberra. Several German military aviation Wings/Units were named after the Baron. Jagdgeschwader 71, the first jet fighter unit established by the post-World War II Germany in June 1959, whose founding commander was the most successful Air Ace in history, Erich Hartmann, was named after Richthofen. In 1968, von Richthofen was inducted into the International Air & Space Hall of Fame. "Red Flag", the US Air Force's counterpart to "Top Gun", was an outcome of "Project Red Baron", which evolved in three phases during the period of the Vietnam War.

Summarise: the Great Air Ace

Manfred von Richthofen, The Red Baron, the German fighter pilot, was the deadliest flying ace of World War I. During a 19-month period between 1916 and 1918, the Prussian aristocrat shot down 80 Allied aircraft and won widespread fame for his scarlet-coloured airplanes and ruthlessly effective flying style. On September 17, 1916, while on patrol over France, Richthofen got the drop on a two-seater British plane and scored his first confirmed kill. "I gave a short series of shots with my machine gun," he later wrote of the dogfight. "I had gone so close that I was afraid I might dash into the Englishman". I never get into an aircraft for fun," Manfred von Richthofen once wrote. "I aim first for the head of the pilot, or rather at the head of the observer, if there is one." It was a maxim that the German aviator followed with ruthless precision. The "Red Baron" inspired both terror and admiration in his Allied adversaries. He also became a potent propaganda symbol in Germany, where he was worshiped as a national hero. German General Erich Ludendorff once remarked that Richthofen "was worth as much to us as three divisions."

Like many pilots, he also had the morbid habit of scrounging souvenirs from the planes he downed. Along with the heads of the animals he killed on hunting trips, his home was decorated with fabric serial numbers, instruments and machine guns looted from Allied wreckage. He even had a chandelier made from the engine of a French plane. Rather than engaging in airborne acrobatics or risky dogfights, he preferred to patiently stalk his enemies, swoop down from high altitude and then blast them out of the sky with pinpoint bursts of machine gun fire. "There is no art in shooting down an aeroplane," he wrote. "The thing is done by the personality or by the fighting determination of the airman." The Circus's "ringmaster," Richthofen became a beloved celebrity. The Red Baron had been the Allied

The Red Baron: The Life and Legacy of Manfred von Richthofen Paperback – October 22, 2014 By Charles River Editors

pilots' most hated adversary, yet in death, he was honoured like a fallen hero. "Anybody would have been proud to have killed Richthofen in action," a correspondent for the British magazine "Aeroplane" later wrote, "but every member of the Royal Flying Corps would also have been proud to shake his hand had he fallen into captivity alive." When Richthofen's body was taken to a British airplane hangar, airmen turned out in droves to pay their last respects. As a sign of respect for the war's most lethal pilot, a wreath was placed on his grave that read: "To Our Gallant and Worthy Foe."

REFERENCES

1. Ace for the Ages: World War I Fighter Pilot Manfred von Richthofen. History Net. https://www.historynet.com/red-baron-world-war-i-ace-fighter-pilot-manfred-von-richthofen.htm
2. Manfred von Richthofen, Wikipedia, https://en.wikipedia.org/wiki/Manfred_von_Richthofen
3. Manfred, baron von Richthofen, German aviator, Britannica, https://www.britannica.com/biography/Manfred-Freiherr-von-Richthofen
4. History.com Authors, Red Baron, History.com, August 21, 2018, https://www.history.com/topics/world-war-i/manfred-baron-von-richthofen
5. The Death of The Red Baron, Frederick Holmes, MA, MD, FACP Professor of Medicine Emeritus and of The History of Medicine University of Kansas School of Medicine, April 08, 2019, http://www.kumc.edu/wwi/biography/red-baron.html
6. Who Killed The Red Baron? German fighter pilot WW1 Manfred von Richthofen, YouTube, January 05, 2020. https://www.youtube.com/watch?v=L_8EtrVgfF4
7. Who's Who – Manfred von Richthofen, firstworldwar.com, https://www.firstworldwar.com/bio/richthofen.htm
8. Jennifer Rosenberg, Biography of Manfred von Richthofen, 'The Red Baron'. ThoughtCo, August 28, 2019, https://www.thoughtco.com/the-red-baron-1779208
9. Don Hollway, The Red Baron, History Magazine, October/November 2015, http://www.donhollway.com/redbaron/?sa=X&ved=0CCoQ9QEwCWoVChMIgK2_iaD9xgIVwr4UCh2cWQA1

4

"Johnnie" Johnson

Highest Scoring Western Fighter Ace in WW II against Germany

Wing Commander Johnnie Johnson with his Spitfire.
Image Source: starduststudios.com

Air Vice Marshal James Edgar "Johnnie" Johnson, CB, CBE, DSO & Bar, DFC & Bar, was a Royal Air Force (RAF) flying Ace of World War II. He was a qualified engineer, and a passionate rugby player. Johnson was interested in aviation. When applied to join the RAF, he was initially rejected, first on social grounds, and then on medical grounds, because of a rugby injury. He was eventually accepted in August 1939. The injury problems, however, returned during his early training and flying career, resulting in him missing the Battle of France and Battle of Britain between May and October 1940. In 1940 Johnson had an operation to reset his collarbone, and began flying regularly. He took part in the offensive against Germany from 1941 to 1944. Johnson was involved in heavy aerial fighting during this period. He became a Group Captain by the end of the war.

Johnson flew nearly 1000 operational sorties and engaged enemy aircraft on 57 occasions. He was credited with 34 individual victories, seven shared, three shared probable, 10 damaged, three shared damaged and one destroyed on the ground. Included in his list of individual victories were 14 Messerschmitt Bf-109s, and 20 Focke-Wulf Fw 190s destroyed, making him the most successful RAF ace against the Fw 190. His score made him the highest scoring Western Allied fighter Ace against the German Luftwaffe. Johnson later served in the Korean War before retiring in 1966 as an Air Vice Marshal. Johnnie Johnson died of cancer in 2001.

Johnnie-Johnson. National Portrait Gallery.
Image Source: npg.org.uk

The Youth

Last of the British Dambusters.
Image Source: BBC Interview

Johnson was born on 9 March 1915 in Barrow upon Soar, Leicestershire. His father, Alfred was a policeman. Johnson's uncle, Edgar Charles Rossell, who had won the Military Cross with the Royal Fusiliers in 1916, paid for Johnson's education. Johnson was nearly expelled from school after refusing punishment for a misdemeanour, believing it to be unjustified. He was very principled and simply dug his heels in. Among Johnson's hobbies and interests were shooting and sports; he shot rabbits and birds in the local countryside. Johnson qualified as a civil engineer at age 22. He became a surveyor, and later an assistant engineer. In 1938, Johnson broke his collarbone playing rugby. The injury was wrongly set and did not heal properly, which later caused him difficulty at the start of his flying career.

Struggle to Join the RAF

Johnson started taking flying lessons at his own expense. In 1937, "Johnnie" Johnson tried to join the Auxiliary Air Force (AAF). On hearing that he came from Melton Mowbray, Leicestershire, the interviewing officer said, "My dear chap, you're just the type. Which hunt do you follow?" When Johnnie said he did not even ride a horse, he was promptly shown the door. Little did that interviewing officer think he had just rejected the man who, in the Second World War, would shoot down more of the enemy than any other pilot in the RAF – and without ever being shot down himself. Johnson felt he was rejected on the grounds of his class status.

The prospect of war increased, and the criteria for applicants changed as the RAF expanded and brought in men from ordinary social backgrounds. Johnson re-applied to the AAF. He was informed that sufficient pilots were already available but there were some vacancies in the balloon squadrons. Johnson rejected the offer. Realising that the AAF was at that time an exclusive club, Johnson then applied to join the Royal Air Force Volunteer Reserve (RAFVR), which was a means to enter the RAF for young men with ordinary backgrounds. All volunteer aircrew were made Sergeant on joining with the possibility of a commission. But, once again he was rejected, because there were too many applicants for vacancies and his shoulder injury made him unsuitable for flight operations. He then joined the Territorial Army as a reserve, to be called in case of war.

Persuasion and War Clouds Supports His joining RAF

With war clouds in the horizon, in August 1939, Johnson was finally accepted by the RAFVR and began training at weekends at an RAF satellite airfield. Johnson trained on the de Havilland Tiger Moth biplane. Upon the outbreak of war in September 1939, Johnson entrained for Cambridge. He arrived at the 2nd Initial Training Wing to begin flight instruction. After many "ifs and buts", Johnson was finally selected for fighter pilot training and given the service number 754750 with the rank of Sergeant. By December 1939, Johnson began his initial training at 22 EFTS (Elementary Flying Training School), Cambridge. On 29 February 1940, Johnson flew solo for the first time. He moved to 5 FTS at Sealand and later to 7 OTU (Operational Training Unit) at RAF Hawarden in Wales. He received his "wings" on 7 August 1940, and was immediately inducted into the General Duties Branch of the RAF as a pilot officer with 55 hours and 5 minutes solo flying.

Broken Shoulder: Tough Days on Spitfire

On 19 August 1940, Johnson flew a Spitfire for the first time, and began operational flying training. During his training flights, he stalled and crashed a Spitfire. Johnson had his harness straps too loose. The shoulder got wrenched revealing that his earlier rugby injury had not healed properly. The Spitfire did a ground loop, ripping off one of the undercarriage legs and forcing the other up through the port main plane. The Commanding Officer (CO) forgave Johnson, for the short airfield was difficult to land on even for an experienced pilot. Johnson was worried that he would

De Havilland Tiger Moth biplane.
Image Source: Wikipedia Commons

1943 Spitfire Mk IX JE-J EN398 personal aircraft of Johnnie Johnson.
Image Source: aviacion.tumblr.com

be under close watch, and could not make another mistake. Johnson packed his injured shoulder with wool, held in place by adhesive tape. He also tightened the straps to reduce vibrations while flying. The measures proved useless and Johnson found he had lost feeling in his right hand. When he dived the pressure changes aggravated his shoulder. He often tried to fly using his left hand only, but Spitfires had to be handled with both hands during anything other than simple manoeuvres. Despite the difficulties with his injuries, on 28 August 1940, he completed the course, and now had over 200 hours in his log book, including 24 on the Spitfire.

Initial Tactical Flying

In August 1940, he was briefly posted to No. 19 Squadron as a probationary Pilot Officer. On 6 September 1940 Johnson was posted to No. 616 Squadron, where he learnt the technique of deflection shooting and how to take a killing shot from line-astern or near line-astern positions. He also learnt that the duty of the No. 2 was not to shoot down enemy aircraft but to ensure the leader's tail was safe. He learnt the importance of correct battle formation and the tactical use of sun, cloud and height. Five days later, Johnson flew an X-Raid patrol in a Spitfire, qualifying for the Battle of Britain Clasp.

Injury Resurfaces: The Shoulder Operation

Johnson's old injury continued to trouble him and he found flying high performance aircraft like the Spitfire extremely painful. RAF medics gave him two options; he could have an operation that would correct the problem, but this meant he would miss the Battle of Britain, or becoming a training instructor flying the light Tiger Moth. Johnson opted for the operation. He was taken off flying duties and sent to a RAF Hospital. He returned to the squadron only on 28 December 1940. He flew a test flight with his CO and was cleared for further flying. And from then on Johnson became a deadly killing machine, not only the master of the Spitfire but also – unlike almost everyone else – a master of accurate deflection shooting, learned against agile rabbits.

Johnnie Johnson
Image Source: ThoughtCo

Second World War: Time for First Action

Johnson was in operational flying in early 1941 in 616 Squadron, which was forming part of the Tangmere Wing. Johnson often found himself flying alongside Wing Commander Douglas Bader and Australian ace Tony Gaze. The only problem was that the great Battle of Britain was over, and "Huns" were hard to find. On 15 January 1941, Johnson, took off as a No.2 in a formation to fly as cover for a convoy off North Cotes. The controller vectored the pair onto an enemy aircraft, a Dornier Do 17. Both attacked the bomber and lost sight of it and each other. Although the controllers intercepted distress signals from the bomber Johnson did not see it crash. The formation was credited with one enemy aircraft damaged. It was the only time Johnson was to engage a German bomber.

Take the Fight to Germans

Johnson flew as a night fighter. Using a day fighter without radar was largely unsuccessful in intercepting German bombers during The Blitz (a German bombing campaign against the United Kingdom in 1940 and 1941). Johnson's only action occurred on 22 February 1941 when he damaged a Messerschmitt Bf 110 flying a Spitfire. Johnson's squadron was moved to RAF Tangmere on the Channel coast. In November 1940 Air Marshal Sholto Douglas became Air Officer Commanding (AOC) RAF Fighter Command. On 8 December 1940 a directive from the Air Staff called for "Sector Offensive Sweeps". It ordered hit-and-run operations over Belgium and France. The operations were to be conducted to harass German air defences. On 10 January 1941 "Circus attacks" were initiated by sending small bomber formations protected by large numbers of fighters. These were designed to draw up the Luftwaffe. These operations became known as the "Circus Offensive". Trafford Leigh-Mallory, AOC 11 Group, outlined distinct mission escort operations for day fighters.

RAF Bombers being escorted by Fighters during "Circus Offensive".
Image Source: Wikipedia

Johnson's First Combat Embarrassment

Johnson's first contact with enemy single-engine fighters was when he led a section behind Bader's patrol section. Johnson spotted three Bf 109s a few hundred feet higher and travelling in the same direction. Johnson, forgetting to calmly report the number, type and position of the enemy, shouted, "Look out Dogsbody" (Bader's call sign). Such a call was only to be used if the pilot in question was in imminent danger of being attacked. The Section broke in all directions and headed to Tangmere singly. The mistake brought an embarrassing rebuke from Bader at the debriefing.

Evolving New Combat Tactics and Formations

Johnson flew various operations over France including ground attack missions, which Johnson hated, as he considered it a waste of pilots. Several successful fighter pilots had been lost in such missions he felt.

During this time, many other pilots also expressed dissatisfaction with the formation tactics being used. After long deliberations, Bader accepted the suggestions and agreed to the use of more flexible tactics to lessen the chances of being taken by surprise. The tactical changes involved operating overlapping line abreast formations similar to the German "finger-four" formation. The formation and tactics were used thereafter by all RAF pilots.

Air Combat Formations.
Image Source: verybrambleberry.com

The First Air Victory

Johnson gained his first air victory on 26 June 1941. Crossing the coast near Gravelines, Bader warned of 24 Bf 109s nearby, southeast, in front of the formation. The Bf 109s saw the British and turned to attack the lower section from the rear. While watching three Bf 109s above him dive to port, Johnson lost sight of his wing leader at 15,000 feet. Immediately a Bf 109E flew in front of him and turned slightly to port at a range of 150 yards. Johnson shot. After being hit, the Bf 109's hood was jettisoned and the pilot bailed out. Several pilots witnessed the victory. He had expended 278 rounds. The Bf 109 was one of five lost by German Fighter Wing 2 that day.

German fighter Bf 109.
Image Source: Pinterest

More Victories Follow

On 1 July 1941 he expended 89 rounds and damaged a Bf 109E. Bader's section was attacked and Johnson out-turned his assailant. Firing, he saw glycol streaming behind it. On 14 July, losing sight of the squadron, Johnson and his wingman proceeded inland at 3,000 feet after spotting three aircraft. Turning in behind them, he identified them as Bf 109Fs. Johnson dived so as to come up and underneath into the enemy's blind spot. Closing to 15 yards, he gave the trailing Bf 109 a two-second burst. The tail was blown off and Johnson's windshield was covered in oil from the Messerschmitt. Johnson saw the other Bf 109s spinning down out of control. Having also lost his wingman, Johnson disengaged. Climbing and crossing

Johnson climbs out of the cockpit before waiting media, at RAF Kings Cliffe, 1941.
Image Source: Wikipedia

the coast at Etaples, Johnson bounced a Bf 109E. Giving chase in a dive to 2,000 feet and firing at 150 yards, he observed something flying off the Bf 109's starboard wing. Johnson could not see any more owing to the oil-covered windscreen and did not make a claim.

Encounter with Adolf Galland

On 21 July, Johnson shared in the destruction of another Bf 109 with Pilot Officer Heppell. Johnson's wingman disappeared during the battle. Sergeant Mabbet was mortally wounded but made a wheels-up landing near St Omer. Impressed with his skilful flying while badly wounded, the Germans buried him with full honours. On 23 July, Johnson damaged another Bf 109. During this battle the famous German Ace Adolf Galland was wounded, but his life was saved by a recently installed armour plate behind his head.

Adolf Galland.
Image Source: flying-tigers.co.uk/ tag/adolf-galland

Mission When Bader Bailed Out

Johnson took part in the 9 August 1941 mission in which Bader was lost over France. On that day Bader had been without his usual wingman Sir Alan Smith who was unable to fly due to a cold. During the sortie, Johnson destroyed a solitary Messerschmitt Bf 109. Johnson flew as wingman to Dundas in Bader's section. As the Wing crossed the coast, around 70 Bf 109s were reported in the area, the Luftwaffe aircraft outnumbering Bader's Wing by 3:1. Spotting a group of Bf 109s 1,000 feet below them, Bader led a bounce on a lower group. The formations fell apart and the air battle became a mass engagement with high risk of collision rather than being shot down. Johnson exited the mass of aircraft and was immediately attacked by three Bf 109s. The closest was 100 yards away. Maintaining a steep, tight, spiral turn, he dived into the cloud and immediately headed for Dover. Coming out of the cloud, Johnson saw a lone Bf 109. Suspecting it to be one of the three that had chased him, he searched for the other two. Seeing nothing, Johnson attacked and shot it down. It was his fourth victory. On 4 September 1941 Johnson was promoted to Flight Lieutenant and awarded the Distinguished Flying Cross (DFC).

Distinguished Flying Cross (DFC).
Image Source: Medals of England

Johnson Becomes a Flying Ace

On 21 September 1941, while escorting Bristol Blenheims to Gosnay, the top cover wings failed to rendezvous with the bombers. Near Le Touquet at around 20,000 feet, Johnson's section was bounced by 30 Bf 109s. Johnson broke and turned in and behind a Bf 109F. Approaching from a quarter astern and slightly below, Johnson fired closing from 200 to 70 yards. The German pilot bailed out. Pursued by several enemy aircraft, Johnson dived to ground level. About 10 miles off Le Touquet, other Bf 109s attacked. Allowing the Germans to close within range, Johnson turned into a steep left-hand turn. It took

German Focke-Wulf Fw 190.
Image Source: YouTube

him onto the tail of a Bf 109. He shot a Bf 109. The two victories made Johnson's total to six destroyed, which now meant he was an official flying Ace. In winter 1941, Johnson and 616 Squadron moved to training duties.

German Focke-Wulf Fw 190 Introduced

RAF Fighter Command resumed its offensive policy in April 1942 when the weather cleared for large-scale operations. Johnnie flew seven sweeps that month. But the situation had now changed. RAF was flying the Spitfire V, which was a match for the Bf 109F. However, the Germans introduced the Focke-Wulf Fw 190. It was faster at all altitudes below 25,000 feet, possessed a faster roll rate, was more heavily armed and could out-dive and out-climb the Spitfire. Only in the turn could the Spitfire outperform the Fw 190. The introduction of this new enemy fighter resulted in heavier casualty rates among the Spitfire squadrons until a new mark of Spitfire could be produced. Johnson claimed a damaged Fw 190 on 15 April 1942 but he witnessed the Fw 190s get the better of the British pilots consistently throughout most of 1942.

Bar to DFC

On 26 June 1942, Johnson was awarded the Bar to his DFC. Meanwhile the Spitfire Mk. IXs began reaching RAF units. On 10 July 1942, Johnson was promoted to the rank of Squadron Leader and given command of 610 Squadron. Johnson preferred the finger-four formation much to the disagreement with his boss Wing Commander Patrick Jameson. On 19 August 1942 was the Dieppe raid. Johnson in his Spitfire VB ran into around 50 Bf 109s and Fw 190s. In a climbing attack Johnson shot down one Fw 190 and shared in the destruction of a Bf 109F.

With His Dog Labrador 'Sally'.
Image Source: Imperial War Museum

Becomes a Wing Commander – Introduces New Tactics

He was frustrated for being given a staff job. But in March 1944, he was switched to command a different Canadian wing in the newly formed 2nd Tactical Air Force. Johnson took command of No. 127 Wing RCAF based at RAF Kenley, and they received the new Spitfire IX, perhaps the answer to the Fw 190. Being a Wing Commander meant his initials could be painted on the machine. After D-Day he organised barrels of beer to be slung under the Spitfires in place of extra fuel tanks, a move welcomed on the dusty front-line airfields of Normandy. His Spitfires now carried JE-J. He was also allotted the call sign "Greycap". Johnson set about changing the Wing's tactical approach. He quickly forced the Wing to abandon the line-astern tactics for the finger-four formation which offered much more safety in combat; enabling multiple pilots to participate in scanning the skies for enemy aircraft so as to avoid an attack, and also being better able to spot and position their unit for a surprise attack upon the enemy. Johnson also abandoned ground attack missions whenever he could. On a fighter sweep, Johnson destroyed an Fw 190 for his eighth victory. Johnson scored more success in July when the USAAF began Blitz Week, a concentrated effort against German targets. Escorting American bombers, Johnson destroyed three Bf 109s and damaged another, the last being shot down on 30 July; his tally stood at 18.

Becomes Highest Scoring Ace

Johnson continued to score regularly. His 22nd & 23rd victories were achieved on 25 April 1944 and Johnson became the highest scoring ace still on operations. After the landings in France on 6 June 1944,

Adolf "Sailor" Malan.
Picture Source: flying-tigers.co.uk

Johnson added further to his tally, claiming another five aerial victories that month. Johnson's Wing was the first to be stationed on French soil following the invasion. With their radius of action now far extended compared to the squadrons still in Britain, the Wing scored heavily through the summer. Johnson had now equalled and surpassed Sailor Malan's record score of 32. However Johnson considered Malan's exploits to be better, and said, Malan had fought with great distinction when the odds were against him.

Last Victory of the War

In September 1944 Johnson's Wing participated in support actions for Operation Market Garden in the Netherlands. On 27 September 1944, Johnson had his last victory of the war, when his flight bounced a formation of nine Bf 109s, one of which Johnson shot down. The Wing rarely saw enemy aircraft for the remainder of the year. Only on 1 January 1945 did the Germans appear in large numbers, during Operation Bodenplatte to support their faltering attack in the Ardennes. He recalled the

Wing Commander Johnnie Johnson, CO No. 144 (Canadian) Wing with his Labrador Retriever "Sally".
Image Source: thisdayinaviation.com

Germans seemed inexperienced and their shooting was "atrocious". Johnson led a Spitfire patrol to prevent a second wave of German aircraft attacking but engaged no enemy aircraft, since there was no follow-up attack. From late January and through most of February, Johnson reduced his flying time.

1944/06/06 Spitfires D Day – Painting by Nicolas Trudgian.
Image Source: pinturas-sgm-aviacion.tumblr.com

Post War: Permanent Commission

On 26 March Johnson was promoted to acting Group Captain, and later took command of No. 125 Wing. Some idea of his character is shown by the fact that in early 1942 he became a Squadron Leader, in 1943 a Wing Commander and in February 1945 a Group Captain. During the last week of the war, Johnson's squadron flew patrols over Berlin and Kiel as German resistance crumbled. After the German capitulation in May 1945, Johnson relocated with his unit to Copenhagen, Denmark. After the war, Johnson was given a permanent commission by the RAF, initially as a Squadron Leader. On promotion to Wing Commander (his wartime rank), he became OC Tactics at the Central Flying Establishment.

Korean War

During an exchange posting to the US Air Force, in 1950 he served in the Korean War flying the Lockheed F-80 Shooting Star, and later flew the North American F-86 Sabres with the US Air Force Tactical Air Command. Johnson did not leave any written record of his experiences but at the end of his tour received the US Air Medal and Legion of Merit.

Lockheed F-80 Shooting Star.
Picture Credit: Lockheed Martin

Further RAF Service

In 1951, Johnson commanded a Wing at RAF Fassberg. In 1952, he was promoted to Group Captain and commanded RAF Wildenrath in West Germany until 1954. From 1954 to 1957 he was deputy director operations (DD (Ops) at the Air Ministry in London. In 1956 his wartime memoir, "Wing Leader" was published. He requested Bader to write the foreword. His old CO wrote back "Dear Johnnie, I did not know you could read and write." That was Johnnie in a nutshell.

On 20 October 1957, Johnson became Commanding Officer of RAF Cottesmore in the UK, a station operating the Victor V bombers. In 1960 he was promoted to Air Commodore, and attended the Imperial Defence College (IDC) course in London and in June 1960 was made a Commander of the Order of the British Empire (CBE) for his work as Station Commander. On 1 October 1963 he was promoted to Air Vice Marshal and served as Air Officer Commanding (AOC) RAF Middle East based at Aden. In 1964 he published his book "Full Circle", a history of air fighting, co-written with Percy "Laddie" Lucas, a former Member of Parliament and Douglas Bader's brother-in-law. In 1965 on retirement from the RAF he was appointed a Companion of the Order of the Bath (CB).

Image Source: rcaf403squadron.wordpress.com

Johnson's Medal Set.
Image Source: thisdayinaviation.com

Post-Retirement Life

Johnson was a Deputy Lieutenant for the County of Leicestershire in 1967. He established the Johnnie Johnson Housing Trust in 1969 and by 2001 the housing association managed over 4,000 properties. After the death of the WW2 RAF fighter pilot Douglas Bader in 1982, Johnson, Denis Ceowley-Milling and Sir Hugh Dundas set up the Douglas Bader Foundation, to continue supporting disabled charities, of which Bader was a passionate supporter. Johnson was also the first to recognise the skills of Robert Taylor, aviation artist, in the 1980s. Depictions of aircraft and battle scenes in print began to become popular and he helped Taylor promote them. The venture was successful and Johnson's sons set up their own distribution networks in the United States and Britain.

Johnson's Uniform.
Image Source: historicflyingclothing.com

'Johnnie' Johnson poses in front of one of the last Lancaster planes at the Rolls Royce Heritage Museum.
Image Source: pippaettore.com

Personal Life

In 1942 he married Paula, and they had two sons, Michael and Christopher. In 1977 he and his wife decided amicably to separate. For a while he lived in Jersey, but he returned to Buxton, Derbyshire.

Johnson spent most of the 1980s and 1990s as a keynote speaker, fundraiser and spending time on his hobbies; travelling, fishing, shooting and walking his dogs. Johnson

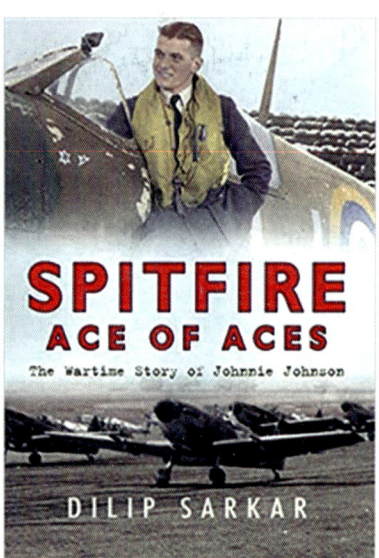

Spitfire Ace of Aces: The Wartime Story of Johnnie Johnson.
Image Source: Amazon.com

appeared on the long–running British television show "This is Your Life" on 8 May 1985, the 40th anniversary of VE Day. Among the program's guests was German fighter ace Waltor Matoni. British wartime propaganda had alleged Johnson had challenged Matoni to a personal duel; a version of events denied by Johnson. The two men arranged to meet after the war but were unable to do so until the TV program. Among other guests was High Dundas, "Nip" Heppel, who flew alongside Johnson on his first operation.

Summary of War Flying and Air Victories

Johnson's wartime record was over 1,000 missions flown, 38 aircraft claimed destroyed with a further seven shared destroyed (three and one shared victories), three probable destroyed, 10 damaged, and one shared, destroyed on the ground. All his victories were fighters. As a Wing Leader, Johnson was able to use his initials "JE-J" in place of squadron code letters. He scored the bulk of his victories flying two Mk IXs. With EN398/JEJ he shot down 12 aircraft, and shared five plus six and one shared damaged. With MK392/JEJ, 12 aircraft plus one shared, destroyed on the ground. His last victory of the war was scored in this aircraft. The ability to verify British claims against the British' main opponents in 1941 and 1942, JG 26 and JG 2, was very limited. Only two of the 30 volumes of War Diaries produced by JG 26 survived the war. Historian Donald Caldwell attempted to use what limited German material was available to compare losses and air victory claims but acknowledges the lack of sources leaves the possibility for error. Officially this remains the highest total of any RAF pilot, though it is widely believed Sqn Ldr St John Pattle exceeded 40 in the turmoil of the Greek campaign. Only one Allied pilot – Richard Bong of the United States Army Air Forces, who shot down 40 Japanese planes – had greater success during the war. The leading American air ace in Europe, Francis Gabreski, shot down 28 German planes.

Richard Bong of the United States Army Air Forces.
Image Source: Wikipedia

George 'Johnny' Johnson, pose holding a Dambusters model at the base.
Image Source: pippaettore.com

Perhaps Johnson's most impressive achievement was that, he was never shot down. Only once was his Spitfire damaged by the enemy. He was awarded the DSO and two bars, the DFC and bar, the Belgian Légion d'Honneur and Croix de Guerre.

Air Chief Marshal Sir Christopher Foxley-Norris, chairman of the Battle of Britain Fighter Association, wrote: Johnnie Johnson's performance was even more credible because he largely missed the Battle of Britain and won his "kills" in fighter-to-fighter combat rather than against heavy bombers. Johnnie's kills were hard-earned, but then Johnnie had the two skills needed to be successful, he was a good shot and a good pilot. Lots of people were good pilots, but Johnnie was also a good shot, gifted in the art of deflection shooting".

He was a hard man, a very tough man, but a very good leader. He was trusted and he looked after his people. But he was intolerant if a man did not come up to scratch. There were some pilots who had to overcome a great deal of fear; but Johnnie did not seem to suffer like that. It was somehow easier for him. He was certainly tough – and demanding, both on and off duty – but then you had to be.

Passes On

Second World War fighter ace credited with more enemy 'kills' than any other British pilot. On 30 January 2001, Johnson, aged 85 years, died from cancer. A memorial service took place on 25 April 2001. The only memorial was a bench dedicated to him at his favourite fishing spot on the estate; the inscription reads "In Memory of a Fisherman".

REFERENCES

1. Dambuster 'Johnny' Johnson recalls World War Two raid, BBC Interview https://www.bbc.com/news/av/uk-42580442
2. https://www.thoughtco.com/air-vice-marshal-johnnie-johnson-2360546
3. Genevieve Hopkins, Aviator Mouse Notes-3, Air Combat Tactics, February 20, 2017 https://verybrambleberry.com/index.php/2017/02/20/aviator-mouse-notes-air-combat-tactics/
4. Wikipedia. https://en.wikipedia.org/wiki/Johnnie_Johnson_(RAF_officer)
5. Johnnie Johnson (RAF officer) https://military.wikia.org/wiki/Johnnie_Johnson_(RAF_officer)
6. Johnnie Johnson, RAF Spitfire Ace, This is Your Life. YouTube. https://www.youtube.com/watch?v=z2ZfHeS6KMU
7. Johnnie Johnson: Kenley's Ace https://www.kenleyrevival.org/content/new-contributions/johnnie-johnson-kenleys-ace
8. Johnnie Johnson, Aces of WW2. https://acesofww2.com/UK/aces/johnson/

5

GERMAN ACE GENERAL ADOLF GALLAND

104 Aerial Victories – all against the Western Allies

Adolf Galland.
Image Source: warrelics.eu

"An excellent weapon and luck had been on my side. To be successful, the best fighter pilot needs both".
—**Adolf Galland**

Adolf Josef Ferdinand Galland was a German Luftwaffe General and flying ace who served throughout the WW II in Europe. Flew 705 combat missions. On four occasions, he survived being shot down, and was credited with 104 aerial victories, all against the Western Allies. He took part in the Spanish Civil War. In 1938 Galland was employed in the Air Ministry writing doctrinal and technical manuals about ground-attack, and also served as an instructor for ground-attack units. Initially in WW II he was tasked for ground attack missions, but later managed to persuade his superiors to allow him to become a fighter pilot.

Galland took part in the Battle of France and Battle of Britain, and later in Northern France. By November 1941, his aerial victories tally had increased to 96, and was given the command of the German Fighter Force, and stayed in that position until January 1945. Galland was forbidden to fly combat missions in this position.

Over subsequent years, Galland had serious disagreements with Reichsmarshall Herman Göring about how best to combat the Allied Air Forces bombing of Germany. The Luftwaffe fighter force was under severe pressure by 1944, and Galland was blamed by Göring for the failure. In early January 1945, Galland was relieved of his command, and put under house arrest following the so-called Fighter Pilot's revolt in which senior fighter pilots confronted Göring about the conduct of the air war.

In March 1945, Galland returned to operational flying and was permitted to form a jet fighter unit. He flew missions over Germany until the end of the war in May. After the war, Galland was employed as a consultant to the Argentine Air Force. Later, he returned to Germany and managed his own business.

Adolf Galland in the Cockpit of a Bf 109.
Image Source: pinterest.ca/Clem_Mx

Early Life and Family of Aces

Galland was born in Westerholt in the Ruhr industrial area, on 19 March 1912 to a family with French ancestry. Galland's father worked as the land manager or bailiff to Count von Westerholt. Adolf's pet name was "Keffer". His two younger brothers also became fighter pilots and aces. Paul claimed 17 victories, before being shot down and killed on 31 October 1942. Wilhelm-Ferdinand, was credited with 55 victories, and was shot down and killed on 17 August 1943.

Ferdinand Wutz Galland.
Image Source: Pinterest (worldwartwo.filminspector.com)

Initial Flying Interest

Galland's lifelong interest in flying started in 1927 when a group of aviation enthusiasts formed an air sports club in a nearby estate. Galland travelled by foot or horse-drawn wagon 30 kilometres until his father bought him a motorcycle to help prepare the gliders for flight. By 19 Galland was a glider pilot. In 1932 he completed pilot training. Under the Treaty of Versailles, Germany was denied an air force. They were however allowed gliders and it became the way for pilots to begin their flying career. The sport became so popular in Germany. The German military also published a magazine, Flight Sport. When he eventually attained his B and C certificates, his father promised to buy him his own glider if he also passed his matriculation examinations, which he succeeded in doing. Galland became an outstanding glider pilot, and he became an instructor.

Begins Flying as a Career

In February 1932, Galland graduated from high school and was among 20 personnel accepted to the aviation school of Germany's national airline, Luft Hansa. Jobs were scarce and life was hard those years in Germany. Adolf applied for German Commercial Flying School which was heavily subsidised by Luft Hansa. He was one of 100 successful applicants out of 4,000. After ten days of further evaluations, he was among just 18 selected for flight training. Galland's first flight was in an Albatros L 101. During training Galland had

Albatros L 101.
Image Source: Wikimedia Commons

Galland with Herman Göring.
Image Source: ww2db.com

two accidents; a heavy landing damaged the undercarriage of his aircraft, and a collision. Worried of being thrown out, in parallel Galland applied to join the German Army. Later he was awarded B1 certificate that allowed him to fly large aircraft over 2,500 kilograms in weight. Meanwhile the Army accepted his application, but the flying school refused to release him. By Christmas 1932, he had logged 150 hours flying and had obtained a B2 certificate.

Early in 1933, Galland was sent to the Baltic Sea training base to train on flying boats. Galland disliked the idea of "seamanship", but logged 25 hours in these aircraft. Later he was sent to Central Airline Pilot School. The group was interviewed by military personnel in civilian clothing. After being informed of a secret military training program being built that involved piloting high performance aircraft, all the pilots accepted an invitation to join the organisation.

Joins the Luftwaffe

In May 1933, Galland was ordered to a meeting in Berlin as one of 12 civilian pilots among 70 airmen who came from clandestine programs. He met Herman Göring for the first time. Galland was impressed by and believed Göring to be a competent leader. In July 1933, Galland travelled to Italy to train with the Italian Air Force. In September 1933, Galland returned to Germany. He was sent to learn instrument flying and piloting heavy transport aircraft, where he flew another 50 hours and also flew Lufthansa airliners. Finally in December 1933, Galland was finally offered the chance to join the new Luftwaffe. As an airline pilot Galland had begun enjoying the easy lifestyle, and visiting exotic places. But he wanted adventure and decided to join the Luftwaffe.

Focke-Wulf Fw 44 biplane.
Image Source: Wikipedia

First Major Crash and Injury

After basic training ground training with the Army, in February 1935 Galland was part of 900 airmen to be inducted to the new ReichsLuftwaffe. Galland was ordered to report to Fighter Wing 2 on 01 April 1935. In October 1935, during aerobatic manoeuvre training, he crashed a Focke-Wulf Fw 44 biplane and was in a coma for three days, other injuries were a damaged eye, fractured skull and broken nose. When Galland recovered, he was declared unfit for flying by the doctors. A friend, Major Rheital, kept the doctor's report secret to allow Adolf to continue flying. The expansion of the Luftwaffe was a priority. Galland's medical report was overlooked. Within a year, Galland showed no signs of injury from his crash.

Second Major Crash and Grounding

In October 1936 he crashed an Arado Ar 68 and was hospitalised again, aggravating his injured eye. It was at this point his previous medical report came to light again and Galland's unfit certificate was discovered. Major Rheital was rumoured to have undergone a court-martial, but the investigators dropped the charges. Galland, however, was grounded. He admitted having fragments of glass in his eye, but convinced the doctors he was fit for flying duty. Galland was ordered to undergo eye tests to validate his claims. Before the testing could begin, one of his brothers managed to acquire the charts. Adolf memorised the charts passing the test and was permitted to fly again.

Arado Ar 68 aircraft. Picture Credit: Klinke & Co.
Source: Wikipedia

Spanish Civil War: Early Tactics and Flying in Swimming Trunks

During the Spanish Civil War, Galland was appointed Captain of a Condor Legion Unit which was sent to support the Nationalist forces. Galland flew ground attack missions in Heinkel He 51s. In Spain Galland first displayed his unique style: flying in swimming trunks with a cigar between his teeth in an aircraft decorated with a Mickey Mouse figure. When asked why, he said "I like Mickey Mouse. And I like cigars, but I had to give them up after the war". Galland flew 300 combat missions in Spain starting 24 July 1937. During his time in Spain, Galland analysed the engagements, evaluated techniques and devised new ground-attack tactics which were passed on to the Luftwaffe. His experiences in pin-point ground assaults were used by Ernst Udet, a proponent of the dive bomber and leading supporter of the Junkers Ju 87 to push for Stuka wings.

Galland and His aircraft decorated with Micky Mouse.
Image Source: flickriver.com (Dreadnought 2003)

Spanish Cross

During his time in Spain, he also developed early gasoline and oil bombs, and proposed quartering of military personnel on trains to aid in quick relocation. Following the Nationalist victory was awarded the Spanish Cross in Gold with Swords and Diamonds'. On 24 May 1938 Galland left Spain. Before leaving he made ten flights in the Bf 109. Deeply impressed with the aircraft, he pushed to change from a strike pilot to a fighter pilot. His colleagues said "Galland was a very good pilot and excellent shot, but ambitious and he wanted to get noticed. He was crazy about hunting anything, from a sparrow to a man."

Spanish Cross in Gold with Swords and Diamonds.
Image Source: Wikipedia

Suggested Operational Aircraft Modifications

On his return to Germany, he was posted to the Ministry of Aviation where he was tasked with preparing recommendations on close air support. Galland favoured the virtually simultaneous attack of the air force before the Army advance, leaving their opponents no time to recover. He reasserted the lessons of WW I, while some others were pessimistic as to whether that kind of coordination was possible. Galland also preferred

Focke-Wulf Fw 190.
Image Source: britannica.com

the Italian suggestion of heavy armament vis-a-vis the light machine guns on early German fighter aircraft. He suggested combining machine guns with cannon. These proved successful on the Bf 109 and Focke-Wulf Fw 190. He also recognised the innovation of drop tanks to extend the range of aircraft. He proposed specialised tactics for escorting bomber fleets. Galland did not subscribe to the prevailing idea in the Luftwaffe, and RAF, that the bomber could get through alone. All of Galland's suggestions were adopted and proved successful in the early campaigns, 1939–41.

Posted for Flight Testing – Unhappy

Unluckily for Galland, his excellence at evaluation earned him a place at training facility where he was asked to test fly prototype reconnaissance and strike aircraft. This was not what he wanted, and he hoped to be returned to a fighter unit to fly the Bf 109. During his time there, he gave positive evaluations on the Focke-Wulf Fw 189 and Henschel Hs 129. He was then given an unwelcome news of posting as a group commander of the 2nd Demonstration Wing. It was not a fighter unit, but a special mix of ground attack aircraft.

Henschel Hs 129.
Image Source: YouTube (Hitler's A-10 – The Henschel Hs 129 Tank Bomber)

Invasion of Poland and the Iron Cross

During the invasion of Poland from 1 September 1939 onward, he flew in a unit equipped with the Henschel Hs 123, nicknamed the "biplane Stuka," in support of German 10th Army. On 1 September, Galland flew a reconnaissance mission and was nearly shot down. He flew many ground attack missions in support of the 1st Panzer Division. The German Army reached near Warsaw by 7 September. Luftwaffe was executing the kind of close air support operations Galland had been advocating. After flying nearly 360 missions in two wars and averaging two missions per day, on 13 September 1939, Galland was awarded the Iron Cross Second Class.

Henschel Hs 123. "Biplane Stuka".
Image Source: Wikimedia Commons

Rheumatism an Excuse to get out of Ground Attack

After the end of the Poland campaign, Galland claimed to be suffering from rheumatism and therefore unfit for flying in open-cockpit aircraft, such as the Hs 123. He tactfully suggested a transfer to a single-engine aircraft type with a closed cockpit. His request was accepted on medical grounds. Galland was removed from his post as a direct ground support pilot. Given his performance with eye specialists, a certain amount of suspicion remained. He was transferred to JG 27 Fighter Wing on 10 February 1940 as the adjutant, and that restricted his flying.

Mölders' Tactics

After his transfer to JG 27, Galland met Mölders again. Due to his injuries, Galland could never match Mölders' sharp eyesight. Mölders, by that time a recognised ace shared his experiences with Galland. These included leadership in the air, tactics and organisation. Mölders was group commander of JG 53. For Galland to gain experience on the Bf 109E, which he lacked, Mölders offered him the chance to fly with his unit. Galland learned Mölders' tactics, such as using spotter aircraft to indicate the position of enemy formation. Galland learned to allow pilots to operate freely in order to seize the initiative. He took the experiences back to JG 27, and its commander Max Ibel, agreed to their implementation.

Col General Ernst Udet with Bomber Pilots Galland and Mölders Photo 1940.
Image Source: akg-images.de

Invasion of Western Europe: First Air Victory

On 10 May 1940, the Wehrmacht (the unified armed forces of Nazi Germany) invaded the Low Countries and France under the codename Fall Gelb. JG 27 supported German forces in the Battle for Belgium. On the third day of the offensive, 12 May 1940, flying a Bf 109, Galland, with Gustav Rodel as his wingman, claimed his first aerial victories, over two Royal Air Force (RAF) Hurricanes. The Hurricanes had been escorting Bristol Blenheim bombers. Galland remembered; "My first kill was child's play. An excellent weapon and luck had been on my side. To be successful, the best fighter pilot needs both"— Galland pursued one of the "scattering" Hurricanes and shot down another at low level. Galland claimed his third Hurricane later that same day. He had long believed that his opponents had been Belgian, not knowing that all of the Belgian Air Force's Hurricanes had been destroyed on the ground in the first two days, without seeing combat.

Bf 109.
Image Source: warhistoryonline.com

Chased and Shot an Aircraft: Ran out of Fuel

On 19 May, Galland shot down a French Potez aircraft. During this flight he ran out of fuel short of the runway and landed nearby, at the base of a hill. With the help of soldiers from a German Flak battery, he pushed the Bf 109 up the hill and then half-flew, half-glided down to an airfield in the valley below. He sent back a can of fuel for his wingman, who had also landed short of the runway. He continued flying and the next day, claimed another three more aircraft, making a total of seven. For this he was awarded the Iron Cross First Class on 22 May.

First Encounter with Spitfire

During the Battle of Dunkirk, after first encounter with the Supermarine Spitfire, Galland was very impressed with these aircraft and their pilots. On

Bristol Blenheim.
Image Source: militaryfactory.com

29 May, Galland claimed a Bristol Blenheim over the sea. Over Dunkirk, the Luftwaffe suffered its first serious rebuff of the war. Galland noted that the nature and style of the air battles should have provided a warning on the inherent weaknesses of the Luftwaffe's force structure.

Battle of Britain

On 6 June 1940, Galland took over the command of a unit under 26th Fighter Wing (JG 26) with Bf 109Es. On 24 July 1940, almost 40 Bf 109s of JG 26 took off for operations over the English Channel. They were met by 12 Spitfires. The Spitfires forced the larger number of Bf 109s into a turning battle that ran down the Germans' fuel. Galland recalled being impressed by the Spitfire's ability to out manoeuvre Bf 109s at low speed and to turn into the Bf 109s within little airspace. Only by executing a "Split S" (a half-roll onto his back, followed by pulling into a long, curving dive) without the float carburettor causing a temporary loss of engine power, could his aircraft escape back to France at low altitude. During the action, two Spitfires were shot down for the loss of four Bf 109s. Galland was shocked by the aggression shown by British pilots who he initially believed to be relatively inexperienced. Galland later said he realised there would be no quick and easy victory.

Supermarine Spitfire.
Image Source: Wikipedia

Knight's Cross of the Iron Cross

As the battles over the Channel continued, Galland shot down Spitfires on 25 and 28 July. On 1 August 1940, Galland was awarded the Knight's Cross of the Iron Cross for his 17 victories by then. The cloudy skies of Britain were a dangerous environment to confront an enemy that had an effective ground control system. Galland resolved to fly higher, where he could see most things and where the Bf 109 performed at its best.

Bomber Bf 110.
Image Source: asisbiz.com

By 15 August, in two weeks' fighting over Britain, Galland had increased his own tally to 21. On this day he claimed three Spitfires. This put him to within three victories of Mölders, who had claimed the highest number of enemy aircraft destroyed and who was wounded and grounded with a damaged knee. Galland and his pilots remained ignorant of the disastrous losses suffered by other German units and the defeat of their attacks by the RAF.

Summoned by Göring: Asked to Fly as AD Escorts

Galland was summoned on 18 August 1940, and missed the intense air battle that day, known as "The Hardest Day". During the meeting, Göring insisted that, in combat, Bf 109 fighters escort Bf 110s, which could not survive against single-engine fighters. As high-scoring aces, both Galland and Mölders shared their concerns that close escort of Bf 110s and bombers robbed fighter pilots of their freedom to roam and engage the enemy on their own terms. They also pointed to the fact that German bombers flew at medium altitudes and low speed, the best height area and speed for the manoeuvrability of the Spitfire. Göring would not move from his position. Galland claimed that fighting spirit was also affected when his pilots were tasked with close-escort missions. The worst disadvantage of this type of escort was not aerodynamic but lay in its deep contradiction of the basic function of fighter aircraft—to use speed and manoeuvrability to seek, find, and destroy enemy aircraft, in this case, those of Fighter Command. The Bf 109s were bound to the bombers and could not leave until attacked, thus giving their opponent the advantage of surprise, initiative, superior altitude, greater speed, and above all fighting spirit, the aggressive attitude which marks all successful fighter pilots.

Hermann Göring (Centre), Werner Mölders (left) and Adolf Galland, operational discussions during the Battle of Britain. (John Frost Newspapers)
Image Source: historynet.com

Becomes Wing Commander

Göring grew frustrated with the lack of aggressiveness of several of his fighter-wing commanders, and on 22 August, he replaced Handrick with Adolf Galland, who became the Wing Commander of JG 26. The pilots were dissatisfied. Galland could not change Göring's mind with respect to the escort missions, but he did take immediate actions to improve pilot morale. He replaced ineffective Group and Squadron commanders with younger, more aggressive, and more successful (aerial victories) officers within the Wing. He also increased the Wing flights from earlier two-aircraft formations to the more lethal four-fighter formations. Galland flew as often as possible and led the most difficult missions in order to encourage his men, and gain respect. From 25 August until 14 September, Galland had victories 23 to 32 shooting down Spitfires and Hurricanes. Göring, worried of growing bomber losses, asked what his fighter pilots needed to win the battle. Werner Mölders replied, the Bf 109 with more powerful engines. Galland said, "I should like an outfit of Spitfires for my squadron." This left Göring speechless with rage. Galland still preferred the Bf 109 for offensive sweeps, but he regarded the Spitfire as a better defensive fighter, owing to its manoeuvrability.

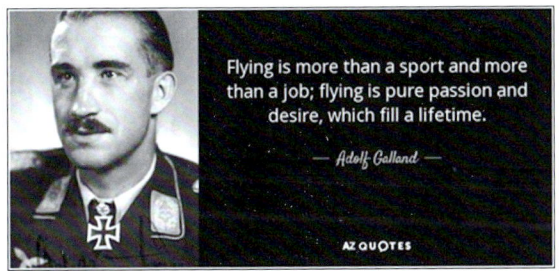

Image Source: azquotes.com

Shooting Down Pilot in Parachute

Göring wanted to know if German pilots had ever thought of killing the enemy pilots while descending in parachute. "I should regard it as murder, Herr Reichsmarschall', Galland told. "I should do everything in my power to disobey such an order" if ever given. "That is just the reply I had expected from you, Galland" said Göring. In practice, this act of mercy was not applied.

Audience with Hitler

On 23 September, Galland became the third member of the *Wehrmacht* to receive the Knight's Cross of the Iron Cross with Oak Leaves for achieving his 39th and 40th aerial victories. On 25 September, he was summoned to Berlin to receive the award from Adolf Hitler. Galland was granted a personal audience with Hitler and during the meeting Galland reported to Hitler that the British had proven tough opponents, and that there were signs of declining morale in

With Adolf Hitler.
Image Source: akg-images.co.uk

the German fighter force in the absence of operational success. Hitler expressed his regret for the war with the "Anglo-Saxons", who he admired, but resolved to fight until total destruction.

German Pilots Fatigue and Low Morale

Hermann Göring.
Image Source alphahistory.com

Morale and exhaustion became a problem in September. The Luftwaffe lacked the pilots and aircraft to maintain a constant presence over Britain. To compensate, commanders demanded three to four sorties per day by the most experienced men. Galland recognised the manifest fatigue of his pilots. By the end of September, Galland noticed that "the stamina of the superbly trained and experienced original cadre of pilots was down to a point where operational efficiency was being impaired. Several factors contributed to this situation. Herman Göring was German political and military leader. One of the most powerful figures in the Nazi Party. A veteran WW I fighter pilot ace, recipient of "The Blue Max". Göring's interference with tactics without regard for the situation or the capabilities of German aircraft; rapid adaptation to German tactics by the British; and the poorer quality replacement pilots to JG 26 Fighter pilots, lost confidence in their aircraft and tactics.

Galland's New Tactics

Galland innovated tactics to improve the situation and found a partial solution to Göring's irrational order to maintain close escort. He developed a flexible escort system that allowed his pilots to constantly change altitude, airspeed, direction, and distance to the bombers during the close-escort missions. The results were better and acceptable to his pilots. By the end of the Battle of Britain, JG 26 had gained a reputation as one of only two fighter Wings that performed escort duties with consistently low losses to the bombers. The fighter-bomber mission was also a problem Galland had to deal with. Göring was committed to fitting one-third of all fighter Wings to use modified Bf 109s to carry bombs. Galland had no choice but to accept the mission but it damaged the morale he had cultivated. Galland's again decided to develop tactics that mixed the bomb-laden Bf 109s with the fighter escort in an effort to deceive the enemy and confound their intercept plans. This tactic slowed down the fighter-bomber losses, but the pilots still felt as though they were being wasted. Galland's leadership still made several errors. Galland did not use training opportunities to improve the bombing accuracy of his pilots. He did not discipline those pilots who were prone to jettison their bombs early. He himself flew escort missions, thus not setting example to his men by flying bombing missions.

Last of Battle of Britain – Leading Fighter Pilots

The Battle of Britain continued with large-scale dogfights well past 31 October, considered by many historians as the end of the campaign. Galland claimed eight victories in October, six Spitfires and two Hurricanes, taking his tally to 50. On 15 November, Galland flew his 150th combat mission. In November, his victories increased to 56, putting him level with the late Helmet Wick who had been shot down and killed on 28 November. On 5 December, Galland recorded his 57th victory. This made him the most successful fighter pilot of the war at that point, putting him ahead of his colleague, friend

Luftwaffe pilots Werner Mölders and Adolf Galland.
Image Source: i.pinimg.com via Pinterest

and rival Werner Mölders. Leading fighter pilots shared special and indefinable qualities in piloting, particularly marksmanship, hunting skills and situational awareness. During the Battle of Britain, Galland accounted for 14% of all JG 26's aerial successes, from a unit of around 120 pilots. Four of the Wing's fighter pilots claimed an astounding 31% of all aircraft shot down.

Channel Front

In March 1941, Göring held a major conference for units in the west. After describing in detail the coming air offensive against Britain, he secretly admitted to Adolf Galland and Werner Mölders that "there's not a word of truth in it." The Luftwaffe was to transfer to the Eastern Front. Approximately two fighter Wings remained in the west for the next year and a half, many of the best fighter crews remained in that theatre. Similarly, the best equipment went to the west, including the Focke-Wulf Fw 190. Just about 180 best in the Luftwaffe aircraft were left with the western fighter forces.

General Adolf Galland in his 109-E4, the only 109 equipped with a cigar lighter.
Image Source: reddit.com. Posted by Epiccow400000 Luftwaffe

Now promoted to Lt Col he continued to lead JG 26 in 1941 against the RAF fighter sweeps across northern Europe. In early 1941, most of the Luftwaffe's fighter units were sent to the Eastern Front or Mediterranean Theatre leaving only JG 26 as the sole single-engine fighters in France. JG 26 were being re-equipped with the new Bf 109F (with 20 mm cannon).

Galland in the Cockpit, on the way to General Theo Osterkamp's birthday get-together.
Image Source: reddit.com (WWII Pics)

Air Combat with Champagne on board

On 15 April 1941, Galland took off with lobster and champagne to celebrate General Theo Osterkamp's birthday in France. He made a detour with his wingman towards England, looking for RAF aircraft. Off the cliffs of Dover, he spotted a group of Spitfires. Galland attacked and claimed two confirmed and one unconfirmed shot down. The actual result was the destruction of one Spitfire; the other two were damaged and force landed with both pilots wounded. During the combat, Galland's undercarriage had dropped causing one of the RAF pilots to claim Galland's aircraft as destroyed, but Galland landed without incident and presented Osterkamp with his gifts. Galland's success that day represented his 60th and 61st aerial victory.

RAF Offensive

RAF mounted a non-stop offensive with Fighter Command over France. Galland intended to engage the British and inflict maximum damage. The fighters were to scramble quickly gain height and make use of the sun and cloud to attack the enemy

Galland and Mölders attending General Theo Osterkamp's birthday.
Image Source: Wikipedia

formation. Under these tactics many JG 26 pilots began to emerge as aces. On 18 June, Galland shot a Spitfire, taking his tally to 67, then the highest against the Western Allies.

Galland Bails out and is Injured

On the morning of 21 June, he shot two Bristol Blenheims but was shot down by a Spitfire. Galland bailed out and tugged at what he thought was his parachute ripcord, but was actually pulling at his parachute release harness. With a "sickening" feeling, he composed himself and pulled the ripcord which opened. Galland was being treated for his wounds in the hospital when he was informed that his 69 victories had now earned him the Knight's Cross of the Iron Cross with Oak Leaves and Swords.

Galland Survives Again

On 2 July 1941, Galland led JG 26 into combat. Galland's fighter was hit by a 20 mm round. The armour plate fitted to the Bf 109 just days earlier saved Galland's life. Wounded in the head he managed to land and was again hospitalised for the second time in a few days. Galland had been shot up and shot down twice in the space of four days.

Galland with Douglas Bader.
Image Source: historiassegundaguerramundial.com

Galland Looks after Douglas Bader

On 9 August 1941, RAF ace Douglas Bader bailed out over St Omer, France. Bader was well known to the Luftwaffe and at the time of his capture had been credited with 22 aerial victories. Galland had claimed two Spitfires on that date. Galland and JG 26 entertained Bader over the next few days. Owing to the significant stature of the prisoner, Galland permitted Bader, under escort, to sit in the cockpit of a Bf 109. Apparently, despite having lost one of his tin legs in the aircraft, Bader, in a semi-serious way, asked if they wouldn't mind if he took it on a test flight around the airfield. Galland replied that he feared Douglas would attempt to escape and they would have to give chase and shoot at each other again, and declined the request.

Posting to High Command – Youngest General

His 96th victim, yet another Spitfire, was claimed on 18 November 1941. It proved to be his last official victory for three years as he was about to be forbidden to fly combat missions. In November 1941, he was chosen by Göring to command Germany's fighter force, in the rank of a Lt General succeeding Mölders who had just been killed in an air crash en route to attend the funeral of Ernst Udet. Galland was not enthusiastic about his promotion, seeing himself as a combat leader and not wanting to be "tied to a desk job". He was the youngest General in the armed forces. Soon afterward, on 28 January 1942, Galland was awarded the Knight's Cross of the Iron Cross with Oak Leaves, Swords and Diamonds, for his service as Commander JG 26.

With Adolf Hitler.
Image Source: hitler-archieve.com

Full Command of Air Ops

Although not keen on a staff position, he planned and executed the German air superiority plan for German Navy's Operation Cerberus, to give air cover to German battleships and heavy cruisers sailing from France to Germany. The operation caught the British off guard. German fighter defences were able to shoot down 43 RAF aircraft with 247 British casualties. The Luftwaffe had prevented any damage on the ships by air attack.

A strong proponent of the day fighter force and the defence of Germany, Galland used his position to improve things. Germany had declared war on the United States on 11 December 1941, and Galland was keen to build up a force that could withstand the resurgence of the Western Allied Air Forces in preparation for what became known as the Defence of the Reich. Galland was outspoken, something that was not often tolerated by Göring. Yet, by earning and cultivating the support of other powerful personalities including Adolf Hitler, Galland was able to survive in his position for three years.

Galland (left) and Mölders.
Picture by Gabe Pecan. Source: Pinterest

Adolf Galland on an inspection in southern Italy, 1943. Note Bf 109 fighter in background.
Image Source: Photo by Ketelhonn, Source: German Federal Archive

Posting to Mediterranean – Major German Failures

The first major crisis for Galland's command occurred in 1943. Galland had been supporting operations in the area, but the Tunisian defeat caused a reorganisation of Axis air forces in the south. Luftflotte (Air Fleet) was divided in two. Wolfram von Richthofen became Luftflotte 2 commander. Galland went to Sicily to control fighter operations under him. Galland was not able to prove himself a capable senior staff officer. Galland's failings delighted Richthofen who was content to allow Galland "enough rope to hang himself", which deflected attention from others.

Upon reaching Sicily, Galland found the state of German air forces shocking. The combat units were exhausted, short of spares, and the 130 fighters on the island were under frequent attack. It was impossible to completely rebuild the squadrons. The resources available could not prevent the Allied air forces acting with impunity. Göring threatened to have one pilot from each unit stand trial by court martial, and if improvements were not forthcoming, they were to be sent as infantry to the Eastern Front. Göring ordered pilots returning without claims and undamaged aircraft suffer court martial for cowardice. Under pressure from Göring, Galland also berated the Wing under him.

Galland interacting with Crew.
Image Source: Pinterest

Adolf Galland with his Focke Wulf Fw 190A6 "White 2" in 1943.
Image Source: asisbiz.com

Considerable reinforcements then arrived. The number of fighters increased from 190 in mid-May to 450 in early July 1943. The movement of additional fighters resulted only in the rise of German losses, which reflected the superiority of Allied production. The weak German bomber force made only a feeble attempt to support the defence of Sicily. Losses were rather high. In the first nine days of July 1943, Galland's command lost approximately 70 fighters. On the fourteenth day he was summoned to Berlin to explain the collapse of air defences on the island. Since the Allied invasion of Sicily, Galland had lost 273 German and 115 Italian aircraft and imposed a cost of only around 100 on Allied air forces.

Conflict with Göring and Failed Leadership

Galland's position brought him into gradual conflict with Göring as the war continued. Galland was often at odds with Göring and Hitler on how to prosecute the air war. In 1942–44, the German fighter forces on all fronts in Europe came under increasing pressure. In the spring of 1943, Galland suggested that the fighter forces defending Germany should be conserved by limiting fighter interceptions, and concentrating on the enemy bombers. Göring found the suggestion unacceptable. He demanded every raid be countered in maximum strength regardless of the size of the Allied fighter escort. According to head of production and procurement, Erhard Milch, "Göring just could not grasp it." The combination of declining production and attrition left Galland with a thin resource-base with which to defend Germany. By early October, American fighters were accompanying bombers as far as Hamburg. When Galland explained this to Göring, he was livid with Galland and the fighter force. He called it the "ranting of a worn-out defeatist".

Göring and Galland.
Image Source: albumwar2.com

Offers to Resign

In October 1943, Galland met Göring and mentioned the need for new and improved interceptor aircraft. Göring wanted heavy cannon-armed fighters (cannons of some 2,000 lb. in weight) to be used en masse. Galland explained that such a weapon could not be used effectively in an aircraft. Galland also asserted that the use of inappropriate weaponry, such as the German heavy-fighter Messerschmitt Me 410, a favourite of Hitler's, had caused heavy losses. Göring disregarded Galland's arguments and continued his frequent attacks on the fighter force, accusing them of cowardice. Galland, as he always did, defended them, risking his career and, near the end of the war, his life in doing so. Galland stated that he could not agree

Göring appreciating Galland's victories.
Image Source: worthpoint.com

to follow Göring's plans and requested to be dismissed from his post and sent back to his unit. Göring accepted, but two weeks later he apologised to Galland and attributed his behaviour to stress. Galland continued in his post.

In November 1943 Galland issued a communique to the fighter forces, announcing the introduction of new weapons, such as heavily armed Fw 190 fighters, to engage and destroy Allied bombers. Göring ordered his units, through Galland, to use ramming methods, and risk sacrificing the pilot. German losses were so heavy that Galland held a special meeting with Division Commanders on 4 November 1943. It was decided the single-engine fighters must engage in protecting the heavier fighters, such as the Bf 110, from enemy escorts, so the latter could attack the bombers. At the end of December, Galland and the staff concluded that their new tactics had failed with high losses. In mid-March 1944, shortages of skilled pilots caused Galland to send a desperate plea asking for volunteer pilots.

Galland inspecting an aircraft.
Image Source: Pinterest (picture uploaded by Natalee).

American air forces continued unrelenting pressure for the duration of the war. A conference between Galland and Göring in mid-May 1944 underlined how enemy air operations were devastating the fighter force. Galland urged all fighter pilots holding short staff positions be transferred immediately to operational units. Transfers were sought from the eastern front of all pilots with more than five aerial victories. Finally, Galland asked flying schools to release 80-plus instructors.

The Disagreements on Me 262 Aircraft

Galland flew the Me 262 aircraft, father-of-all-jet-planes, in May 1943 and became an enthusiastic supporter of the aircraft as the saviour of the fighter force. Galland failed to appreciate the difficulties involved in transferring a design into production, especially under the circumstances. There were also problems with the engines and series production was difficult. By spring 1944, the Me 262 was sufficiently ready for operational service. By this time, Galland faced rivalries amongst the Luftwaffe Command over how best to employ the aircraft. Galland thought the bomber corps should be disbanded and its pilots converted onto fighters. Göring opposed. Galland did not give up. He made repeated appeals for the Me 262 to be used as a fighter aircraft. Even this was difficult, as Hitler had taken personal control of turbo-jet production and checked where each batch of the aircraft were being deployed.

Father of Jet Planes. German Me 262.
Image Source: warhistoryonline.com

It was not until September 1944 that Hitler rescinded his directive that the Me 262 be used as a fighter-bomber. Galland decided to test the Me 262 against high-flying Allied reconnaissance aircraft. He selected a highly decorated pilot. Hitler heard of the experiment and ordered a stop to it. Galland persisted

Mustang P 51.
Image Source: eskipaper.com

with the experiments and ordered operations to be continued. They achieved isolated successes until the ace pilot Thierfelder was shot down and killed by a P-51 Mustang on 18 July 1944. On 20 August, Hitler finally agreed to allow one in every 20 Me 262 to go into fighter service, which allowed Galland to build an all jet unit. The unit struggled into November 1944 without much success and high losses. Galland himself flew on unauthorised interception flights to experience the combat pressures of the pilots, and witnessed USAAF bombers being escorted by large numbers of P-51 Mustangs.

Dismissal and Relieved of Command

Despite Göring's apology after their previous dispute, the relationship between the two men did not improve. Göring's influence was in decline by late 1944 and he had fallen out of favour with Hitler. Göring became increasingly hostile to Galland, blaming him and the fighter pilots for the situation. In 1944, the situation worsened. By the spring of 1944, the Luftwaffe could not effectively challenge the Allies over France or the Low Countries. Operation Overlord, the Allied invasion of German Occupied Europe took place in June 1944. In the previous four months 1,000 German pilots had been killed.

Adolf Galland inspects a BK 50 armed Me 410B-2/U4 Hornisse unit.
Image Source: i.pinimg.com through Pinterest

Galland reported that the enemy outnumbered his fighters between 6:1 and 8:1, and the standard of Allied fighter pilot training was "astonishingly high".

To win back some breathing space for his force and German industrial targets, Galland formulated a plan which he called the "Big Blow". It called for the mass interception of USAAF bomber formations by approximately 2,000 German fighters. Galland hoped that the German fighters would shoot down some 400–500 bombers. Acceptable losses were to be around 400 fighters and 100–150 pilots. Galland's staff could muster 3,700 aircraft of all types by 12 November 1944, with 2,500 retained for this specific operation. Hitler rejected Galland's plan. Hitler distrusted Galland's theory, and believed him to be afraid, and stalling for time. The Führer was not willing to have German resources sit idle on airfields to wait for an improvement in flying conditions. Göring and Hitler handed over the forces pooled by Galland to Peltz whom they had now appointed commander responsible for virtually all fighter forces in the West. Peltz and his fighter Staff Officer, who were vociferous opponents of Galland, eventually engineered his dismissal. Whether the "Big Blow" operation would have worked is a matter of academic debate. Historians remained divided, with some believing it was a lost opportunity while others think it would have had much less impact than Galland estimated. The operation never took place. Instead, the fighter force was committed to the disastrous operation designed to support German forces during the Battle of Bulge. Galland's influence on matters was now virtually nil. SS General Heinrich Himmler was the most powerful man after Hitler at that time. Himmler, whose relationship with Göring was poor, took the opportunity to exploit the dissent in the Luftwaffe. It was also an opportunity for the SS to seize control of the Luftwaffe and for Himmler to oust Göring from power. Göring, for his part, offered no support to Galland. On 13 January 1945, Galland was finally relieved of his command.

Fighter Pilots Revolt

On 17 January, a group of senior pilots took part in a "Fighter Pilots Revolt". Galland's high standing with his fighter pilot peers led to a group of the most decorated Luftwaffe combat leaders loyal to Galland (including Johannes Steinhoff and Gunter Lutzow) confronting Göring with a list of demands for the survival of their service. Göring initially suspected Galland had instigated the unrest. Heinrich Himmler had wanted to put Galland on trial for treason himself, and the SS and Gestapo had already begun investigations. The more politically acceptable Austrian fighter Ace, Gordon Gollob, a national socialist supporter, succeeded Galland on 23 January. Although professional contemporaries, Gollob and Galland had a mutual dislike. Much earlier Gollob had started to gather evidence to use against Galland, detailing false accusations of his gambling, womanising, and alleged private use of Luftwaffe transport aircraft. The official reason for Galland being relieved of command was his ill health. For his own safety, Galland went to a retreat in the Harz Mountains. He was to keep the government informed of his whereabouts, but was effectively under house arrest. Hitler, who liked Galland, learned of the revolt and ordered that "all this nonsense" was to stop immediately. Hitler had been informed by Galland's close friends. After Hitler's intervention, Göring contacted Galland and invited him to Carinhall, the country residence of Göring. In light of his service to the fighter arm, he promised no further action would be taken against him and offered command of a unit of Me 262 jets. Galland accepted.

Fighter Pilots Revolt. Galland with Göring.
Image Source: Military Wikia

As Germany Collapsed: Command of a New Unit

On 24 February 1945 a new flying unit Jagdverband 44 (JV 44) was formed. The commander of this unit had the disciplinary powers of a Divisional Commander. It was to be directly under Berlin. Galland was given sixteen operational Me 262s and fifteen pilots. Galland quickly got the unit going. Göring showed sympathy for Galland's efforts. Galland requested that all experienced fighter pilots flying with Bf 109 or Fw 190 units should be made to join the Me 262 unit. Galland believed he could get 150 jets in action against the USAAF fleets. The general chaos and impending collapse prevented his plans from being realised. On 31 March 1945, Galland flew 12 operational jets to Munich and on 5 April, began operations. The Me 262s started destroying American aircraft. On 21 April, the unit was visited by Göring for the final time. Göring confessed to Galland that his assertions about the Me 262 and the use of bomber pilots with experience as jet fighter pilots had been correct. Göring said, "I envy you Galland, for going into action. I wish I were a few years younger and less bulky. I would have gladly put myself under your command." On 21 April, Galland was credited with his 100th aerial victory. He was the 103rd and last Luftwaffe pilot to achieve the century mark. On 26 April, Galland claimed his 103rd and 104th aerial victories. His last.

German Surrender: Offer to Join Americans against Soviet Union

By late April, the war was effectively over. On 1 May 1945, Galland attempted to make contact with US Army forces to negotiate the surrender of his unit. The act itself was dangerous. SS forces roamed the countryside and towns executing anyone who was considering capitulation. The Americans requested that Galland fly his

Galland surrenders to the Americans.
Image Source: warfarehistorynetwork.com

unit and Me 262s to a USAAF controlled airfield. Galland declined citing poor weather and technical problems. In reality, Galland was not going to hand over Me 262 jets to the Americans. Galland had harboured the belief that the Western Alliance would soon be at war with the Soviet Union, and he wanted to join American forces and to use his unit in the coming war to free Germany from Communist occupation. Galland made his whereabouts known to the Americans, and offered his surrender once they arrived at the hospital where he was being treated. Galland then ordered his unit to destroy their Me 262s. At the time of his surrender, Galland had filed claims for 104 Allied aircraft shot down. His claims included seven with the Me 262. On 14 May 1945, Galland was flown to England and interrogated by RAF personnel about the Luftwaffe, its organisation, his role in it and technical questions. Galland returned to Germany on 24 August and was imprisoned. On 7 October, Galland was returned to England for further interrogation. He was eventually released on 28 April 1947.

Self-appraisal and Introspection

Galland did not pretend to have been error free. After the war, he was candid about his own mistakes. As the Air Force Fighter Commander, production and aircraft procurement were not his responsibility but Galland identified four major mistakes during the war, and accepted partial responsibility for the first three.

Making of Movie "Battle of Britain". German Technical advisor Adolf Galland and British Technical advisor Stanford Tuck. Both were Aces of respective air forces.
Image Source: daveswarbirds.com

(a) Fighter pilots received no instrument training until very late in the war, after the training courses had already been curtailed because of fuel shortages. Galland also did not make sure all-weather flying was incorporated into pilot training, which was of great importance for an effective air defence force.

(b) Attrition by 1942 had created a shortage of experienced combat leaders. No special training was made available for combat role. Galland set up a course in late 1943, but it only lasted for a few months. Galland thought they could learn the skills while on operations, as he himself had. This was unreasonable.

(c) The Me 262, while not a war winner, might have extended the "Defence of the Reich" campaign. The problems with the engines, failures of production priorities and Hitler's meddling are well known, but the long delay between operational testing, tactical and doctrinal development and training were largely Galland's fault.

(d) The German pilots were increasingly lacking in quantity and quality. Galland recognised this but could not correct it without stepping outside his own authority.

Years in Argentina

After release, Galland travelled to Schleswig-Holstein in the north of Germany, now in Denmark to join Baroness Gisela von Donner, an earlier acquaintance, on her estate and lived with her three children. During this time, Galland found work as a forestry worker. There he convalesced and came to terms with his career and Nazi war crimes. Galland began to hunt for

Galland in Cockpit. British Gloster Meteor. In Argentina 1948-55.
Image Source: adolphgallandinargentina.blogspot.com

the family, and traded the kills in the local markets to supplement meagre meat rations. Soon Galland rediscovered his love of flying. Kurt Tank, the designer of the Fw 190, had been asked to work for the British and Soviets, and had narrowly avoided being kidnapped by the latter. Tank, through a contact in Denmark, informed Galland about the possibility of the Argentinian government employing him as a test pilot for Tank's new generation of fighters. Galland accepted and flew to Argentina. He settled with Gisela in Buenos Aires. Galland enjoyed the slow life. His time there, aside from work commitments, was taken up with Gisela and the active Buenos Aires nightlife. Soon, he took up gliding again. Galland spoke fluent Spanish, which helped in his instruction of new pilots. He flew the British Gloster Meteor which was a contemporary of the Me 262. He claimed that if he could have fitted the Meteor engines to the Me 262 airframe he would have had the best fighter in the world. Galland continued training, lecturing and consulting for the FAA until 1955. In between he kept returning to Europe to test fly new types. For his services to Argentina, Galland was awarded a pilot's wings badge and the title of the Honorary Argentine Military Pilot.

Adolf Galland (in Cockpit), Stanford Tuck and Douglas Bader (bandage on forehead).
Image Source: Flickr (By Andrew Hurdle)

Return to Germany: Denied Job in West German Air Force

In 1955 Galland left South America. By that time, he had begun writing his autobiography, "The First and the Last" that was published in 1954. It was a best-seller in 14 languages and sold three million copies. It was well received by the RAF and USAF. Galland returned to Germany and was approached to join the armed forces of West Germany which was to join NATO. But the chief of staff of the USAF, got inputs that Galland had alleged "strong neo-Nazi leanings", and was known to have served with the Perón dictatorship, which was not on good terms with the United States. Galland was disapproved of for the position of chief of staff to the German Air Force. Americans also suspected that Galland's rapid promotions were due to his association with Hitler rather than his merits. RAF had a different view because of his close association with Jewish pilots who had served in the RAF.

Movie Chief Technical Advisor. Real General Adolf Galland (Left), Shaking-Hands with Actor Hein Riess Playing Reichmarschall Hermann Goering (Right).
Image Source: mediastorehouse.com

Own Aircraft Consultancy

In 1957, Galland moved to Bonn and began his own aircraft consultancy. Galland worked hard and continued flying, taking part in national air shows. In 1956, he was appointed honorary chairman of the Association of German Fighter Pilots. He came into contact with contemporaries in Britain and America. In 1961, he joined the Gerling Group of Cologne who contracted Galland to help develop their aviation business. With business going well, Galland bought his own aircraft, Beechcraft Bonanza, on 19 March 1962, his 50th birthday, and named it *Die Dicke* (Fatty).

Other Engagements

In 1969, he served as technical adviser for the film Battle of Britain in which the character Major Falke was based on Galland. Galland was upset about the director's decision not to use the real names. While making the film, Galland was joined by his friend, British fighter Ace, Robert Stanford Tuck. In 1973, Galland appeared in the British television documentary series The World at War. In 1974, he was part of the remaining German General Staff that took part in the Operation Sea Lion war-game at Sandhurst in the United Kingdom, replicating the planned German invasion of Britain in 1940 (which the German side lost). In 1975, he was a guest at the RAF Museum Hendon, during the unveiling of the Battle of Britain Hall, where he was entertained

Galland chats with U.S. Air Force pilots at Cologne's Flying Day of Nations in June 1956. (Ullstein Bild via Getty Images)

by Prince Charles. In 1980, Galland's eyesight became too poor for him to fly and he retired as a pilot. However, he continued to attend numerous aviation events, including being a periodic guest of the US Air Force for their annual "Gathering of Eagles" program at the Air Command and Staff College at Maxwell, Alabama, USA. In October 1980, he was returned the two Merkel shotguns stolen by American soldiers after his capture in 1945. Galland had located them before and had tried to buy them back, only to be turned down, as they would be worth more after his death. Towards the end of the 1980s, Galland's health began to fail.

Three Marriages and Last Days

Baroness Gisela von Donner had refused to marry Galland as the restrictions imposed upon her by her former husband's will would deny her the wealth and freedom she had enjoyed. She left for Germany in 1954. Galland married Sylvinia von Donhoff on 12 February 1954. However, she was unable to have children and they divorced on 10 September 1963. On 10 September 1963, Galland married his secretary, Hannelies

Ladwein. They had two children: a son and a daughter. The RAF ace and good friend Robert Stanford Tuck was the godfather of his son Andreas. Galland's marriage to Hannelies did not last and on 10 February 1984, he married his third wife, Heidi Horn, who remained with him until his death. By the 1980s, Galland was regularly attending the funerals of friends like Tuck, and Douglas Bader. In early February 1996, Galland was taken seriously ill. He had wanted to die at home and so was released from hospital and returned to his own house. With his wife Heidi, son and daughter present, he was given the last rites. Adolf Galland died in the morning of Tuesday, 9 February 1996. His body was buried at St Laurentius Church, Oberwinter on 21 February.

Adolf "Dolfo" Joseph Ferdinand Galland (19 March 1912 – 9 February 1996). Grave at Oberwinter.
Image Source: tracesofwar.com

Aerial Victory Claims

With his slicked-back black hair and matching moustache, broken nose and perennial cigar, Lieutenant General Adolf Galland was the personification of the Luftwaffe fighter arm during World War II. His Messerschmitt 109s bearing the incongruous Mickey Mouse emblem became iconic images for generations of historians, artists and modellers. Yet those were superficial manifestations of his personality; the man beneath the image was far more intriguing. Researchers have confirmed records for 100 aerial victory claims, plus nine further unconfirmed claims, all of which claimed on the Western Front. This figure of confirmed claims includes two four-engine bombers and six victories with the Me 262 jet fighter.

REFERENCES

1. Adolf Galland, Britannica, https://www.britannica.com/biography/Adolf-Galland
2. Barrett Tillman, Adolf Galland: The Luftwaffe's Fighter General, HistoryNet, https://www.historynet.com/adolf-galland-luftwaffes-fighter-general.htm
3. Colin D. Heaton, Interview with World War II Luftwaffe General and Ace Pilot Adolf Galland, originally appeared in the January 1997 issue of "World War II". HistoryNet. https://www.historynet.com/interview-with-world-war-ii-luftwaffe-general-and-ace-pilot-adolf-galland.htm
4. Adolf Galland, Wikipedia https://en.wikipedia.org/wiki/Adolf_Galland
5. Adolf Galland, Military Wiki Org, https://military.wikia.org/wiki/Adolf_Galland
6. David Binder, Adolf Galland, Top Aviator for the Nazis, Is Dead at 83, The New York Times, February 14, 1996.
7. Grave Adolf Galland, Traces of War, https://www.tracesofwar.com/sights/90163/Grave-Adolf-Galland.htm

6

Fighter Ace Lydia Litvyak "White Lily"

Highest Aerial Victories by a Female Fighter Pilot

Lydia Litvyak.
Image Source: warhistoryonline.com

Lydia Vladimirovna Litvyak also known as Lilya, was a fighter pilot in the Soviet air Force during World War II. Historians' estimates for her total solo victories vary from five to twelve, and two to four shared kills in her 66 combat sorties. In about two years of operations, she was the first female fighter pilot to shoot down an enemy aircraft, and among the first two female fighter pilots who have earned the title of fighter ace and the holder of the record for the greatest number of kills by a female fighter pilot. She was herself shot down near Orel during the Battle of Kursk as she attacked a formation of German aircraft.

Early Years: Harsh Reality

Lydia Litvyak was born in Moscow, on 18 August 1921, into a Russian Jewish family. Her mother Anna Vasilievna Litvyak was a shop assistant, her father Vladimir Leontievich Litvyak worked as a railwayman, train driver and clerk. During the "Great Purge" or the "Great Terror", a campaign of political repression which occurred from 1936 to 1938, he was arrested as an "enemy of the people" and disappeared. Raised at the outset of one of the most exciting era of aviation, the roaring 20's, Lydia grew up reading about aviation, and watching military pilots practice nearby. Lydia became interested in aviation at an early age. At 14, she enrolled in a flying club. She performed her first solo flight at 15 and later graduated from Kherson military flying school. She became a flight instructor at the Kalinin aero club. By the time the German-Soviet war broke out, she had already trained 45 pilots.

Lydia Litvyak.
Image Source: warlordgames.com

Marina Raskova, Hero of the Soviet Union and founder of the 46th Taman Guards Night Bomber Aviation Regiment.
Image Source: warhistory.com

Women's Regiment

After Germany attacked the Soviet Union in June 1941, Litvyak tried to join a military aviation unit, but was turned down because of lack of experience. After deliberately exaggerating her pre-war flight time by 100 hours, she joined the all-female 586th Fighter Aviation Regiment of the Air Defence Force, which was formed by Marina Raskova. According to some stories, Marina Raskova, a famous Soviet navigator, used her personal connections with Joseph Stalin to found the three first combat regiments for female pilots. Thanks to Raskova, by the time Lydia made it past the disapproving recruiters and red tape, she was able to join the all-female 586th Fighter Regiment. She trained there on the Yakovlev Yak-1 aircraft. What the Axis powers were not expecting were the 800,000 Russian women who volunteered for front-line action. Among these women was Lydia Litvyak who would become the holder of the greatest number of kills by a female fighter pilot.

Lydia Litvyak.
Image Source: warhistoryonline.com

Marina Mikhaylovna Raskova

Marina Raskova was the first woman in the USSR to achieve the diploma of professional air navigator. Raskova went from a young woman with aspirations of becoming an opera singer to a military instructor, the first female navigator in the Soviet Union. She was navigator to many record-setting as well as record-breaking flights, as well as founding and commanding officer of the 587th Dive Bomber Regiment which was renamed the 125th M.M. Raskova Borisov Guards Dive Bomber Regiment in her honour. Raskova founding three female air regiments, one of which was would eventually fly over 30,000 sorties in WW II and produce at least 30 Heroes of the Soviet Union.

First four Soviet female pilots.
Image Source: disciplesofflight.com

"Night Witches": The All-Female Soviet Night Bomber Aviators

"Night Witches" was a World War II German nickname for the all-female military aviators of the 588th Night Bomber Regiment, known later as the 46th "Taman" Guards Night Bomber Aviation Regiment, of the Soviet Air Forces. Though women were initially barred from combat, Major Marina Raskova used her position and personal contacts with the Chairman of the Council of People's Commissars, Joseph Stalin, to obtain permission to form female combat units. On 8 October 1941, Order number 0099 specified the creation of three women's regiments—all personnel from technicians to pilots would be entirely composed of women. This included the 588th Regiment. The regiment, formed by Major Marina Raskova and led by Major Yevdokiya Bershanskaya, comprised primarily female volunteers in their late teens and early twenties. An attack technique of the night bombers involved idling the engine near the target and gliding to

Lydia Litvyak.
Image Source: theparisreview.org

the bomb-release point, with only wind noise left to reveal their presence. German soldiers likened the sound to broomsticks and named the pilots "Night Witches".

During the yearlong training, the women aviators were sorted by ability levels to form the three all-female regiments: the 586 Fighter Aviation Regiment, the 587 Bomber Aviation Regiment, and the 588 Night Bomber Aviation Regiment. The most-skilled aviators became fighter pilots and, to the ire of their male counterparts, were issued brand-new Yakovlev Yak-1s. The middle-tier pilots were assigned to the bomber regiment, and the lowest-scoring pilots were assigned to fly night bombers and were issued a plane that no one else wanted to fly: the Polikarpov Po-2, a 1928 trainer constructed from wood and canvas with no heat, an open cockpit, and a 100-horsepower engine. Litvyak was among the best and therefore was chosen as a day fighter pilot.

Assigned to Men's Regiment

Litvyak flew her first combat flights in the summer of 1942 over Saratov. In September, she was assigned to the 437 Fighter Regiment, a men's regiment fighting over Stalingrad. On 10 September she moved along with Yekaterina Budanova, and two other women pilots, and accompanying female ground crew, to the regiment airfield, at Verkhnaia Akhtuba, on the east bank of the Volga River, near Volgograd. But when they arrived the base was empty and under attack. They were then moved to Srednaia Akhtuba. Here, flying a Yak-1, Cowl number "32" on the fuselage, she would achieve considerable success. Boris Yeremin (later Lt General of aviation), then a regimental commander in the division to which she and another female pilot Budanova were assigned, saw her as "a very aggressive person" and "a born fighter pilot".

Yekaterina Budanova and Lydia Litvyak.
Image Source: war history.com

World War II: First Women to Score Aerial Victory

In the 437th Fighter Regiment, Litvyak scored her first two kills on 13 September, three days after her arrival and on her third mission to cover Stalingrad, becoming the first woman fighter pilot to shoot down an enemy aircraft. That day, four Yak-1s, with Major S. Danilov in the lead—attacked a formation of Junkers Ju 88s escorted by Messerschmitt Bf 109s. Her first kill was a Ju 88 which fell in flames from the sky after she fired several bursts. Then she shot down a Bf 109 G-2 "Gustav" on the tail of her squadron commander, Raisa Beliaeva. The Bf 109 was piloted by a decorated pilot from the 4th Air Fleet, and an 11-victory ace, Staff Sergeant Erwin Maier of the 2nd Staffel of JG 53. Maier parachuted from his aircraft, was captured by Soviet troops, and made to meet the Russian ace who had shot him down. When he was taken to Litvyak, he thought he was being made the butt of a Soviet joke. It was not until Litvyak described each manoeuvre of the fight to him in perfect detail that he knew he had been shot down by a woman pilot. But according to some other authors, the first air victory by a female pilot was achieved by Lieutenant Valeriya Khomyakova of the 586th Regiment when she shot down the Ju 88 flown by Oblt. Gerhard Maak of 7/KG76 on the night of 24 September 1942.

A wrecked Junkers Ju 88 – the first "victim" of Litvyak.
Image Source: warhistory.com

Shoots a Knight's Cross Holder

On 14 September, Litvyak shot down another Bf 109. Her victim was probably Knight's Cross holder and "71 kill ace" Lt. Hans Fuss (Adj.II./JG-3), injured in aerial combat with a Yak-1 on 14 September 1942 in Stalingrad area, when his G-2 fuel tank was hit, his plane somersaulted during the landing when he ran out of fuel flying back to base. He was critically injured, lost one leg, and died of his wounds on 10 November 1942. On 27 September, Litvyak scored an air victory against a Ju 88, the gunner having shot up the regiment commander, Major M.S. Khovostnikov. Possibly Ju 88A-4 "5K + LH" of Iron Cross holder Oberleutnant (Lieutenant) Johann Wiesniewski, 2/KG 3, killed in Action (KIA) with all crew members. Some historians credit it as her first kill.

Bf 109.
Image Source: Wikipedia

Change of Unit to Stay on Yak-1s

Female pilots Litvyak, Beliaeva, Budanova and Kuznetsova stayed in the 437th Regiment for a short time only, mainly because it was equipped with LaGG 3s (Lavochkin-Gorbunov-Gudkov) rather than Yak-1s, that the women were trained on. So the four women were moved to the 9th Guards Fighter Regiment. From October 1942 till January 1943, Litvyak and Budanova served, still in the Stalingrad area, with this famous unit, commanded by Lev Shestakov, Hero of the Soviet Union.

Lev Lvovich Shestakov.
Image Source: alchetron.com

Lev Lvovich Shestakov

Lev Shestakov was a Soviet military aviator and the Red Air Force's leading ace in the Spanish war. Flying Polikarpov I-16s he claimed eight solo victories and 31 collaborative (shared) victories gained in 90 sorties during the Spanish Civil War. Shestakov joined 69th Fighter Aviation Regiment in September 1939, and was at the time one of the most famous Soviet aces. His unit operated in the Odessa front. He became the Regiment Leader on 16 July 1941. During the battle for Odessa 69th Fighter Aviation Regiment pilots achieved 94 air victories. The losses inflicted on the Romanian Air Force above Odessa in 1941 by Shestakov's fighter pilots compelled the Romanian High Command to withdraw its entire air force from the Eastern Front. At the end of 1941, 69th Fighter Aviation Regiment received the LaGG-3 to replace the outdated I-16 and relocated to the Stalingrad area. Shestakov eventually flew more than 200 missions during the war, took part in 32 aerial combats, and was credited with 26 kills before being killed in action on 13 March 1944. His 26 victories in WW II raised his career total to 65, including shared. His decorations included Hero of the Soviet Union, and Order of Lenin, among others.

Lavochkin, Gorbunov, Gudkov (LaGG). LaGG-3 aircraft.
Image Source: flugzeuginfo.net

The Free Hunter Concept

In January 1943, the 9th was re-equipped with the Bell P-39 Airacobras and Litvyak and Budanova were moved to the 296th Fighter Regiment (later designated again as the 73rd Guards Fighter Aviation Regiment) of Nikolai Baranov, of the 8th Air Army, so that they could still fly the Yaks. On February 23, she was awarded the Order of the Red Star, and made a junior lieutenant, and selected to take part in the elite air tactic called "okhotniki' (free hunter), where pairs of experienced pilots searched for targets on their own initiative. Twice, she was forced to land due to battle damage.

Order of the Red Star.
Image Source: Wikimedia Commons

Wounded For the First Time

On 22 March she was wounded for the first time. That day she was flying as part of a group of six Yak fighters when they attacked a dozen Ju 88s. Litvyak shot down one of the bombers but was in turn attacked and wounded by the escorting Bf 109s. She managed to shoot down a Messerschmitt and to return to her airfield and land her plane, but was in severe pain and losing blood. Later she recovered.

Lydia Litvyak.
Image Source: Pinterest (Upload by Natalee)

In Love with Her Leader

While in 73rd Regiment, she often flew as wingman of Captain Aleksey Solomatin, a flying ace with a claimed total of 39 victories (22 shared). On 21st May, while training a new flyer, Solomatin was killed in front of the entire regiment in Pavlonka when he flew into the ground. Litvyak was devastated by the crash and wrote a letter to her mother describing how she realized only after Solomatin's death that she had loved him. Senior Sergeant Inna Pasportnikova, Litvyak's mechanic during the time she flew with the men's regiment, narrated in 1990 that after Solomatin's death, Litvyak wanted nothing but to fly combat missions, and she fought angrily and desperately.

Alexei Solomatin.
Image Source: livelongandprosper28.wordpress.com

Shooting a Manned Observation Balloon

Litvyak scored against a difficult target on 31 May 1943, a manned German artillery observation balloon. German artillery used the balloons as an observation post for guiding artillery. Shooting the balloons had been attempted by other Soviet airmen but all had failed due to a dense protective belt of anti-aircraft fire defending the balloon. Litvyak volunteered to take out the balloon but was not allowed initially. She insisted and described her tactics to the commander. She would attack it from the rear after flying in a wide circle around the perimeter of the battleground and over German-held territory. The tactic worked. The hydrogen-filled balloon caught fire under her stream of tracer bullets and was destroyed.

German artillery observation balloon.
Image Source: flightlineweekly.com

Flight Commander and Lots More Action

On 13 June 1943, Litvyak was appointed flight commander of the 3rd Aviation Squadron within the 73rd Guards Fighter Aviation Regiment. Litvyak had another victory on 16 July 1943. That day, six Yaks encountered 30 German Ju 88 bombers with six escorts. The female ace downed a bomber and shared a victory with a comrade, but her fighter was hit and she had to make a belly landing. She was wounded again but refused to take medical leave. She shot down one Bf 109 on 19 July 1943, probably 6-kill ace Uffz. Helmuth Schirra, 4/JG-3 (KIA, Luhansk area).

Litvyak (far left). Part of the Men's Regiment.
Image Source: aviationoiloutlet.com

Another Bf 109 kill followed two days later on 21 July 1943, possibly Bf 109G-6 of Iron Cross holder and 28-kill expert Lt. Hermann Schuster 4./JG-3(KIA, near Pervomaysk, Luhansk area).

Last Mission

On August 1, 1943, Litvyak did not come back to her base at Krasnvy Luch. It was her fourth sortie of the day, escorting a flight of Ilyushin IL-2 ground-attack aircraft. As the formation was returning to base near Orel, a pair of Bf 109 fighters dove on Litvyak while she was attacking a large group of German bombers. Soviet pilot Ivan Borisenko recalled, "Lily just didn't see the Messerschmitt 109s flying cover for the German bombers. A pair of them dove on her and when she did see them she turned to meet them. Then they all disappeared behind a cloud." Borisenko, involved in the dogfight, saw her the last time, through a gap in the clouds, her Yak-1 pouring smoke and pursued by as many as eight Bf 109s. Borisenko descended to see if he could find her. No parachute was seen, and no explosion. She never returned from the mission. Litvyak was just 21 years old. Soviet authorities suspected that she might have been captured, a possibility that prevented them from awarding her the title of Hero of the Soviet Union.

Krasnyi Luch wall of Honour to the Heroes of War and Labour. Litvyak took off for her last mission from an airfield close to this city, where a museum dedicated to her is located.
Image Source: Wikiwand

Who Shot Her

One of the two German pilots is believed to have shot down Litvyak. Iron Cross holder and 30-kill expert Fw Hans-Jorg Merkle, or Knight's Cross holder and future 99-kill expert Lt Hans Schleef. Merkle is the only pilot that claimed a Yak-1 near Dmitryevka on 1 August 1943, his 30th victory. Dmitryevka is where she was last seen and was reportedly buried. This occurred before being rammed and killed by his own victim. Luftwaffe combat report of collision also mentioned 3 km east of Dmitryevka. While Schleef claimed a LaGG-3, often confused in combat with Yak-1s by German pilots, kill on the same day, in the South-Ukraine area where Litvyak's aircraft was at last found.

Search and Recognition

Hero of the Soviet Union.
Image Source: pinpng.com

In an attempt to prove that Litvyak had not been taken captive, senior Sergeant Inna Pasportnikova, Litvyak's mechanic, embarked on a 36-year search for the Yakovlev Yak-1 crash site assisted by the public and the media. For three years she was joined by relatives who together combed the most likely areas with a metal detector. In 1979, after uncovering more than 90 other crash sites, 30 aircraft and many lost pilots killed in action, "the searchers discovered that an unidentified woman pilot had been buried in the village of Dmitrievka, Shakhterski district. It was then assumed that it was Litvyak and that she had been killed in action after sustaining a mortal head wound. Pasportnikova said that a special commission was formed to inspect the exhumed body and it concluded the remains were those of Litvyak. On 6 May 1990, USSR President Mikhail Gorbachev posthumously awarded her the title Hero of the Soviet Union. Her final rank was senior lieutenant, as documented in all Moscow newspapers of that date.

Death Controversy

There is still some dispute about the official version of Litvyak's death. Yekaterina Valentina Vaschenko, the curator of the Litvyak museum in Krasnyi Luch has stated that the body was disinterred and examined by forensic specialists, who determined that it was indeed Litvyak. However, Kazimiera Janina Cottam claims,

Image Source: kjohnsonnz.blogspot.com

on the basis of evidence provided by Yekaterina Polunina, chief mechanic and archivist of the 586th Fighter Regiment in which Litvyak initially served, that the body was never exhumed and that verification was limited to the comparison of a number of reports. Cottam, an author, and researcher focusing on Soviet women in the military, concludes that Litvyak made a belly-landing in her stricken aircraft, was captured and taken to a prison of war camp. In her book published in 2004, Polunina lists the evidence that led her to conclude that Litvyak was pulled from the downed aircraft by German troops and held prisoner for some time.

Gian Piero Milanetti, the author of a recent book about Soviet aviatrixes, wrote that an airwoman parachuted in the approximate location of the alleged crash landing of Litvyak's aircraft. No other Soviet airwomen operated in that area, so Milanetti believes the pilot was Litvyak, probably captured by the enemy. Russian aviation historian, Anatoly Plyac, former KGB major, told Milanetti: "Litvyak survived and was taken prisoner. A television broadcast from Switzerland was seen in 2000 by Raspopova, a veteran of the women's night bomber regiment. It featured a former Soviet woman fighter pilot who Raspopova thought may have been Litvyak. This veteran was wounded twice. Married outside of the Soviet Union, she had three children. Raspopova promptly told Polunina what she inferred from the Swiss broadcast.

Russian Blondes

Lidya Litvyak and Roza Shanina were famously called the "Beautiful and Lethal". Roza was a Soviet sniper during World War II who was credited with fifty-nine confirmed kills. Shanina volunteered for the military after the death of her brother in 1941 and chose to be a marksman on the front line. Praised for her shooting accuracy, Shanina was capable of precisely hitting enemy personnel and making doublets (two target hits by two rounds fired in quick succession).

Number of Kills and Awards

There is no consensus among historians about the number of aerial victories scored by Litvyak. Russian historians Andrey Simonov and Svetlana Chudinova were able to confirm five solo and three team shoot-downs of enemy aircraft plus the destruction of the air balloon with archival documents. Various other tallies are attributed to her, including eleven solo and three shared plus the balloon, as well as eight individual and four team kills. Anne Noggle credits her with twelve individual and two

Image Credit: flightlineweekly.com

team shoot-downs, Pasportnikova stated in 1990 that the tally was eleven solo kills plus the balloon, and an additional three shared. Polunina has written that the kills of famous Soviet pilots, including those of Litvyak and Budanova, were often inflated; and that Litvyak should be credited with five solo aircraft kills and two group kills, including the observation balloon. What comes out clear is that she had between 5 and 12 kills. Litvyak was awarded the Order of the Red banner, Order of the Red Star, and was twice honoured with the Order of the Patriotic War. She was posthumously awarded the title Hero of the Soviet Union.

Lidiya Litvyak, a rebellious and romantic woman.
Image Source: oddfibulae.com

Strong Character Yet believed in Luck

Litvyak displayed a rebellious and romantic character. Returning from a successful mission, she would "buzz" the aerodrome and then indulge in unauthorised aerobatics, knowing that it enraged her commander. Litvyak could also be superstitious. Paspotnikova testified, "She never believed that she was invincible. She believed that some pilots had luck on their side and others didn't. She firmly believed that, if you survived the first missions, the more you flew and the more experience you got your chances of making it would increase. But you had to have luck on your side".

Femininity and Fashion

Despite the predominantly male environment in which she found herself, she never renounced her femininity and would carry on dyeing her hair blonde, sending her friend Inna Pasportnikova to the hospital to fetch hydrogen peroxide for her. She would fashion scarves from parachute material, dyeing the small

Lidia Vladimirovna Litvyak "White Lily".
Image Source: findagrave.com

pieces in different colours and stitching them together and would not hide her love of flowers, which she picked at every available occasion, favouring red roses. She would make bouquets and keep them in the cockpit, which were promptly discarded by the male pilots who shared her aircraft.

Yak-1B, White 23, 73rd GvIAP. Summer 1943, flown by Lydia Litvyak.
Image Source: forum.worldofwarplanes.com/

Fiancé and Love

Her comrade Solomatin is believed to have been her fiancé, and after his death, she wrote to her mother, "You see, he was not my type, but his insistence and his love for me convinced me to love him.....and now, it seems I will never meet someone like him ever again". The novel Vernis iz Poleta ("Return from Flight") by Natalya Kravtsova fictionalizes the death of Solomatin, stating that he was killed when he ran out of ammunition while battling with a German Bf 109 fighter plane over his own airfield. In this version, Litvyak and others at the airfield watched the fight and witnessed his death.

White Lily and Red Rose

Litvyak was called the "White Lily of Stalingrad" in Soviet press releases. The white lily flower may be translated from Russian as Madonna lily. She has also been called the "White Rose of Stalingrad" in Europe and North America since reports of her exploits were first published in English.

Dramatization and Books "Call Sign, White Lily"

Litvyak is the major character in Mary Ann Cook's romanticized novel The White Rose, a fictional account of her wartime experiences. A heavily fictionalized Litvyak (called Natasha in the book) is the main character of Belinda Alexandra's novel "Sapphire Skies". Perhaps the most detailed work of literary fiction about Litvyak, her life, times and loves, was written by an American, M.G. Crisci, with no Russian ancestry, in cooperation with Valentina Vaschenko, the curator of the Lilya Litvyak Museum and School in Krasny Luch, Eastern Ukraine. The book entitled "Call Sign, White Lily," also contains never-before-seen photographs contributed by the museum. Litvyak posed as a model for Sanya V. Litvyak, a character in Strike Witches. In Zap Comix (February 1985), graphic artist Spain Rodriguez dramatizes the fighter pilot's story in a ten-page narrative, "Lily Litvak, the Rose of Stalingrad." The work is also included in Rodriguez's 1995 Fantagraphics Books anthology, "My True Story." On March 22, 2019, director Andrei Chaliop announced a film about Lydia Litvyak to be filmed in conjunction with director Kim Druzhinin and produced by the 28 Panfilov studio. In July 2019 the Israeli Metal Band Desert Released a song named "Fortune favours the brave" about Lily and her bravery in their album with the same name.

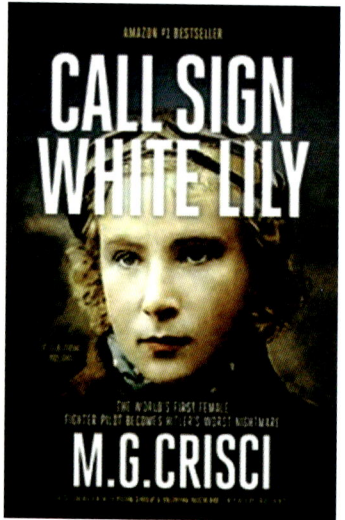
Image Source: books.google.co.in

Image Source: Amazon.com

Stage Play

The play "White Rose" by Scottish playwright Peter Arnott portrays Litvyak's imagined political thoughts, with her character discussing war and Soviet women's resistance against Nazism. It was first performed on 22

May 1985 at the Edinburgh Festival. There has been a production at the Tron Theatre in Glasgow in February–March 2013.

REFERENCES

1. Lydia Vladimirovna Litvyak, Wikipedia, https://en.wikipedia.org/wiki/Lydia_Litvyak
2. David T. Zabecki, First Female Ace: Lydia Litvyak, History Net, July 2020, https://www.historynet.com/first-female-ace-lydia-litvyak.htm
3. Lydia Litvyak, World War II Database, https://ww2db.com/person_bio.php?person_id=433
4. Claudia Mendes, Guest Author, The Highest-Scoring Female Fighter Ace Ever: The Short but Daring Life of Lydia Litvyak, War History online, November 13,2018, www.warhistoryonline.com/instant-articles/life-of-lydia-litvyak.html
5. Pilot Profile: Lydia Litvyak, the World's First Female Fighter Ace By Aviation Oil Outlet on Apr 13th 2018, https://aviationoiloutlet.com/blog/lydia-litvyak-first-female-fighter-ace/
6. Edward White, The Short, Daring Life of Lilya Litvyak, the Paris Review, October 6, 2017, https://www.theparisreview.org/blog/2017/10/06/short-daring-life-lilya-litvyak-white-rose-stalingrad/
7. Carly Courtney, The First Female Flying Ace: Lydia Litvyak, January 17, 2017, https://disciplesofflight.com/first-female-ace-lydia-litvyak/
8. The story of Lydia Litvyak, YouTube, https://www.youtube.com/watch?v=w1ogph2KHA0
9. Lydia Vladimirovna "Lilya" Litvyak, Find A Grave, https://www.findagrave.com/memorial/195310349/lydia-vladimirovna-litvyak
10. Lydia Litvyak, https://alchetron.com/Lydia-Litvyak

7

"Triple Ace" American Fighter Pilot "Robin Olds"

World War II and Vietnam War

Robin Olds.
Image Source: af.mil

R obin Olds was a "triple ace" American fighter pilot with a combined total of 17 victories in World War II and the Vietnam War. He retired in 1973 as a Brigadier General of the United States Air Force (USAF) after 30 years of service. The son of Army Air Force Major General Robert Olds, he was educated at West Point and saw upbringing in the early years of the United States Army Air Corps. Olds epitomized the youthful World War II fighter pilot. He remained in the service as it became the USAF, despite often being at odds with its leadership, and was one of its pioneer jet pilots. Rising to the command of two fighter wings, Olds is regarded among aviation historians, and his peers, as the best wing commander of the Vietnam War, for both his air-fighting skills and his reputation as a combat leader.

Olds was promoted to brigadier general after returning from Vietnam but did not hold another major command. The remainder of his career was spent in non-operational positions, such as Commandant of Cadets at the USAF Academy and a posting in the Air Force Inspector General's Office. His inability to rise higher as a general officer is attributed to both his maverick views and his penchant for drinking.

Robin's Father Major General Robert Olds.
Image Source: af.mil

Olds had a highly publicized career and life, including marriage to Hollywood actress Ella Raines. As a young man, he was also recognised for his athletic prowess in both high school and college, being named an All-American lineman in college football. Robin's expression about fighter pilots is summed up in his quote "There are pilots and there are pilots; with the good ones, it is inborn. You can't teach it. If you are a fighter pilot, you have to be willing to take risks."

Young Days: Fascinated About Aviation

Robin was born in Honolulu, Hawaii, on July 14, 1922, into an army family and spent much of his youth and did his schooling in Hampton Virginia. His father was Captain (later Major General) was an instructor pilot in France during WW I, and former aide to Brigadier General Billy Mitchell,

Major Robin Olds, United States Army Air Forces. 1946.
Image Source: Life Magazine

who was a leading advocate of strategic bombing in the Air Corps. His mother, died when Robin was four and he was raised by his father. Olds was the eldest of four brothers. Growing up primarily at Langley Field (airbase on the east coast), Virginia, Olds virtually made daily contact with the small group of officers who would lead the U.S. Army Air Forces (USAAF) in World War II. One neighbour was Major Carl Spaatz, destined to become the first Chief of Staff of the US Air Force (USAF). As a result, Olds was imbued with an unusually strong dedication to the air service, and conversely, with a low tolerance for officers who did not exhibit the same. On November 10, 1925, his father appeared as a witness on behalf of Billy Mitchell during Mitchell's court-martial in Washington D.C. He brought three-year-old Robin with him to court, dressed in an Air Service uniform, and posed with him for newspaper photographers before testifying.

First Flight

Olds first flew at the age of eight, in an open cockpit biplane operated by his father. At the age of 12, Olds made up his mind to join the U.S. Military Academy at West Point. His goals were to become an officer and a military aviator, and become a football player. His father was made commander of the pioneer B-17 Flying Fortress 2nd Bombardment Group at Langley Field on March 1, 1937, and promoted to Lieutenant Colonel. Olds attended Hampton High School where he was elected president of his class three successive years and played varsity high school football on a team that won the state championship of Virginia in 1937. Olds was aggressive, even mean, as a player.

Olds' Football Position: Tackle.
Years: 1941-1942.
Image Source: footballfoundation.org

Enters West Point

Instead of joining college after graduating in 1939, Olds enrolled at Millard Preparatory School for West Point in Washington D.C. When Germany invaded Poland, Olds attempted to join the Royal Canadian Air Force but was thwarted by his father's refusal to approve his enlistment papers. After completing Millard Prep, he applied for the U.S. Military Academy at West Point. He passed the West Point entrance examination and was accepted into the Class of 1944 on June 1, 1940. One month after he entered the academy the Japanese attacked Pearl Harbour, Olds was sent to the Spartan School of Aeronautics in Tulsa, Oklahoma, for flight training. This training ended a year later by Christmas 1942.

Football Hall of Fame

Olds played football on a freshman squad. Olds played on the varsity college football team in both 1941 and 1942. At 6 foot 2 inches in height and weighing 92 kg, he played tackle on both offense and defence. In 1942 he was named by 'Collier's Weekly' as its "Lineman of the Year" and by Grantland Rice as "Player of the Year." Olds was also selected as an All-American as the cadets compiled a 6–3 record, beating most major opponents. In the Army-Navy game of 1942, Olds had both upper front teeth knocked out when he received a forearm blow to the mouth while making a tackle. In 1985 Olds was enshrined in the College Football Hall of Fame.

West Point Military Academy.
Representative Image Source: britannica.com

Days at West Point

Olds developed ambivalent feelings about West Point, admiring its dedication to "Duty, Honour, Country", but disturbed by the tendency of many tactical officers to distort the purpose of its Honour Code. In March 1943, Olds was braced by an officer upon returning from leave in New York City, and compelled on penalty of an honour violation to admit he had consumed alcohol. The infraction reduced him in rank from cadet captain to cadet private, characterized by Olds in his memoirs as "only the second cadet in the history of West Point to earn that dubious honour." He walked punishment tours until the day of his graduation in June. The incident left its mark on Olds such that when he became Commandant of Cadets at the Air Force Academy, use of the Honour Code as an instrument for integrity rather than as a tool for petty enforcement of discipline became a point of emphasis in his administration. During his Academy years Olds also acquired a strong contempt for alumni networking, commonly called "ring knocking", to the degree that he went out of his way to conceal his West Point background.

Curriculum Cut Short Due War

By an Act of Congress on October 1, 1942, during Olds' second year, the academy began a three-year curriculum for the duration of the war for cadets entering after July 1939. Cadets applying to the Air Corps were classified as Air Cadets, with a modified curriculum that provided flying training but removed military topography and graphics from the ground syllabus for pilots. Olds' class was given an abridged second class course of study until January 19, 1943, when it began an abridged first class course. Olds completed basic and advanced flying training at Stewart Field, New York. 208 cadets including Olds completed the course, while five classmates died in accidents. Olds received his pilot's wings on May 30, 1943, and graduated on June 1 as a member of the Class of June 1943, 194th in general merit of 514 graduates.

29 October 1941, the Stewart Field lands were made part of West Point.
Image Source: military.wikia.org

World War II. Flying P-38 Lightning: First Kill

Second Lieutenant Olds completed fighter pilot training at the operational training unit on Lockheed P-38 Lightning. Olds was promoted to first Lieutenant on December 1, 1943. Olds logged 650 hours of flying time during training, including 250 hours in the P-38. His unit the 479th fighter group moved to Scotland on May 14, 1944. The 479th began combat on May 26, flying bomber escort missions and attacking transportation targets in occupied France in advance of the invasion of Normandy. Olds flew a new P-38J that he nicknamed Scat II. Olds' crew chief, T/Sgt. Glen A. Wold, said that he showed an immediate interest in aircraft maintenance and learned emergency servicing under Wold. He also insisted his aircraft be waxed to reduce air resistance and helped his maintenance crew carry out their tasks. On July 24 Olds was promoted to Captain and became a flight and later squadron leader. Following a low-level bridge-bombing mission to Montmirail, France, on August 14, Olds shot down his first German aircraft, a pair of Focke-Wulf FW 190s.

Lieutenant Robin Olds with "SCAT II," a Lockheed P-38J-15-LO Lightning.
Image Source: Imperial War Museum

Dead Stick Shoot

On an escort mission on August 23, his flight was on the far left of the group's line abreast formation and encountered 40–50 Messerschmitt Bf 109s in a loose formation of three large Vics. Olds turned his flight left and began a ten-minute pursuit in which they climbed to altitude above and behind the Germans. Undetected by the Germans, Olds and his wingman jettisoned their fuel drop tanks and attacked. Just as Olds began firing, both engines of his P-38 quit from fuel exhaustion. In the excitement of the attack, he had neglected to switch to his internal fuel tanks. He continued attacking in "dead-stick mode", hitting his target in the

Lockheed P-38 Lightning.
Image Source: warhistoryonline.com

fuselage and shooting off part of its engine cowling. After fatally damaging the Bf 109 he dived away and restarted his engines. Despite battle damage to his own plane, including loss of a side window of its canopy, Olds shot down two during the dogfight and another on the way home to become the first ace of the 479th FG.

He made a total of eight claims while flying the P-38 (five of which are sustained by the Air Force Historical Research Agency) and was originally credited as the top-scoring P-38 pilot of the European Theatre of Operations.

Messerschmitt Bf 109.
Image Source: Wikipedia

P-51 D 'SCAT VI' Mustang Robin Olds, 434th FS, 479th FG.
Image Source: digitalcombatsimulator.com

P-51 Mustang Pilot

The 479th FG converted to the P-51 Mustang in mid-September. On his second transition flight, at the point of touchdown during landing, Olds learned a lesson in "false confidence" when the powerful torque of the single-engine fighter forced him into a ground loop after the Mustang veered off the runway. Olds shot down an Fw 190 in his new "P-51 Scat VI" on October 6 during a savage battle near Berlin in which he was nearly shot down by his own wingman. He completed his first combat tour on November 9, 1944, accruing 270 hours of combat time and six kills.

Second Tour in Europe

After returning to the United States for a two-month leave, Olds began a full second tour at Wattisham on January 15, 1945. He was assigned duties as operations officer of the 434th Fighter squadron. Promoted to Major on February 9, 1945, Olds claimed his seventh victory southeast of Magdeburg, Germany, and the same day, downing another Bf 109. On February 14, he claimed three victories, two Bf 109s and an Fw 190, but one of the former was credited only as a "probable".

Focke-Wulf Fw 190.
Image Source: Pinterest (Uploaded by Madi H)

Final WW II Kill

His final World War II aerial kill occurred on April 7, 1945, when Olds in 'Scat VI' led the 479th Fighter Group on a mission escorting B-24s bombing an ammunition dump in Germany. The engagement marked the only combat appearance of "Sonderkommando Elbe", a German Air Force squadron formed to ram Allied bombers. Olds noticed contrails popping up above a bank of cirrus clouds, of aircraft flying above and to the left of the bombers. For five minutes these bogeys paralleled the bomber stream while the 479th held station. Turning to investigate, Olds saw pairs of Me 262s turn towards and dive on the Liberators. After damaging one of the jets in a chase meant to lure the fighter escort away from the bombers, the Mustangs returned to the bomber stream. Olds observed a Bf 109 attack the bombers and shoot down a B-24. Olds pursued the Bf 109 through the formation, and shot it down.

Messerschmitt Me 262.
Image Source: asisbiz.com

Strafing Credits

Olds achieved the bulk of his strafing credits the following week in attacks on German airdromes on April 13, and an airfield in Austria on April 16, when he destroyed six German planes on the ground. He later reflected on the hazards of such missions: "I was hit by flak as I was pulling

Airmen First. Celebrating with his ground Crew.
Image Source: medium.com/@the.colosseum.blog/brig-gen-robin-olds

out of a dive-strafing pass on an airfield called Tarnewitz, up on the Baltic. Five P-51s made a pass on the airdrome that April day. I was the only one to return home. When I tested the stall characteristics of my wounded bird over our home airfield, I found it quit flying at a little over 175 mph (282 km/h) indicated and rolled violently into the dead wing (the right flap had been blown away and two large holes knocked in the same wing). What to do? Bailout seemed the logical response, but here's where sentiment got in the way of reason. That Scat VI airplane had taken me through a lot and I was damned if I was going to give up on her…why the bird and I survived the careening, bouncing and juttering ride down the length of the field, I guess I'll never know."

Robin Olds as depicted in a new documentary on Uflytv.
Image Source: Uflytv.

P-80 Shooting Star aircraft.
Image Source: Wikipedia

Command of the Squadron: "Ace on Two Types"

Olds was given command of his squadron on March 25, less than two years out of West Point and at only 22 years of age. By the end of his combat service he was officially credited with 13 German planes shot down and 11.5 others destroyed on the ground. Olds became an ace on both of his combat tours and was twice awarded the Silver Star, for the mission of August 25, and for the achievements of himself and his squadron during his combined tours. As recognized by the American Fighter Aces Association, Olds was the only pilot to "make ace" in both the P-38 (5 victories) and the P-51 (8 victories).

Post WW II Assignments: Professional vs. Career Struggle

Returning to the United States after the war, Olds was assigned at West Point as an assistant football coach for Red Blaik. Apparently resented by many on the staff for his rapid rise in rank and plethora of combat decorations, Olds was transferred in February 1946 to the 412th Fighter Group to fly the P-80 Shooting Star. He then began a career-long professional struggle with superiors he viewed as more promotion, than warrior-minded.

Olds with the P-80 Demonstration team.
Image Source: the rake.com

First Jet Aerobatics Team and Demonstration Flights

In April 1946, he and Lieutenant Colonel John C. "Pappy" Herbst formed what he believed was the Air Force's first jet aerobatics demonstration team. In late May, the 412th was ordered to undertake Project Comet a nine-city transcontinental mass formation flight. Olds and Herbst performed a two-ship aerobatic routine that thrilled the crowds at every stop. In June, Olds was one of four pilots who participated in the first one-day, dawn-to-dusk, transcontinental round trip jet flight from March Field, California to Washington, D.C. The jet demonstration performances with Herbst ended tragically on July 4, 1946, when Herbst crashed at the Del Mar Racetrack after his aircraft stalled during an encore of their routine finale in which the P-80s did a loop while configured to land. Later that same year Olds took second place in the "Thompson Trophy" race (Jet Division) of the Cleveland National Air Races in Ohio. In this first "closed course" jet race, six P-80s competed against each other on a three pylon course 30 miles in length.

USAF/RAF Exchange Program to England

Olds went to England under the USAF/Royal Air Force (RAF) Exchange Program in 1948. Flying the Gloster Meteor jet fighter, he commanded No.1 Squadron at RAF Station Tangmere between October 20, 1948 and September 25, 1949, becoming the first foreigner to command an RAF unit in peacetime.

Missed Service in Korean War: Wanted to Resign

Strong and Clear Views. Olds making a point.
Image Source: worldwarwings.com

End of exchange program, on November 15, 1949, Olds returned to become operations officer of the 94th Fighter Squadron flying F-86 Sabres. Later Olds was assigned to command the 71st Fighter Squadron which soon joined the Air Defence Command based in Pennsylvania. As a result, he missed service in the Korean War despite repeated applications for a combat assignment. Discouraged and at odds with the Air Force, in which he was seen as an iconoclast, Olds reportedly was in the process of resigning when he was talked out of it by a mentor, Major General Frederic H. Smith Jr., who brought him to work at Eastern Air Defence Command headquarters in New York.

Promoted Colonel: Unenthusiastic Staff Appointments

Olds was promoted to Lt Colonel on February 20, 1951, and Colonel on April 15, 1953, while just thirty years of age and just short of ten years after his graduation from West Point. Olds served unenthusiastically in several staff assignments until returning to flying in 1955. At first on the command staff of the 86th Fighter-Interceptor Wing, in West Germany. Olds then commanded its Sabre-equipped 86th Fighter-Interceptor Group from October 8, 1955, to August 10, 1956. He then was made chief of the Weapons Proficiency Center in Libya, in charge of all fighter weapons training for the USAF Europe until July 1958.

Deputy Chief, Air Defence Division

Olds had administrative and staff duty assignments at the Pentagon between 1958 and 1962 as the Deputy Chief, Air Defence Division, Headquarters USAF. In this assignment he prepared a number of papers, iconoclastic at the time, which soon became prophetic, including identifying the need for upgraded conventional munitions (foretelling the "bomb shortage" of the Vietnam War), and the lack of any serious tactical air training in conventional warfare. From November 1959 to March 1960, his section worked intensely to develop a program reducing the entire structure of the ADC with the purpose of generating $6.5 billion for classified funding to develop the SR-71 Blackbird. Following his Pentagon assignment, Olds attended the National War College in Washington D.C. graduating in 1963.

SR-71 Blackbird.
Image Source: nationalinterest.org

Commander 81st Tactical Fighter Wing: Near Court Martial

Olds next became commander of the 81st Tactical Fighter Wing (TFW), at RAF Bentwaters, England on September 8, 1963. It was an F-101 Voodoo fighter-bomber wing. The 81st TFW was a major combat unit in USAF Europe, having both a tactical nuclear and conventional bombing role supporting NATO. Olds commanded the wing until July 26, 1965. As his Deputy Commander of Operations Olds brought with him Colonel Daniel "Chappie" James Jr., whom he had met during his Pentagon assignment and who would go on to become the first African-American 4-star Air Force General. James and Olds worked closely together for a year as a command team and developed both a professional and social relationship which was later renewed in combat. Olds formed a demonstration team for the F-101 using pilots of his Wing, without command authorization, and performed at an Air Force open house at Bentwaters. He asserted that his superior at Third Air Force attempted to have him court-martialled, but the commander of USAFE, General Gabriel P. Disosway, instead authorised his removal from command of the 81st TFW, cancellation of a recommended Legion of Merit award, and transfer to the headquarters of the 9th Air Force in South Carolina.

A pair of F-101Bs.
Image Source: U.S. Air Force via historynet.com

Command in Thailand: Vietnam War – "Blackman and Robin"

On September 30, 1966, Olds took command of the 8th Tactical Fighter Wing based at Ubon Royal Thai Air Force Base. A lack of aggressiveness and sense of purpose in the Wing had led to the change in command (Olds' predecessor had flown only 12 missions during the 10 months the Wing had been in combat). The 44-year-old Colonel also set the tone for his command stint by immediately placing himself on the flight schedule as a rookie pilot under officers junior to himself, then challenging them to train him properly because he would soon be leading them. Olds' Vice Commander was Col. Vermont Garrison, an Ace in both World War II and Korea, and in December Olds brought in Daniel James Jr., to replace an ineffective Deputy Commander for operations, creating arguably the strongest and most effective tactical command triumvirate of the Vietnam War. The Olds-James combination became popularly nicknamed "Blackman and Robin". James was named 8th TFW Vice Commander in June 1967, succeeding Garrison, who had completed his tour. Olds took to the air war over North Vietnam in an F-4C Phantom he nicknamed "Scat XXVII", in keeping with his previous combat aircraft that all carried the "Scat" name.

Colonel Robin Olds, 8th Tactical Fighter Wing, with SCAT XXVII, his McDonnell F-4C-24-MC Phantom II, 64-0829, at Ubon RTAFB, May 1967.
Image Source: thisdayinaviation.com

MiG Killer: Operation Bolo

After suggesting the idea to 7th Air Force Commander, himself a former Commander of the 8th TFW, Olds was directed to plan a mission designed to draw the North Vietnamese MiG 21s into an aerial trap, called "Operation Bolo". Multirole fighters flew a mission

Col Robin Olds painting a victory star on his F 4 Phantom II on May 4, 1967 after shooting a MiG 21.
Image Source: Pinterest.co.uk (Photo by aeroman3)

along flight paths typically used by the U.S. bombers during Rolling Thunder. The ruse drew an attack by Vietnamese MiG 21s. This worked and MiGs started getting shot up. In October 1966, strike force F-105 Thunderchiefs were equipped with QRC-160 radar jamming pods whose effectiveness virtually ended their losses to SAMs. As a result, SAM attacks shifted to the Phantoms, which were unprotected because of a shortage of pods. To protect the F-4s, they would penetrate to the edge of SAM coverage. MiG interceptions increased as a result, primarily by MiG-21s using high-speed hit-and-run tactics against bomb-laden F-105 formations. The Bolo plan was to equip the F-4s with jamming pods, using the call signs and communications code-words of the F-105 wings, and flying their flight profiles through northwest Vietnam, and entice the MiG-21s into intercepting Phantoms configured for air-to-air combat. The first mission was flown on January 2, 1967. The deceptive strike force began arriving over the target area, five-minute intervals separating the flights of F-4s. Leading the first flight, Olds overflew the primary MiG-21 base at Phúc Yên and was on a second pass when GCI controlled MiGs finally began popping up through the cloud base. The F-4s claimed seven MiG-21s destroyed, almost half of the 16 then in service with the VPAF without loss to USAF aircraft. Olds himself shot down one of the seven, for which he and the other aircrew were awarded Silver Stars. Follow-up interceptions over the next two days by MiGs against RF-4C reconnaissance aircraft led to a similar mission on a smaller scale on January 6, with another two MiG-21s shot down. VPAF fighter activity diminished to almost nothing for 10 weeks thereafter.

Silver Star Medal.
Image Source: afpc.af.mil

More Kills for Olds: "Triple Ace"

On May 4, Olds destroyed another MiG-21 over Phúc Yên. Two weeks later, on May 20, he destroyed two MiG 17s in what one of his pilots described as a "vengeful chase" after they shot down his wingman during a large dogfight, bringing his total to 16 confirmed kills (12 in World War II and four in Vietnam) and making him a triple ace. Olds states that following the shoot down of his fourth MiG, he intentionally avoided shooting down a fifth, even though he had at least ten opportunities to do so, because he had learned in the middle of June that Seventh Air Force, at the direction of Secretary of the Air Force, would immediately relieve him of command to return to the United States as a publicity asset if he did. He was awarded a fourth Silver Star for leading a three-aircraft low-level bombing strike on March 30, 1967, and the Air Force Cross for an attack on the Paul Doumer Bridge in Hanoi on August 11, one of five awarded to Air Force pilots for that mission.

The Final Mission and Flying Summary

He flew his final combat mission over North Vietnam on September 23, 1967. His 259 total combat missions included 107 in World War II and 152 in Southeast Asia, 105 of those over North Vietnam. Scat XXVII (F-4C-24-MC 64-0829) was retired from operational service and placed on display at the National Museum of the USAF, Wright-Patterson AF Base, Ohio.

"Wolfpack" aviators of the 8th Tactical Fighter Wing carry their Commanding Officer, Colonel Robin Olds, following his return from his last combat mission over North Vietnam, on 23 September 1967.
Image Source: Wikipedia

Olds' Moustache: Showing the Middle Finger

Olds was known for the extravagantly waxed (non-regulation) handlebar moustache he began sporting in Vietnam. It was a common superstition among airmen to grow a "bulletproof moustache", but Olds also used his as "a gesture of defiance". The kids on the base loved it. Almost everybody grew a moustache. Olds started the moustache in the wake of the success of Operation Bolo and let it grow beyond regulation length because "It became the middle finger I couldn't raise in the PR photographs. The moustache became my silent last word in the verbal battles…with his higher headquarters on rules, targets, and fighting the war." Returning home, however, marked the end of this flamboyance. When he reported to his first interview with USAF Air Chief, General John McConnell walked up to him, stuck a finger under his nose and said, "Take it off." Olds replied, "Yes, sir."

Olds' Moustache.
Image Source: therake.com

For his part, Olds was not upset with the order, recalling: "To tell the truth, I wasn't all that fond of the damned thing by then, but it had become a symbol for the men of the 8th Wing. I knew McConnell understood. During his visits to Ubon over the past year he had never referred to my breach of military standards, just seemed rather amused at the variety of 'staches sported by many of the troops. It was the most direct order I had received in twenty-four years of service." The incident with the moustache is given credit as the impetus for a new Air Force tradition, "Moustache March", in which, every March, aircrew, aircraft maintainers, and other airmen worldwide show solidarity by a symbolic, albeit good-natured "protest" for one month against Air Force facial hair regulations, to honour Air Force legend, Robin Olds.

Dog Fight Advocate: But No Gun Pods

We weren't allowed to dogfight. Very little attention was paid to strafing, dive-bombing, rocketry, stuff like that. It was thought to be unnecessary. Yet every confrontation America faced in the Cold War years was a 'bombs and bullets' situation, raging under an uneasy nuclear standoff. "The Vietnam War" proved the need to teach tactical warfare and have fighter pilots. It caught us unprepared because we weren't allowed to learn it or practice it in training.

Olds' MiG scoreboard on splitter vane of a McDonnell F-4C Phantom II.
Image Source: Wikipedia

Olds often lamented the lack of an internal gun in the F-4C he flew during his tour in Vietnam, but would not allow his fighters to be equipped with the gun pods then available. While he knew that he would be capable of effectively using them, he was also aware that none of his pilots were trained in the use of a gun or dogfighting. He also reasoned that the drag of the pod would both degrade the performance characteristics of the F-4 while not gaining it any advantage against the more manoeuvrable MiG-17s and MiG-21s, result in unnecessary losses strafing worthless targets, and reduce the number of bombs carried by the Phantoms, the delivery of which was the 8th's primary mission.

Brigadier General Robin Olds.
Image Source: thisdayinaviation.com

TV Episodes

Operation Bolo, and Olds' P-38 dogfights were recreated using computer animation in the episode "Air Ambush", of The History Channel "Dogfights" series, the first telecast on November 10, 2006. His fourth MiG kill in Vietnam was recreated in the season 2 episode "No Room for Error". Olds, then 84 years old, appeared as a commentator.

Air Force Academy 1967–71

After relinquishing command of the 8th TFW on September 23, 1967, Olds reported for duty to the USAF Academy Colorado, in December 1967. He served as commandant of cadets for three years and sought to restore morale in the wake of a major cheating scandal. Olds was promoted to Brigadier General on June 1, 1968, with seniority dating from May 28.

Director of Aerospace Safety

In February 1971 he began his last duty assignment as director of aerospace safety in the Office of the Inspector General, Headquarters USAF, and after December 1971 as part of the Air Force Inspection and Safety Center, a newly activated separate operating agency. Olds oversaw the creation of policies, standards, and procedures for Air Force accident prevention programs, and dealt with work safety education, workplace accident investigation and analysis, and safety inspections

At Oval Office with President Lyndon Baines Johnson (LBJ).
Image Source: firstaerosquadron.com

1971 Inspector General Tour

Air Force Inspector General and Olds' West Point classmate Lt Gen Louis L. Wilson Jr. sent Olds to Southeast Asia in the autumn of 1971 to determine the state of readiness of Air Force pilots. Olds toured USAF bases in Thailand (flying several unauthorized combat missions in the process) and brought back a blunt assessment. Air Force pilots, he reported to the Air Force Chief of Staff, Gen John D.Ryan (a former Strategic Air Command (SAC) general and bomber pilot often at odds with the tactical fighter community), who Olds thought "…couldn't fight their way out of a wet paper bag." Olds wrote "there is a systemic lack of interest by the USAF in air-to-air combat training for fighter crews. He warned that losses would be severe in any resumption of aerial combat". Olds recalled that Ryan expressed surprise at this assessment and reflected his disagreement.

Brigadier General Robin Olds retired as the director of aerospace safety in the Air Force Inspection and Safety Center.
Picture Source: af.mil

Leaves Air Force When Refused another Tour to Vietnam

When Operation Linebacker began in May 1972, American fighter jets returned to the offensive in the skies over North Vietnam for the first time in nearly four years. Navy and Marine Corps fighters, reaped the benefits of their TOPGUN program, immediately enjoyed considerable success. In contrast by June, as Olds had predicted, the Air Force's fighter community was struggling with a nearly 1:1 kill-loss ratio. To the new Inspector

General, Olds offered to take a voluntary reduction in rank to colonel so he could return to operational command and straighten out the situation. Olds decided to leave the Air Force when the offer was refused (he was offered another inspection tour instead) and he retired on June 1, 1973.

Robin Olds – WWII and Vietnam ACE.
Image Source: duotechservices.com

Awards and Decorations

Olds' awards and decorations included, USAF Command pilot badge, Air Force Cross, Air Force Distinguished Service Medal with one bronze oak leaf cluster, Silver Star with three oak leaf cluster, Legion of Merit, Distinguished Flying Cross with Valour device and silver oak leaf cluster, Air Medal with four silver oak leaf cluster, Vietnam Air Gallantry Cross, among many others.

Air Force Cross Citation

Colonel Robin Olds, U.S. Air Force, Date Of Action: August 11, 1967, read "The President of the United States of America, ... takes pleasure in presenting the Air Force Cross to Colonel Robin Olds (AFSN: 0-26046), United States Air Force, for extraordinary heroism in military operations against an opposing armed force while serving as Strike Mission Commander in the 8th Tactical Fighter Wing, Ubon Royal Thai Air Base, Thailand, against the Paul Doumer Bridge, a major north-south transportation link on Hanoi's Red River in North Vietnam, on 11 August 1967. On that date, Colonel Olds led his strike force of eight F-4C aircraft against a key railroad and highway bridge in North Vietnam. Despite intense, accurately directed fire, multiple surface-to-air missile attacks on his force, and continuous harassment by MiG fighters defending the target, Colonel Olds, with undaunted determination, indomitable courage, and professional skill, led his force through to help destroy this significant bridge. As a result the flow of war materials into this area was appreciably reduced. Through his extraordinary heroism, superb airmanship, and aggressiveness in the face of hostile forces, Colonel Olds reflected the highest credit upon himself and the United States Air Force."

Olds married Hollywood actress (and pin-up girl) Ella Raines in Beverly Hills on February 6, 1947.
Image Source: Pinterest (Brenda Bennett)

Hollywood Actress, Wife, Ella Raines.
Image Source: Universal Pictures

Personal Life: Marriage to Hollywood Actress

Olds was briefly a stepbrother of the famous author Gore Vidal after Olds' father married for the fourth time in June 1942, to Nina Gore Auchincloss. His father died of pneumonia on April 28, 1943, after hospitalization at the age of 46, just prior to Olds' graduation from West Point. In 1946, while based at March Field, Olds met Hollywood actress (and pin-up girl) Ella Raines on a blind date in Palm Springs. They married in Beverly Hills on February 6, 1947, and

had three children. Most of their 29-year marriage, marked by frequent extended separations and difficult homecomings, was turbulent because of a clash of lifestyles, particularly her refusal to ever live in government housing on military bases. Robin Olds and Ella Raines separated in 1975 and divorced in 1976. Robin married Abigail Morgan Sellers Barnett in January 1978, and they divorced after fifteen years of marriage.

Retired Life: Alcohol

After his retirement at Steamboat Springs, Colorado, Olds pursued his love of skiing and served on the city's planning commission. He was active in public speaking, making 21 events as late in his life as 2005 and 13 in 2006. Olds' fondness for alcohol was well known. John Darrell Sherwood, in his book "Fast Movers: Jet Pilots and the Vietnam Experience", posits that Olds' heavy drinking hurt his post-Vietnam career. On July 12, 2001, Olds was arrested for driving under the influence of alcohol and resisting arrest near his home in Steamboat Springs. Olds, briefly hospitalized during the incident for facial cuts, pleaded guilty in return for charges of weaving and felony vehicular eluding being dropped. Olds was placed on one year probation, and ordered to pay almost $900 in fines and costs, attend an alcohol education course, and perform 72 hours of community service.

L to R: National Aviation Hall of Fame enshrinees Joe Kittinger, Dick Rutan, and Robin Olds shared the experience of being combat fighters in Vietnam.
Image Source: airportjournals.com

National Aviation Hall of Fame

Days later, on July 21, 2001, Olds was enshrined at Dayton, Ohio, in the National Aviation Hall of Fame Class of 2001, along with test pilot Joe H, Engle of Marine Corps, and ace Marion E, Carl, and Albert Lee Ueltschi. He became the only person enshrined in both the National Aviation Hall of Fame and the College Football Hall of Fame.

Book by his daughter Christina Olds and Ed Rasimus

Dies of Cancer: Made a Class Exemplar

Gravestone, Section 6, Row D, Grave 34, United States Air Force Academy Cemetery, Colorado Springs, El Paso County, Colorado, USA

In March 2007 Olds was hospitalized in Colorado for complications of Stage 4 prostate cancer. On the evening of June 14, 2007, he died from congestive heart failure in Steamboat Springs, Colorado, a month before his 85th birthday. He was honoured with a flyover and services at the USAF Academy, where his ashes are housed, on June 30, 2007. Olds is remembered as the Class Exemplar of the Academy Class of 2011, which had begun Basic Cadet Training, the first step towards becoming Air Force officers, two days before Olds' funeral. Cadets choose a class exemplar who becomes the class' honorary namesake. The exemplar is typically a deceased former member of the Air Force or Army

Air Force, with a few notable exceptions like the Wright Brothers and Neil Armstrong. The tradition began with the Class of 2000. The selection of the class exemplar is celebrated with a class-wide dinner.

REFERENCES

1. Robin Olds, Wikipedia, https://en.wikipedia.org/wiki/Robin_Olds
2. Walter J. Boyne, The Robin Olds Factor, Air Force Magazine, June 1, 2008, https://www.airforcemag.com/article/0608olds/
3. Author Christina Olds, Book, Fighter Pilot: The Memoirs of Legendary Ace Robin
4. Christian Barker, A REAL MAVERICK: ROBIN OLDS, The Rake, www.therake.com/stories.icons/real-maverick-robin-olds/
5. Stephen Losey, Mustache March: Robin Olds' mustache is just a sliver of his story, The Air Force Times, March 4, 2019, https://www.airforcetimes.com/news/your-air-force/2019/03/04/mustache-march-olds-mustache-is-just-a-sliver-of-his-story/
6. Olds, Robin, Military Combat, The National Aviation Hall of fame, Enshrined 2007, https://www.nationalaviation.org/our-enshrinees/olds-robin/
7. Legend Robin Olds' Advice On How To Be A Badass Fighter Pilot, https://worldwarwings.com/legend-robin-olds-advice-on-how-to-be-a-badass-fighter-pilot/
8. Anthony B. Carr, Brig. Gen. Robin Olds: American Warrior-Scholar, August 20, 2018, https://medium.com/@the.colosseum.blog/brig-gen-robin-olds-warrior-scholar-total-badass-a6f98bf23c4e
9. Blake Stilwell, We are the Mighty, This is how triple-ace Robin Olds achieved his perfect victory over Vietnam, April 02, 2018, https://www.wearethemighty.com/articles/bolo-was-triple-ace-robin-olds-perfect-victory-over-vietnam/
10. Robin Olds, Find a Grave, https://www.findagrave.com/memorial/19948151/robin-olds

8

Hans-Joachim Marseille

"Triple Ace" – 17 Victories in a Day

Hans-Joachim Marseille.
Image Source: reddit.com

"As long as I look into the muzzles, nothing can happen to me. Only if he pulls lead am I in danger".

– **Hans Joachim Marseille**

To be a Triple-ace in a day, the pilot must have destroyed 15 enemy aircraft in a single day. This has been achieved by only five pilots ever. All from the Luftwaffe. Hans-Joachim Marseille, also called "Stern von Afrika" (Star of Africa) by the Germans. This German Luftwaffe fighter pilot, flying ace of World War II was known for his aerial battles during the North African Campaign and his Bohemian lifestyle. One of the most successful fighter pilots, Marseille claimed all but seven of his 158 victories against the British Commonwealth's Desert Air Force over North Africa, flying the Messerschmitt Bf 109 fighter through his entire combat career. No other pilot claimed as many Western aircraft as Marseille.

Marseille, belonged to the French Huguenot ancestry (Huguenots were French Protestants who held to the Reformed, or Calvinist, tradition of Protestantism). He joined the Luftwaffe in 1938, at the age of 20, having graduated from one of the Luftwaffe's fighter pilot schools just in time to participate in the Battle of Britain. A charming person, he had such a busy night life that sometimes he was too tired to be allowed to fly the next morning. As a result of poor discipline, he was transferred to JG 27 (Fighter Wing 27), which relocated to North Africa in April 1941.

Image Source: alchetron.com

Young Marseille.
Image Source: facebook.com/
pages/category/Public-Figure/
Hans-Joachim-Marseille

Under the guidance of his new commander, who recognised the latent potential of the young officer, Marseille quickly developed his abilities as a fighter pilot. He reached the zenith of his fighter pilot career on 1 September 1942, when during the course of three combat sorties, in a single day, he claimed 17 enemy fighters shot down and earning him the Knight's Cross with Oak Leaves, Swords, and Diamonds. Only 29 days later, Marseille was killed in a flying accident, when he was forced to abandon his fighter due to engine failure. After he exited the smoke-filled cockpit, Marseille's chest struck the vertical stabilizer of his aircraft. The blow either killed him instantly or incapacitated him so that he was unable to open his parachute.

Youth and Family

Hans-Joachim "Jochen" Walter Rudolf Siegfried Marseille was born to Charlotte Marie Johanna Pauline Gertrud Riemer and Hauptmann Siegfried Georg Martin Marseille, a family with paternal French ancestry, in Berlin Charlottenburg on 13 December 1919. As a child, he was physically weak, and he nearly died from a serious case of influenza. His father was an Army officer during World War I, and later left the armed forces to join the Berlin police force. Hans-Joachim also had a younger sister, Ingeborg. Many years later, while on sick leave in

A School Picture.
Image Source: ww2gravestone.com

Athens at the end of December 1941, he was summoned to Berlin by a telegram from his mother. Upon arriving home, he learned his sister had been killed by a jealous lover while living in Vienna, Hans-Joachim reportedly never recovered emotionally from this blow.

Troubled Childhood: Parents Separate

When Marseille was still a young child his parents divorced and his mother subsequently married a police official named Reuter. Marseille initially assumed the name of his stepfather at school (a matter he had a difficult time accepting) but he reverted to his father's name of Marseille in adulthood. A lack of discipline gave him a reputation as a rebel, which plagued him early on in his Luftwaffe career. Marseille also had a difficult relationship with his natural father, whom he refused to visit in

Young Days.
Image Source: Pinterest (Pasha Idris)

Hamburg for some time after the divorce. Eventually he attempted a reconciliation with his father, who subsequently introduced him to the nightlife that initially hampered his military career during his early years in the Luftwaffe. However, the rapprochement with his father did not last and he did not see him again. After initial years of schooling in Berlin, between April and September 1938, he served in the Reich Labour Service.

Joins Luftwaffe

Marseille joined the Luftwaffe on 7 November 1938 as an officer candidate and received his basic training. On 1 March 1939 Marseille was transferred to the LKS 4 Air War School. Among his classmates was Werner Schröer, a German WW II fighter ace credited with shooting down 114 enemy aircraft. Werner served in the Luftwaffe from 1937, initially as a member of the ground staff, until the end of World War II in Europe on 8 May 1945. Interestingly Schröer was the second most successful claimant of air victories after Marseille in the Mediterranean. Marseille completed his training at a Fighter Pilot School in Vienna. One of his instructors' was the Austro-Hungarian World War I ace Julius Arigi, a WW I ace with a total of 32 credited victories. He was Austro-Hungary's most highly decorated ace. Marseille graduated with an outstanding evaluation on 18 July 1940 and was assigned to a unit having air defence duties from the outbreak of war until the fall of France. On 10 August 1940 he was assigned to the Instructional Squadron 2, based in Calais-Marck, to begin operations against Britain and again received an outstanding evaluation this time by commander Herbert Ihlefeld (130 enemy aircraft shot down in over 1,000 combat missions), himself an Ace.

Early days.
Image Source: facebook.com/pages/category/Public-Figure/Hans-Joachim-Marseille

Battle of Britain: First Engagement – First Victory

In his first dogfight over England on 24 August 1940, Marseille engaged in a four-minute battle with a skilled opponent while flying Messerschmitt Bf 109 (E-3 W.Nr. 3579). He defeated his opponent by pulling up into a tight chandelle, to gain an altitude advantage before diving and firing. The British fighter was struck in the engine, pitched over, and dove into the English Channel. This was Marseille's first victory. Marseille was then engaged from above by more Allied fighters. By pushing his aircraft into a steep dive, then pulling up meters above the water, Marseille escaped from the machine gun fire of his opponents. "Skipping away over the waves, I made a clean break, no one followed me and I returned to an alternative airbase" he later said. The act was not praised by his unit. Marseille was reprimanded when it emerged he had abandoned his wingman, and "Staffel" to engage the opponent alone. In so doing, Marseille had violated a basic rule of air combat. Reportedly, Marseille did not take any pleasure in this victory and found it difficult to accept the realities of aerial combat.

Messerschmitt Bf 109 (E-3 W.Nr. 3579).
Image Source: bigginhillheritagehangar.co.uk/messerschmitt-me109

Gets Shot and bails out over the Sea: Dismissed for Squadron

While returning from a bomber-escort mission on 23 September 1940 flying BF 109 No. 5094, his engine failed 10 miles (16 km) off the coast after combat damage sustained over Dover. Various Royal Air Force (RAF) pilots had claimed to have shot him. No. 5094 aircraft was also claimed destroyed by Robert

Heinkel He 59 SAR plane.
Image Source: commons.wikimedia.org

Stanford Tuck (an RAF fighter pilot, test pilot, and ace with 29 victories). Tuck had pursued a Bf 109 to that location and whose pilot was rescued by a German naval aircraft. Marseille is the only German airman known to have been rescued by a Heinkel He 59 on that day and in that location. Although Marseille tried to radio his position, he bailed out over the sea. He paddled around in the water for three hours before being rescued by the float plane. Exhausted and suffering from exposure, he was sent to a field hospital.

When he returned to duty, he received a stern rebuke from his Commander, Herbert Ihlefeld. In this engagement, Marseille had abandoned his leader Adolf Buhl, who was shot down and killed. During his rebuke, his Commander tore up Marseille's flight evaluations. Other pilots also voiced their dissatisfaction concerning Marseille. Because of his alienation of other pilots, his arrogant and unapologetic nature, Ihlefeld eventually dismissed Marseille from LG 2.

German Bf 109 fighter after force-landing on a French beach, 1940-1941; reportedly of Hans-Joachim Marseille's fighter that crashed on 28 Sep 1940.
Source: Federal Archives via ww2db.com

Another Combat: Another Rebuke – Passed over for Promotion

In another mission Marseille had once ignored an order to turn back from a fight when outnumbered by two to one, but seeing an Allied aircraft closing on his wing leader, Marseille broke formation and shot the attacking aircraft down. Expecting congratulations when he landed, his Commander was critical of his actions, and Marseille received three days of confinement for failing to carry out an order. Days later, Marseille was passed over for promotion and was now the sole "Fähnrich" (junior most rank) in the "Geschwader" (Squadron). This was a humiliation for him, suspecting that his abilities were being suppressed so the Squadron leaders could take all the glory in the air.

He had girlfriends everywhere.
Image Source: deviantart.com

Wrote-Off Four Aircraft

Shortly afterwards, in early October 1940, after having claimed seven aerial victories all of them while flying with the LG 2 squadron, and flying with likes of Johannes "Macky" Steinhoff (Ace with 176 victories) and Gerhard "Gerd" Barkhorn (Ace with 300 victories). He had written off four aircraft as a result of operations during this period. Marseille was transferred to LG 4 squadron under the same wing JG 52. Steinhoff, later recalled: "Marseille was extremely handsome. He was a very gifted pilot, but he was unreliable. He had girlfriends everywhere, and they kept him so busy that he was sometimes so worn out that he had to be grounded. His sometime irresponsible way of conducting his duties was the main reason I fired him. But he had an irresistible charm." "Telling Marseille that he was grounded was like telling a small child that it could not go out and play. He sometimes acted like one too." said Werner Schroer.

Punishment for Insubordination: Move to New Wing

The Playboy.
Image source: facebook.com/pages/category/Public-Figure/Hans-Joachim-Marseille

As punishment for "insubordination", rumoured to be his penchant for American jazz music, womanizing and an overt "playboy" lifestyle – and inability to fly as a wingman, Steinhoff transferred Marseille to JG 27 on 24 December 1940. His new Group Commander, Eduard Neumann, later recalled, "His hair was too long and he brought with him a list of disciplinary punishments as long as your arm. He was tempestuous, temperamental and unruly. Thirty years later, he would have been called a playboy." Nevertheless, Neumann quickly recognised Marseille's potential as a pilot. He stated in an interview: "Marseille could only be one of two, either a disciplinary problem or a great fighter pilot." JG 27 was soon relocated to North Africa.

Arrival in North Africa: Force Landing in Desert – Hitch Hiked to Airbase

Marseille's unit briefly saw action during the invasion of Yugoslavia, deployed at Zagreb on 10 April 1941, before transferring to Africa. On 20 April on his flight from Tripoli to his front airstrip, Marseille's Bf 109 E-7 (1259) developed engine trouble and he had to make a forced landing in the desert short of his destination. His squadron departed the scene after they had ensured that he had got down safely. Marseille continued his journey, first hitchhiking on an Italian truck, then, finding, this too slow; he tried his luck at an airstrip in vain. Finally, he made his way to the General in charge of a supply depot on the main route to the front, and convinced him that he should be available for

Bf 109.
Image Source: Pinterest (gde-fon.com)

operations the next day. Marseille's character appealed to the General and he put at his disposal his own chauffeur-driven Opel Admiral. "You can pay me back by getting fifty victories, Marseille!" were his parting words. He caught up with his squadron on 21 April.

Initial Victories in North Africa: And is shot down Twice

In Bf 109 Cockpit.
Image Source: ww2db.com

Marseille scored two more victories on 23 and 28 April, his first in the North African campaign. However, on 23 April, Marseille himself was shot down during his third sortie of that day by a Free French pilot, James Denis, flying an RAF Hawker Hurricane. Marseille's Bf 109 E-7 (5160) received almost 30 hits in the cockpit area, and three or four shattered the canopy. As Marseille was leaning forward the rounds missed him by inches. Marseille managed to crash-land his fighter near Tobruk. Just a month later, the same James Denis shot down Marseille again on 21 May 1941. Marseille had engaged Denis, but overshot his target. A dogfight ensued, in which Denis once again bested Marseille. His Bf 109 E-7 (567) came down in the vicinity of Tobruk behind German lines. In a post-war account, Denis wrote that he waited for Marseille to close on him while he feigned ignorance, then skidded (side slipped) forcing the faster German to over shoot. Marseille was lucky. Bullets passed in front of his face and behind his head. 30 hits were counted after Marseille crash-landed.

Downs a Bristol Blenheim

In between the battles with Denis, Marseille downed a Bristol Blenheim on 28 April. Blenheim (T2429), of No.45 Squadron RAF, piloted by Pilot Officer B. C. de G. Allan, crashed killing all five men aboard. Jan Yindrich, a Polish army soldier, witnessed the attack, later said: "when a Blenheim came roaring down over our heads at about 50 feet, there was a terrific rattle of machine gun fire and at first I thought the Blenheim had made a mistake and was firing at us or choosing an awkward spot to clear his guns. Bullets whistled around, so we dived into the trench. A Messerschmitt, hot on the tail of the Blenheim, was responsible for the bullets. The Blenheim roared down the "Wadi", out to sea, trying to escape from the Messerschmitt, but the Messerschmitt was too close. The Blenheim fell out of the sky and crashed into the sea. The plane disappeared completely not leaving a trace. The Messerschmitt banked and flew inland again."

Bristol Blenheim.
Image Source: militaryfactory.com

Low Kill Rate and Four More Crashes

His boss Neumann encouraged Marseille to self-train to improve his abilities. By this time, he had crashed or damaged another four Bf 109 E aircraft, including an aircraft he was ferrying on 23 April 1941. Marseille's kill rate was low, and he went from June to August without a victory. He was further frustrated after damage forced him to land on two occasions: once on 14 June 1941 and again after he was hit by ground fire over Tobruk and was forced to land blind. His tactic of diving into opposing formations often found him under fire from all directions, resulting in his aircraft frequently being damaged beyond repair. Consequently, even Neumann grew impatient with him.

Neumann (right) pins the German Cross in Gold on the "Star of Africa" on Marseille.
Image Source: heatonlewisbooks.com

Marseille Introspects: Creates Unique Self-Training Program

Marseille persisted, and created a unique self-training program for himself, both physical and tactical, which resulted not only in outstanding situational awareness, marksmanship and confident control of the aircraft, but also in a unique attack tactic. He now preferred a high angle deflection shooting attack and shooting at the target's front from the side, instead of the common method of chasing an aircraft and shooting at it directly from behind. Marseille often practiced these tactics on the way back from missions with his comrades and became known as a master of deflection shooting.

Regular Victories Now On: Flies to Pick Downed Pilots-Penance

As Marseille began to claim Allied aircraft regularly, interestingly on occasion he organised the welfare of the downed pilot personally, driving out to remote crash sites to rescue downed Allied airmen. On 13 September 1941 Marseille shot down Pat Byers of the Royal Australian Air Force (RAAF). Marseille flew to Byers' airfield and dropped a note informing the Australians of his condition and treatment. He returned several days later to second the first note with

New Tactics.
Image Source: weaponsandwarfare.com

93

Marseille with Hawker Hurricane Mk IIB of 274 RAF Squadron, North Africa – 30 March 1942.
Image Source: samilhistory.com

news of Byers' death. Marseille repeated these sorties after being warned by Neumann that Göring had forbade any more flights of this kind. After the war, Marseille's JG 27 comrade Werner Schroer stated that Marseille attempted these gestures as "penance" for a group that "loved shooting down aircraft" but not killing a man; "we tried to separate the two. Marseille allowed us that escape, our penance I suppose."

Claims Four Hurricanes in a Day

Finally, on 24 September 1941, his deflection shooting practice came to fruition, with his first multiple victory sortie, claiming four Hurricanes of South African Air Force (SAAF). These victories represented his 19–23rd victory. Marseille became known amongst his peers for accounting for multiple enemy aircraft in a sortie. By mid-December, he had reached 25 victories, and was awarded the German Cross in Gold. His Squadron was rotated to Germany in November/December 1941 to convert to the Bf 109 F-4, the variant that was described as the "experts mount."

German Cross in Gold.
Image Source: epicartifacts.com

Personal Fitness Training

"Marseille was the unrivalled virtuoso among the fighter pilots of WW II. His achievements had previously been regarded as impossible and they were never excelled by anyone after his death," said Adolf Galland about him later. Marseille always strove to improve his abilities. He worked to strengthen his legs and abdominal muscles, to help him tolerate the extreme 'g' forces of air combat. Marseille also drank an abnormal amount of milk and shunned sunglasses, in the belief that doing so would improve his eyesight.

Marseille doing air combat briefing.
Image Source: Wikipedia

The "Lufbery circles"

To counter German fighter attacks, the Allied pilots flew "Lufbery circles", a defensive air combat tactic evolved in WW I. It involved forming a horizontal circle in the air when attacked, in such a way that the armament of each aircraft offers a measure of protection to the others in the circle. It complicates the task of an attacking fighter – the formation as a whole has far fewer "blind spots". The tactic was effective and dangerous as a pilot attacking this formation could find himself constantly in the sights of the opposing pilots. Marseille often dived at high speed into the middle of these defensive formations from either above or below, executing a tight turn and firing a two-second deflection shot to destroy an enemy aircraft.

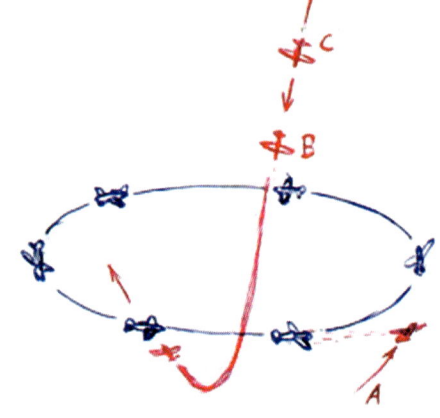

Lufbery circles.
Image Source: suptg.thisisnotatrueending.com

Unorthodox Combat: One man Show

Marseille's successes had begun to become readily apparent by early 1942. He claimed his 37–40th victories on 8 February 1942 and 41–44th victories four days later which earned him the Knight's Cross of the Iron Cross that same month for 46 victories. Marseille attacked under conditions many considered unfavourable, but his marksmanship allowed him to make an approach fast enough to escape the return fire of the two aircraft flying on either flank of the target. Marseille's excellent eyesight made it possible for him to spot the opponent before he was spotted, allowing him to take the appropriate action and manoeuvre into position for an attack. He was credited with outstanding situational awareness. In combat, Marseille's unorthodox methods led him to operate in a small leader/wingman units, which he believed to be the safest and most effective way of fighting in the high-visibility conditions of the North African skies. Marseille "worked alone" in combat keeping his wingman at a safe distance so he would not collide or fire on him in error.

Hans-Joachim Marseille in Berlin, Germany.
Image Source: ww2db.com.

Own Special Tactics: Appreciation from other Aces

In a dogfight, particularly when attacking Allied aircraft in a Lufbery circle, Marseille would often favour dramatically reducing the throttle and even lowering the flaps to reduce speed and shorten his turn radius, rather than the standard procedure of using full throttle throughout. Emil Josef Clade, who himself was a German flying Ace and figured in German civilian aviation after the war, had said that none of the other pilots could do this effectively, preferring instead to dive on single opponents at speed so as to escape if anything went wrong. Clade said of Marseille's tactics: "Marseille developed his own special tactics, which differed significantly from the methods of most other pilots. (When attacking a Lufbery circle) he had to fly very slowly. He even took it to the point where he had to operate his landing flaps as not to fall down, because, of course, he had to fly his curve (turns) more tightly than the upper defensive circle. He and his fighter were one unit, and he was in command of that aircraft like no-one else."

Painting of Marseille's Bf 109F-4 over North Africa.
Image Credit: ww2f.com/threads/gallantry-over-north-africa-hans-joachim-marseille-the-star-of-africa

Friedrich Körner (36 victories) also recognised this as unique: "Shooting in a curve (deflection shooting) is the most difficult thing a pilot can do. The enemy flies in a defensive circle. It means they are already lying in a curve and the attacking fighter has to fly into this defensive circle. By pulling his aircraft right around, his curve radius must be smaller, but if he does that, his target disappears in most cases below his wings. So he cannot see it anymore and has to proceed simply by instinct." The attack was, however, carried out at close-range; Marseille dived from above, climbed underneath an opponent, fired as the enemy aircraft disappeared under his own, and then used the energy from the dive to climb and repeat the process.

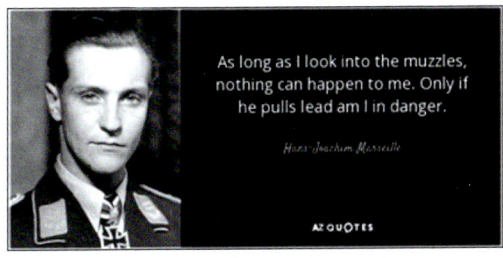
Shooting in a curve (deflection shooting).
Quote Source: azquotes.com

A popular Squadron Commander.
Image Source: facebook.com/pages/category/Public-Figure/Hans-Joachim-Marseille

Promotions and Added Responsibility

His success as a fighter pilot, finally, also led to promotions and more responsibility as an officer. 1 May 1942 saw him receive an unusually early promotion to Lieutenant and then Captain on 8 June 1942. He then got command of a squadron under the famous Fighter Wing JG 27.

Style and Idea of Air Combat

In a conversation with his friend, Air Ace, and sister Squadron Commander Hans-Arnold Stahlschmidt, Marseille commented on his style, and his idea of air-to-air combat: "I often experience combat as it should be. I see myself in the middle of a "British swarm", firing from every position and never getting caught. Our aircraft are basic elements, which have got to be mastered. You've got to be able to shoot from any position. From left or right turns, out of a roll, on your back, whenever. Only this way can you develop your own particular tactics. Attack tactics that the enemy simply cannot anticipate during the course of the battle – a series of unpredictable movements and actions, never the same, always stemming from the situation at hand. Only then can you plunge into the middle of an enemy swarm and blow it up from the inside."

Style and Idea of Air Combat.
Image Source: facebook.com/pages/category/Public-Figure/Hans-Joachim-Marseille

German crewmen Hoffmann and Berger cleaning the machine gun barrel of Bf 109 fighter W. Nr. 8673, which belonged to Hans-Joachim Marseille, Martuba, Libya, March 1942.
Image Source: Federal Archives via ww2db.com

Shot Down: Narrow Escape

Marseille had a narrow escape on 13 May 1942, when his Bf 109 was damaged during a dogfight with 12 Curtiss Kittyhawks Mk I, of RAAF. With a wingman, Marseille bounced the Kittyhawks. After he downed one of the Australian pilots, Marseille's Bf 109 took hits in the oil tank and propeller. Marseille nevertheless managed to shoot down another Kittyhawk, before nursing his overheating aircraft back to base. The repairs to Marseille's Bf 109 took two days. The aerial victories were recorded as numbers 57–58.

Mercy Mission and Letter of Regret

Weeks later, on 30 May, Marseille performed another mercy mission after witnessing his 65th victory. Pilot Officer Graham George of RAF struck the tail plane of his fighter during bailout and fell to his death when the parachute did not open. After landing he drove out to the crash site. The P-40 had landed over Allied lines but they found the dead pilot within German territory. Marseille marked his grave, collected his papers and verified his identity, then flew to Buckland's airfield to deliver a letter of regret. Buckland died two days before his 21st birthday.

Curtiss P-40 E Kittyhawk Mk I, Graham George Buckland, 30 May 1942.
Image Source: centurabooks.com

High Proportion of Victories: Five in Six Minutes – More Peer Praise

His attack method to break up formations, which he perfected, resulted in a high proportion of victories, and in rapid, multiple victories per attack. On 3 June 1942, Marseille attacked alone a formation of 16 Curtiss P-40 fighters and shot down six aircraft, five of them in six minutes, including three Aces: Robin Pare (six victories), Cecil Golding (6.5 victories), and Andre Botha (five victories). This success rose his score further, recording his 70 to 75th victories. Marseille was awarded the Knight's Cross of the Iron Cross with Oak Leaves on 6 June 1942. His wingman Rainer Pöttgen, nicknamed Fliegendes Zählwerk (the "Flying Counting Machine"), said of this fight: "All the enemy were shot down by Marseille in a turning dogfight. As soon as he shot, he needed only to glance at the enemy plane. His pattern of gunfire, began at the front, the engine's nose, and consistently ended in the cockpit. How he was able to do this, not even he could explain. With every dogfight he would throttle back as far as possible; this enabled him to fly tighter turns. His expenditure of ammunition in this air battle was 360 rounds (60 per aircraft shot down)."

Hans-Joachim Marseille (on wing) with Willy Messerschmitt (right) after trying out the new "Gustav" (Bf-109G) at the Messerschmitt plant, Augsburg, Germany, early 1942. Image Source: Federal Archives via ww2db.com

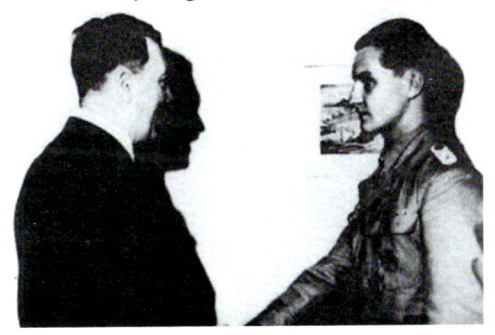

Adolf Hitler and Hans-Joachim Marseille at Wolfsschanze headquarters, Rastenburg, Germany, 28 Jun 1942.
Image Source: Federal Archives via ww2db.com

Most Amazing and Ingenious Combat Pilot

Werner Schröer (114 Victories), did, however, place Marseille's methods into context: "He was the most amazing and ingenious combat pilot I ever saw. He was also very lucky on many occasions. He thought nothing of jumping into a fight outnumbered ten to one, often alone, with us trying to catch up to him. He violated every cardinal rule of fighter combat. He abandoned all the rules."

100th Victory: Oak Leaves and Swords

On 17 June 1942, Marseille claimed his 100th aerial victory. He was the 11th Luftwaffe pilot to achieve the century mark. Marseille then returned to Germany for two months leave and the following day was awarded the Knight's Cross of the Iron Cross with Oak Leaves and Swords. On 6 August, he began his journey back to North Africa accompanied by his fiancée Hanne-Lies Küpper.

Knight's Cross of the Iron Cross with Oak Leaves and Swords.
Image Source: military.wikia.org

Highest Italian Military Award for Bravery: Missing in Italy

On 13 August, he met Benito Mussolini in Rome and was presented with the highest Italian military award for bravery, the Gold Medal of Military Valour (Medaglia d'Oro al Valour Militare). While in Italy Marseille disappeared for some time prompting the German authorities to compile a missing persons report, submitted by the 'Gestapo' head in Rome. He was finally located. According to rumours he had run off with an Italian girl and was eventually

Night Life.
Image Source: facebook.com/pages/category/Public-Figure/Hans-Joachim-Marseille

persuaded to return to his unit. Unusually, nothing was ever said about the incident and no repercussions were visited upon Marseille for this indiscretion.

Returns to Combat: 17 Victories in a Day – 54 in a Month

Leaving his fiancée in Rome, Marseille returned to combat duties on 23 August. 1 September 1942 was Marseille's most successful day, claiming to destroy 17 Allied aircraft (nos. 105–121), and September would see him claim 54 victories, his most productive month. The 17 aircraft claimed included eight in 10 minutes; as a result of this feat, he was presented with a Volkswagen Kübelwagen by a Italian air Force squadron, on which his Italian comrades had painted "Otto" (Otto = eight). 17 was the most aircraft from Western Allied air forces shot down by a single pilot in one day. Only one pilot, Emil "Bully" Lang, on 4 November 1943, would better this score, against the Soviet Air Force on the Eastern Front. Lang had 72 victories in a three-week period, among them an unsurpassed total of 18 on 3 November 1943. The post-war analysis shows that the actual results of the day were probably eight to nine destroyed by Marseille with three or four more damaged.

Presented with a Volkswagen Kübelwagen painted "Otto".
Image Source: facebook.com/pages/category/Public-Figure/Hans-Joachim-Marseille

On Cover of Magazine "Der Adler" (The Eagle).
Image Source: rommelsriposte.com

Famous Through Propaganda

On 3 September 1942 Marseille claimed six victories (nos. 127–132) but was hit by fire from the British-Canadian ace James Edwards. "Der Adler" (The Eagle) a biweekly propaganda magazine published by the Luftwaffe, also reported his actions in volume 14 of 1942. Marseille was made famous through propaganda that treated fighter pilots as superstars and continued to do so after his death. He regularly signed postcards with his image. Aside from Der Adler, his exploits were published in many newspapers and magazines. Three days later Edwards likely killed Günther Steinhausen, a German Ace with 40 victories, and friend of Marseille. The next day, 7 September 1942, another close friend, German Ace with 59 victories, Hans-Arnold Stahlschmidt was posted missing in action. These personal losses weighed heavily on Marseille's mind along with his family tragedy. It was noted he barely spoke and became more morose in the last weeks of his life. The strain of combat also induced consistent sleepwalking at night and other symptoms that could be construed as Post Traumatic Stress Disorder (PTSD). Marseille never remembered these events.

Fractures Arm: Continues to Fly and Score

Marseille flew Bf 109 E-7 aircraft and Bf 109 F-4/Z aircraft. Marseille continued scoring multiple victories throughout September, including seven on 15 September (nos. 145–151). Between 16 and 25

September, Marseille failed to increase his score due to a fractured arm, sustained in a force landing soon after the 15 September mission. As a result, he had been forbidden to fly by Eduard Neumann. But the same day, Marseille borrowed the Machhi C-202 of the Italian ace Tenente Emanuele Annoni, from neighbouring Italian squadron for a test flight. But the one-off flight ended in a wheels-up landing, when the German ace accidentally switched the engine off, as the throttle control in Italian aircraft was opposite to that of the German aircraft. The event was photographed.

Messerschmitt Bf 109 F-4, Hans Joachim Marseille.
Colourised Source: samilhistory.com

Better and More Western Aircraft – More Combat Strain

Marseille had nearly surpassed his friend Hans-Arnold Stahlschmidt's score of 59 victories in just five weeks. However, the massive material superiority of the Allies meant the strain placed on the outnumbered German pilots was now severe. At this time, the strength of German fighters was 112 (65 serviceable) aircraft against the British muster of some 800 machines. Marseille was becoming physically exhausted by the frenetic pace of combat. After his last combat on 26 September, Marseille was reportedly on the verge of collapse after a 15-minute battle with a formation of Spitfires, during which he scored his seventh victory of that day.

Messerschmitt Bf 109F4 (Yellow 14) Hans Joachim Marseille Tail No. 8693 at Martuba on 06 Feb 1942.
Picture Credit: samilhistory.com

The Toughest Adversary

Of particular note was Marseille's 158th claim. After landing in the afternoon of the 26 September 1942, he was physically exhausted. Several accounts allude to his Squadron members being visibly shocked at Marseille's physical state. Marseille, according to his own post-battle accounts, had been engaged by a Spitfire pilot in an intense dogfight that began at high altitude and descended to low-level. Marseille recounted how both he and his opponent strove to get onto the tail of the other. Both succeeded and fired but each time the pursued managed to turn the table on his attacker. Finally, with only 15 minutes of fuel remaining, he climbed into the sun.

Artist Impression. BF 109 vs. Spitfire.
Image Credit: americanartpublishing.com

The RAF fighter followed and was caught in the glare. Marseille executed a tight turn and roll, fired from 100 metres range. The Spitfire caught fire and shed a wing. It crashed into the ground with the pilot still inside. Marseille wrote, "That was the toughest adversary I have ever had. His turns were fabulous... I thought it would be my last fight". Unfortunately the pilot and his unit remain unidentified.

Reluctant to Use New Aircraft

The two missions of 26 September 1942 had been flown in Bf 109, in one of which Marseille had shot down seven Allied aircraft. The first six of these machines were to replace the Group's Bf 109Fs. All had been allocated to Marseille's unit. Marseille had previously ignored orders to use these new aircraft because of its high engine failure rate, but on the orders of General Field Marshal Albert Kesselring, head of Luftflotte 2.

Marseille reluctantly obeyed. One of these machines, WK-Nr. 14256, was to be the final aircraft Marseille flew.

Refuses to Accompany Erwin Rommel to Berlin

Hans-Joachim Marseille with Erwin Rommel and others, Libya, 16 Sep 1942.
Image Source: Federal Archives via ww2db.com

Over the next three days Marseille's Squadron was rested and taken off flying duties. On 28 September Marseille received a telephone call from General Field Marshal Erwin Rommel asking to return with him to Berlin. Hitler was to make a speech at the Berlin Sportpalast on 30 September and Rommel and Marseille were to attend. Marseille rejected this offer, citing that he was needed at the front and had already taken three months' vacation that year. Marseille also said he wanted to take leave at Christmas, to marry his fiancée Hanne-Lies Küpper.

Bailout and the Fatal Fall

On 30 September 1942, Marseille was leading his flight on a "Stuka" escort mission covering the withdrawal of the group. Marseille's flight was vectored onto Allied aircraft in the vicinity but the opponent withdrew and did not take up combat. While returning to base, his new Messerschmitt Bf 109 G-2/trop's cockpit began to fill with smoke; blinded, he was guided back to German lines by his wingmen, Jost Schlang and Lt Rainer Pöttgen. Upon reaching friendly lines, "Yellow 14" had lost power and was drifting lower and lower. Pöttgen called out after about 10 minutes that they had reached the White Mosque of Sidi Abdel Rahman and were thus within friendly lines. At this point, Marseille deemed his aircraft no longer flyable and decided to bail out, his last words to his comrades being "I've got to get out now, I can't stand it any longer". Eduard Neumann was personally directing the mission from the command post: "I was at the command post and listening to the radio communication between the pilots. I realised immediately something serious had happened; I knew they were still in flight and that they were trying to bring Marseille over the lines into our territory and that his aircraft was emitting a lot of smoke."

Marseille's aircraft crash site, called Piramide Memorial Hans Joachim Marseille (Bf 109 G-2/trop WK-Nr. 14256).
Image Source: tracesofwar.com

His flight, which had been flying a tight formation around him, peeled away to give him the necessary room to manoeuvre. Marseille rolled his aircraft onto its back, the standard procedure for bail out, but due to the smoke and slight disorientation, he failed to notice that the aircraft had entered a steep dive at an angle of 70–80 degrees and was now travelling at a considerably faster speed of about 640 km/h. He worked his way out of the cockpit only to be carried backward by the slipstream. The left side of his chest struck the vertical stabilizer of his fighter, which either killed him instantly or rendered him unconscious to the point that he could not deploy his parachute. He fell almost vertically, hitting the desert floor. As it transpired, a gaping 40 cm (16 in) hole had been made in his parachute and the canopy spilled out. After recovering the body, the parachute release handle was still on "safe," suggesting Marseille had not attempted to open it. Whilst the body was checked, a regimental doctor noted Marseille's wristwatch had stopped at exactly 11:42 am. The doctor had been the first to reach the crash site, having been stationed just to the rear of the forward mine defences. He had also witnessed Marseille's fatal fall.

Autopsy Report

The autopsy report stated: "The pilot lay on his stomach as if asleep. His arms were hidden beneath his body. As I came closer, I saw a pool of blood that had issued from the side of his crushed skull; brain matter was exposed. I then noticed the awful wound above the hip. With certainty this could not have come from the fall. The pilot must have been slammed into the airplane when bailing out. I carefully turned the dead pilot over onto his back. Opened the zipper of his flight jacket, saw the Knight's Cross with Oak Leaves and Swords (Marseille never actually received the Diamonds personally) and I knew immediately who this was. The paybook also told me. I glanced at the dead man's watch. It had stopped at 11:42."

Funeral and Inquiry

Marseille lay in state in the Station sick bay, his comrades coming to pay their respects throughout the day. Marseille's funeral took place on 1 October 1942 at the Heroes Cemetery in Derna, Libya, with Field Marshal Albert Kesselring and Eduard Neumann delivering a eulogy.

An enquiry into the crash was hastily set up. The commission's report concluded that the crash was caused by damage to the differential gear, which caused an oil leak. Then a number of teeth broke off the spur wheel and ignited the oil. Sabotage or human error was ruled out. The aircraft, W. Nr. 14256, had recently been ferried in. The mission that ended in its destruction was its first mission. There was no fire and a glycol leak responsible for the engine failure. Fire was ruled out, for Marseille could have spoken for nine minutes without fatigue in smoke caused by a fire.

Image Source: findagrave.com/memorial/81058613/hans_joachim-marseille

Fighter Wing Hits Low Morale: Marseille's Leadership Style

The African Star.
Image Source: 3squadron.org.au

JG 27 was moved out of Africa for about a month because of the impact Marseille's death had on morale. The deaths of two other German aces, three weeks earlier had reduced spirits to an all-time low. One biographer suggests these consequences were a result of the command style of Marseille. The more success Marseille had, the more his Squadron relied on him to carry the greater share of aerial victories claimed by the unit. So his death, when it came, was something which JG 27 had seemingly not prepared for.

Historians Hans Ring and Christopher Shores also point to the fact that Marseille's promotions were based on personal success rates more than any other reason, and other pilots did not get to score air victories, let alone become experts themselves. They flew in support as the "maestro showed them how it was done", and often "held back from attacking enemy aircraft to help him build his score still higher". As a result there were no

other "Expert" to step into Marseille's shoes if he was killed. Eduard Neumann explained: "This handicap that very few pilots scored was partially overcome by the morale effect on the whole fighter wing of the success of pilots like Marseille. In fact most of the pilots in Marseille's squadron acted in secondary role as escort to the 'master'." Allied fighter pilot, Pilot Officer Bert Houle of RAF said, "He was an extremely skilled pilot and a deadly shot. It was a helpless feeling to be continually bounced, and to do so little about it."

Victory Claims and Controversies

Marseille flew his first combat mission on Wednesday, 13 August 1940, and claimed his first aerial victory on 24 August 1940. In little more than two years, he amassed another 157 aerial victories. His 158 aerial victories were claimed in 382 combat missions. The German Federal Archives still hold records for 109 of Marseille's aerial victories. A further biographer of Marseille, Walter Wübbe, has made an attempt to link these records to Allied units, squadrons and when possible even to individual pilots, in order to verify the claims as much as possible. Some serious discrepancies between Allied

Marseille's life collage by: *deviantart.com*

squadron records and German claims have caused some historians and Allied veterans to question the accuracy of Marseille's official victories. Attention is often focused on the 26 claims made by *JG 27* on 1 September 1942, of which 17 were claimed by Marseille alone. A USAF historian, Major Robert Tate states: "for years, many British historians and militarists refused to admit that they had lost any aircraft that day in North Africa. Careful review of records however do show that the British (and South Africans) did lose more than 17 aircraft that day, and in the area that Marseille operated." Tate also reveals 20 RAF single-engine fighters and one twin engine fighter were destroyed and several others severely damaged, as well as a USAAF P-40 shot down. However, overall Tate reveals that Marseille's kill total comes close to 65–70 percent corroboration, indicating as many as 50 of his claims may not have actually been kills. Tate also compares Marseilles rate of corroboration with the top six P-40 pilots. There are others who have corroborated 70% to 80% as correct. Some have concluded that Marseille had developed such a supreme confidence in his ability his mentality dictated, "If I fire at it, it must go down." They estimate two-thirds to three-quarters of his claims were aircraft that were destroyed, crash-landed or at least were heavily damaged.

Marseille in the Media

In his short life span, Marseille appeared many times in the German propaganda newsreel. The first time on 17 February 1942 when General Adolf Galland visited the airbase in North African Desert. On 1 July 1942 when Marseille received the Knight's Cross of the Iron Cross with Oak Leaves and Swords from Adolf Hitler. On 9 September 1942 announcing Marseille's 17 aerial victories on 1 September 1942 and that he had been awarded the Diamonds to his Knight's Cross. His last appearance was on 30 September 1942 showing Marseille visiting Erwin Rommel. He was often on the front page in print media. In 1957, a German film, Der Stern von Afrika (The Star of Africa), a fictionalised account of Marseille's wartime service was made.

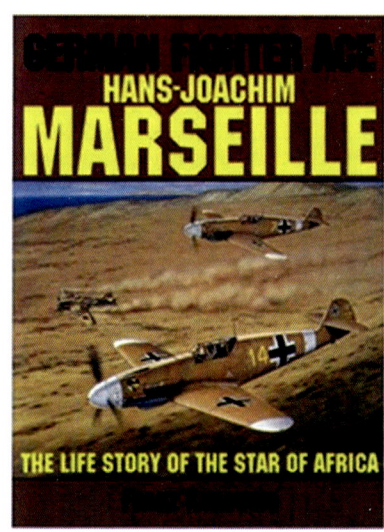

The Book by Franz Kurowski.
Image Source: Amazon.com

Portrait of Hauptmann Hans-Joachim Marseille, mid-Sep 1942; note Knight's Cross of the Iron Cross with Oak Leaves and Swords.
Image Source: Federal Archives via ww2db.com

Apolitical to the Core

It is clear that Marseille was totally an apolitical soldier, despite the prevailing political situation in the Third Reich. Several biographies of Marseille have described his disdain for authority and for the National Socialist movement in general. Some biographers, describe him as "openly anti-Nazi." When Marseille first met Hitler in 1942, he did not form a positive impression. After returning to Africa, Eduard Neumann recalled, "After his first visit with Hitler, Marseille returned and said that he thought 'the Führer was a rather odd sort'." On the visit, Marseille also said some unflattering things about Hitler and the Nazi Party. Several senior officers, including Adolf Galland overheard his remarks during one of the award ceremonies. When asked if he would join the Nazi Party and within earshot of others, Marseille responded, "that if he saw a party worth joining, he would consider it, but there would have to be plenty of attractive women in it." The remarks visibly upset Hitler, who was left "puzzled" by Marseille's behaviour.

Plays American Jazz in Presence of Hitler: Angers Him

At the home of Willy Messerschmitt, industrialist and designer of the Messerschmitt Bf 109 fighter, Marseille played American Jazz on Messerschmitt's piano in front of Adolf Hitler, and Commander-in-Chief of the Luftwaffe, Hermann Göring, head of the SS, Heinrich Himmler and Propaganda Minister Joseph Goebbels. Hitler purportedly left the room. Later that month Marseille was invited to another party function, despite his earlier stunt. Marseille overheard a conversation which mentioned crimes against the Jews and other people. When Marseille returned to his unit, he reportedly asked his friends Franzisket, Clade and Schröer whether they had heard what was happening to Jews and if perhaps something was underway that they did not know about. His friends noticed a change in Marseille's attitude toward his nation's cause. He never spoke of this with his comrades again.

Marseille played American Jazz on Messerschmitt's piano in front of Adolf Hitler.
Image Source: samilhistory.com

"Where I Go, Mathias Goes"

With Mathias. Personal helper.
Image Credit: samilhistory.com

In 1942, Marseille befriended a South African 'Black' Army prisoner of war, Corporal Mathew Letulu, nicknamed Mathias. Marseille took him as a personal helper rather than allow him to be sent to a prisoner of war camp in Europe. Over time, Marseille and Mathias became inseparable. Blacks were looked down upon by Nazis as part of racial theories. This was again an anti-Nazi trait. Marseille was concerned how Mathias would be treated by other units of the Wehrmacht and once remarked "Where I go, Mathias goes." Marseille secured promises from his senior commander Neumann that if anything should happen to him Mathias was to be kept with the unit. Mathias duly remained with JG 27 until the end of the war and attended post-war reunions until his death in 1984.

Memorials: Grave has one Word "Undefeated"

A wartime pyramid was constructed by Italian engineers at the site of Marseille's fall but over time it decayed. On 22 October 1989, Eduard Neumann and other former JG 27 personnel, in co-operation with the Egyptian government, erected a new pyramid. In the weeks following Marseille's death 3/JG 27 was renamed as the "Marseille Staffel" His grave bears a one-word epitaph: "Undefeated". It is understood that Marseille's remains were brought from Derna and reinterred in the memorial gardens at Tobruk. They are now in a small clay coffin (sarcophagus) bearing the number "4133". The tail rudder of his second to last Messerschmitt Bf 109 F-4/trop (Werknummer 8673) now bearing 158 victory marks is on display at Luftwaffe Museum in Berlin. It had initially been given to his family as a gift by Hermann Goring and was donated to the museum.

Marseille Pyramid seen from West, Sidi Abd el-Rahman, Egypt.
Image Source: wikimedia.org

REFERENCES

1. Hans-Joachim Marseille, Wikipedia, https://en.wikipedia.org/wiki/Hans-Joachim_Marseille
2. Katie Serena, Hans-Joachim Marseille Was Hitler's Star Flying Ace – And A Brazen Anti-Nazi, June 6, 2020 https://allthatsinteresting.com/hans-joachim-marseille
3. "Our Adversary" Hauptmann Hans-Joachim MARSEILLE, 3 Squadron RAAF, https://www.3squadron.org.au/subpages/marseille.htm
4. Hans Joachim Marseille Rumba Azul, YouTube, https://www.youtube.com/watch?v=gOECEkUHR10
5. Hans-Joachim Marseille, World War II Database, https://ww2db.com/person_bio.php?person_id=688
6. Hans Joachim Marseille, German Fighter Ace befriends a Black South African POW & defies the Nazi status quo! https://samilhistory.com/tag/hans-joachim-marseille/
7. Not Crusader – Report on the Crash of Hans-Joachim Marseille, Rommelsriposte, May 29, 2018 https://rommelsriposte.com/2018/05/29/not-crusader-report-on-the-crash-of-hans-joachim-marseille/

9

Charles B. DeBellevue

Top American Ace of the Vietnam War – A Non-Pilot

Charles B. DeBellevue.
Image Source: Wikipedia

While Aces are generally thought of exclusively as fighter pilots, but such status has been also accorded to gunners on bombers or reconnaissance aircraft, and to observers in two-seater fighters such as the early Bristol F-2b, and Navigators/weapons officers in aircraft like the F-4 Phantom. Because pilots often teamed with different air crew members, an observer or gunner might be an Ace while his pilot is not, or vice versa. Observer Aces constitute a sizable minority in many lists. Charles George Gass, who tallied 39 victories, was the highest scoring observer Ace in World War I.

In World War II, United States Army Air Forces B-17 tail gunner S/Sgt. Michael Arooth (379th Bomber Group) was credited with 17 victories. The Royal Air Force's leading bomber gunner, Wallace McIntosh, was credited with eight kills, including three on one mission. Flight Sergeant F. J. Barker scored 13 victories while flying as a gunner in a Boulton Paul Defiant turret fighter piloted by Flight Sergeant E.R. Thorne.

Gunner Ace S/Sgt. Michael Arooth.
Image Source: fold3.com

Observor Ace Charles George Gass.
Image Source: airwar19141918.wordpress.com

Captains Steve Ritchie (left) and Charles DeBellevue.
Image Source: wikiwand.com

With the advent of more advanced technology, a third category of Ace appeared. Charles B. DeBellevue became not only the first U.S. Air Force (USAF) weapon systems officer (WSO) to become an Ace but also the top American Ace of the Vietnam War, with six victories. Close behind with five were fellow WSO Jeffrey Feinstein and Radar Intercept Officer William P. Driscoll. Colonel Charles Barbin DeBellevue is a retired officer of the USAF. In 1972, DeBellevue became one of only five Americans to achieve flying Ace status during the Vietnam War, and the first as an Air Force WSO, an integral part of two-man aircrew, with the emergence of air-to-air missiles as the primary weapons during aerial combat. He was credited with a total of six MiG kills, the most earned by any U.S. aviator during the Vietnam War, and is a recipient of the Air Force Cross.

DeBellevue.
Image Source: Pinterest (military.com)

Early Years: Joins USAF

DeBellevue (left), with Captain Steve Ritchie.
Image Source: alchetron.com

DeBellevue was born in New Orleans on August 15, 1945, and grew up in Louisiana. After applying unsuccessfully to the USAF Academy, he attended and graduated from the University of Louisiana at Lafayette (then named the University of South-Western Louisiana), in 1968. Upon graduation, he was commissioned as a second lieutenant through the Air Force Reserve Officers Training Corps (AFROTC) program at the university. Accepted into Undergraduate Pilot Training (UPT), he failed to complete the course, but subsequently applied for and was accepted into Undergraduate Navigator Training (UNT) at Mather Air Base California in July 1969. He completed F-4 combat crew training, and was assigned to the 335th Tactical Fighter Squadron (TFS) as a McDonnell Douglas F-4D weapon systems officer (WSO).

Vietnam War: First MiG Kill

In October 1971, DeBellevue was sent to the famed 555th "Triple Nickel" Tactical Fighter Squadron, of the 432nd Tactical Reconnaissance Wing, at Udorn Royal Thai Air Force Base, Thailand. Flying in a F-4D as the WSO with pilot Captain Steve Ritchie on May 10, 1972, he and Ritchie scored the first of the four Mikoyan MiG 21 kills they would achieve together. Both DeBellevue and Ritchie, along with Captain Jeffrey Feinstein of the 13th Tactical Fighter Squadron, 432nd Tactical Reconnaissance Wing, would become the only USAF "Aces" during the Vietnam War. May 10, 1972 was the same day that Cunningham and Driscoll scored their third, fourth and fifth aerial victories, becoming the Navy's only aces of the war.

Image Source: shop.aviationmilitary.com

An advantage that the "Triple Nickel Squadron" pilots and WSOs had over other U.S. aircrews was that eight of their F-4D Phantoms had the top-secret APX-80 electronic set installed, known by its code-name "Combat Tree". Combat Tree could read the IFF signals of the transponders built into the MiGs so that North Vietnamese GCI radar couldn't discriminate its aircraft from that of the Americans. Displayed on a scope in the WSO's cockpit, Combat Tree gave the Phantoms the ability to identify and locate MiGs when they were still beyond visual range (BVR).

MiGCAP

An air-to-air right side view of two F-4 Phantom II aircraft.
Image Source: nara.getarchive.net

MiGCAP was the term used primarily during the Vietnam War. A MiGCAP was directed specifically against MiG aircraft. MiGCAP during Operation Linebacker became highly organized and meant, an ingress MiGCAP of 2–3 flights (8–12 fighters) that preceded the first supporting forces such as chaff bombers or SAM suppressors and remained until they departed the hostile zone. A target area MiGCAP of at least 2 flights that immediately preceded the actual strikers; and an egress MiGCAP of 1 or 2 flights that arrived on station at the projected exit point ten minutes prior to the earliest egress time. All egress MiGCAP flights were fully fuelled from tankers and relieved the target area CAP.

The First Major Day of Air Combat

Painting of Phantom Vs. MiG 21 in Vietnam:
Image Source: Pinterest

Ritchie and DeBellevue's assignment on May 10, 1972, the first major day of air combat in Operation Linebacker, was as element leader (Oyster 03) of one of two flights of the F-4D MiGCAP for the morning strike force. Oyster flight had three of its Phantoms equipped with Combat Tree IFF interrogators, and two days previously its flight lead, Major Robert Lodge, and his WSO Captain Roger Locher had scored their third MiG kill to lead all USAF crews then flying in Southeast Asia. At 0942 h, forewarned 19 minutes earlier by the EC-121 (AEW&C) Callsign "Disco" over Laos and then by "Red Crown", the US Navy radar picket ship USS Chicago, Oyster flight engaged an equal number of MiG-21s head-on, scattering them. Oyster flight shot down three and nearly got the fourth, but fell victim to a MiG tactic dubbed "Kuban tactics" after those of the Soviet World War II ace Pokryshkin, in which a GCI-controlled flight of MiG-19s trailed so that they could be steered behind the American fighters manoeuvring to attack the MiG-21s. The F-4 flown by Lodge and Locher was shot down. Major Lodge was killed, Capt. Locher ejected and was rescued three weeks later. Almost simultaneously Ritchie and DeBellevue rolled into a firing position behind the remaining MiG-21 of the original four, with a radar lock, launched two Sparrows and scored a kill with the second. Their first.

July 8, 1972, MiG Kills 2 and 3

Ritchie/DeBellevue's Double MiG Killer F-4E.
Image Source: internetmodeler.com

USAF strike and chaff forces suffered a severe series of losses to MiGs between June 24 and July 5 (seven F-4s) without killing a MiG in return. As a counter-measure, 7th Air Force added a second EC-121 to its airborne radar coverage, positioning it over the Gulf of Tonkin. On July 8, 1972, Ritchie and DeBellevue were leading Paula flight, in gun-equipped F-4Es instead of the Combat Tree F-4Ds they usually flew, on a MiGCAP to cover the exit of the strike force. While they were west of Phu Tho and south of Yen Bai, the EC-121 vectored

them to intercept MiG-21s returning to base after damaging one of the US chaff escorts. The MiGs were still approximately 4 miles away and Ritchie turned the flight south to cross the Black River. As they closed, "Disco" gave them a warning that the MiG return had "merged" with the Paula flight's return on his screen. Ritchie reversed course, observed the first MiG at his 10 o'clock position, and turned left to meet it head-on. When Ritchie passed the first MiG-21, he recalled the engagement of May 10 and waited to see if there was a trailing MiG. When he observed the second MiG, which he also passed head-on, he reversed hard left to engage. The MiG turned to its right to evade the attack, an unusual manoeuvre, and Ritchie used a vertical separation move to gain position on its rear quarter. DeBellevue obtained a solid bore-sight radar lock on it while at the MiG's 5 o'clock, and fired from the edge of the flight envelope of AIM-7s. Both missiles struck home.

The first MiG had also turned back and was attacking the last F-4 in Ritchie's flight from behind, an often fatal consequence to US aircraft employing the then-standard "fluid four" tactical formation. Ritchie made a hard turn across the curving intercept of the MiG, again coming out at its 5 o'clock, and the MiG, apparently perceiving the threat, broke hard right and dove away. Ritchie fired an AIM-7 from inside its minimum range and at the limit of its capability to turn. Expecting the Sparrow to miss, he was trying to switch to a gun attack in the relatively unfamiliar F-4E he was flying that day when the missile exploded the MiG, 1 minute and 29 seconds after the first kill.

Steve Ritchie (left) and Chuck DeBellvue (right).
Image Source: Pinterest (History – War & Military, tujifulgueras)

Competition to Become USAF's first Vietnam "Ace"

A competition to become the Air Force's first Vietnam "Ace" developed between Ritchie and Captain Jeffrey S. Feinstein, a WSO in another one of the 432nd's squadrons, the 13th TFS, who scored his 3rd and 4th kills on July 18 and July 29. Each had a claim denied by Seventh Air Force's Enemy Aircraft Claims Evaluation Board, Ritchie and DeBellevue for a claim of a MiG-21 on June 13, and Feinstein for a claim June 9.

August 28, 1972, MiG Kill 4

Ritchie's final victory (his 5th making him an "ace") with DeBellevue (his 4th) came on August 28, 1972, while leading Buick flight, a MiGCAP for a strike north of Hanoi. During the preceding month, 7th Air Force had instituted daily centralized mission debriefings of leaders and planners from all fighter wings called "Linebacker Conferences". Ritchie had just started his flight of Combat Tree Phantoms on its return to base (Ritchie and DeBellevue were flying F-4D AF Serial No. 64-7463, in which they had scored their first kill). Red Crown, now the USS Long Beach, alerted the strike force to "Blue Bandits" (MiG-21s) 30 miles southwest of Hanoi, along the route back to Thailand. Approaching the area of the reported contact at 15,000 feet, Ritchie recalled recent Linebacker Conference information that MiGs had returned to using high altitude tactics and suspected the MiGs were high. Buick and Vega flights, both of the MiGCAP, flew toward the reported location.

28 August 1972: Steve Ritchie and Charles DeBellevue after Ritchie's fifth.
Image Source: retireenews.org

DeBellevue picked up the MiGs on the Phantom's onboard radar and using Combat Tree, discovered that the MiGs were ten miles behind Olds flight, another flight of MiGCAP fighters returning to base. Ritchie called in the contact to warn Olds flight. Ritchie, concerned that MiGs might be at an altitude above them, made continuous requests for altitude readings to both Disco and Red Crown. He received location, heading, and speed data on the MiGs (now determined to be returning north at high speed to their base) but not altitude as Buick flight closed to within 15 miles of the MiGs. DeBellevue's radar then painted the MiGs dead ahead at 25,000 feet, and Ritchie ordered the flight to light afterburners. DeBellevue warned Ritchie they were closing fast and were in range. About the same time Ritchie saw the MiGs himself headed in the opposite direction.

Fifth star. Admiring the painting of their fifth aerial kill on their F-4 Phantom after returning from a mission August 28, 1972, making them the first Air Force Aces in Vietnam, (Left to Right) Capt. Charles D. DeBellevue, weapons systems operator (WSO); Sgt Reggie Taylor, crew chief, Capt. Richard S. Ritchie, pilot, and SSgt. Frank Falcone, assistant crew chief. (U.S. Air Force photo by A1C Larry E. Groom) AFA LIBRARY

Attacking in a climbing curve behind the MiG-21's with his AIM-7 guidance radar locked on, Ritchie was given continuous range updates by DeBellevue. With his Phantom barely making enough speed to overtake the targets, Ritchie launched two Sparrows from over four miles away. The firing parameters of the two shots were out of the missiles' performance envelope, an attempt to influence the MiGs to turn and thus shorten the range. Both shots not only missed but failed to influence the opponents. Moments later, tracking one MiG visually by the contrail it was making, Ritchie fired his remaining two Sparrows, also at long range. The first missed, but the MiG made a hard turn and actually shortened the range, and was destroyed by the second. Short on fuel, Ritchie elected not to try to pursue the second MiG-21. That was Ritchie's final victory (his 5th making him an "Ace")

Vietnamese MiG 19 pilots briefing tactics.
Image Source: nationalmuseum.af.mil

September 9, 1972, "Ace Day", MiG Kills 5 and 6

During Linebacker strikes on September 9, 1972, a flight of four F-4Ds on MiGCAP west of Hanoi shot down three MiGs. Following his fifth kill, Steve Ritchie had been removed from active combat. Two were MiG-19s downed by the new team of Capt. John A. Madden, Jr. and his WSO Capt. DeBellevue. For Madden, the victories constituted his first and second MiG kills, but for DeBellevue they were numbers five and six, moving him up as the leading MiG destroyer of the war and elevating him to "Ace" status. When DeBellevue acquired the MiGs on the radar, the flight manoeuvred to attack. Madden and DeBellevue made the first move. They got a visual on the MiG about 5 miles out on final approach with his gear and flaps down. Getting a lock on him, they fired missiles but they missed. They were coming in from the side-rear and slipped up next to that MiG no more than 500 feet apart. "He got a visual on us, snatched up his flaps, and hit afterburner, accelerating out. It became obvious we weren't going to get another shot at the MiG", says DeBellevue.

An autographed picture of DeBellevue.
Source: Pinterest (GP1011 saved to American Heroes)

DeBellevue Explains the Two Engagements

DeBellevue describes the next two engagements. "We acquired the MiG's on radar and positioned as we picked them up visually. We used a slicing low-speed yo-yo to a position behind the MiG-19's and started turning hard with them. We fired one AIM-9L missile which detonated 25 feet from one of the MiG-19's. We switched the attack to the other MiG-19 and one turn later we fired an AIM-9 at him. I observed the missile impact the tail of the MiG. The MiG continued normally for the next few seconds, then began a slow roll and spiralled downward, impacting the ground with a large fireball."

Madden and DeBellevue returned to their base thinking they had destroyed only the second MiG-19. Only later did investigation reveal that they were the only aircrew to shoot at a MiG-19 which crashed and burned on the runway at Phuc Yen that day. That gave them two MiG-19 kills for the day and brought DeBellevue's total to six MiG kills, the most earned during the war.

Capt. John A. Madden, Jr.
Picture Source: mholloway63.wordpress.com

Flying Summary and Honour

During his combat tour, DeBellevue logged 550 combat hours while flying 220 combat missions, 96 of which were over North Vietnam. His skill as a weapon systems officer was recognized when he and the other two Air Force "Aces", Ritchie and Feinstein, received the 1972 "Mackay Trophy". The Mackay Trophy is awarded yearly by the USAF for the "most meritorious flight of the year" by an Air Force person, persons, or organization. The trophy is housed in the Smithsonian Institution's National air and Space Museum. The award is administered by the U.S. National Aeronautics Association. The award was established on 27 January 1911 by Clarence Mackay, who was then head of the Postal Telegraph-Cable Company and the Commercial Cable Company. Originally, aviators could compete for the trophy annually under rules made each year or the War Department could award the trophy for the most meritorious flight of the year. DeBellevue also received the Veterans of Foreign Wars' Armed Forces Award and the Eugene M. Zuckert Achievement Award.

The 1972 Mackay Trophy goes to three Vietnam aces: Captains Richard S. "Steve" Ritchie, Charles B. DeBellevue, and Jeffery S. Feinstein.
Image Source: afhistory.org

MiG Credits: Pilots, Aircraft and Weapons

The six MiG kills credited to DeBellevue in 1972 included 4 MiG 21s and two MiG 19s. All the MiG 21s were shot using AIM-7 missile and the MiG 19s using AIM-9L. Flights when MiG 21s were shot, the pilot was Capt. Richard S. Ritchie. For remaining two the pilot was Capt. John A. Madden, Jr. The aircraft flown for these six kills included F-4D, AF Serial No. 66-0267, which is now on display at the Homestead Air Reserve Base, Florida; F-4D, AF Serial No. 66-7463, had six confirmed MiG kills, is now on display at the USAF Academy; F-4E, AF Serial No. 67-0362 was sold to Israel as a part of Operation Nickel Grass. Operation Nickel Grass was a strategic airlift conducted by the U.S. to deliver weapons and supplies to Israel during the 1973 Yom Kippur War.

Post-Vietnam War: Made a Pilot

The night of his fifth and sixth victories, DeBellevue was given transfer papers while being toasted at the military officer's club. USAF had a policy of removing aces from combat. He was ordered by the USAF to enter pilot training at Williams AFB, Arizona, in November 1972, or accept his discharge from service. His stated desire to train Weapons System Officers. But the USAF felt that the highest ranking ace of the Vietnam War would not be a non-pilot. He was sent for pilot training, and after pinning on his new pilot wings, he returned to the F-4 as a pilot assigned to the 49th Tactical Fighter Wing at Holloman AFB, New Mexico. This was followed by assignments in Alaska, Japan, and many other airbases in USA. DeBellevue was the last American ace on active duty when he retired from active duty as a full colonel, while serving as commander of Air Force Reserve Officer Training Corps (AFROTC) Detachment 440 at the University of Missouri in January 1998 after 30 years of military service.

The MiG killers head for a pre-mission briefing at Udorn. Captains DeBellevue and Ritchie (front row), and Lt Col Baily and Capt. Feinstein (back row).
Picture Source: wikiwand.com

Pilot DeBellevue.
Picture Source: veterantributes.org

Honours and Decorations

On May 20, 2015, DeBellevue was one of 77 American flying Aces to receive the Congressional Gold Medal in a ceremony in Washington D.C. The Congressional Gold Medal is the highest honour Congress can bestow on behalf of the American people. He was a recipient of the Air Force Cross, Vietnam Gallantry Medal, Distinguished Flying Cross among many others.

DeBellevue's Medal Ribbons.
Image Source: Wikipedia

Congressional Gold Medal.
Image Source: Wikipedia

McDonnell F-4D-29-MC Phantom II 66-0267, flown by Madden and DeBellevue, 9 September 1972, on display at the main gate, Homestead AFB, Florida. (© Europix)

Air Force Cross Citation

Date of Action: September 9, 1972

The President of the United States of America, authorized by Title 10, Section 8742, United States Code, takes pleasure in presenting the Air Force Cross to Captain Charles B. DeBellevue (AFSN: 0-3210693), United States Air Force, for extraordinary heroism in military operations against an opposing armed force as an F-4D Weapon Systems Officer in the 555th Tactical Fighter Squadron, Udorn Royal Thai Air Force Base, Thailand, in action on 9 September 1972. On that date, while protecting a large strike force attacking a high priority target deep in hostile territory, Captain DeBellevue engaged and destroyed a hostile aircraft. Through superior judgment and use of aircraft capabilities, and in complete disregard for his own safety, Captain DeBellevue was successful in destroying his fifth hostile aircraft, a North Vietnamese MiG-19. Through his extraordinary heroism, superb airmanship, and aggressiveness in the face of the enemy, Captain DeBellevue reflected the highest credit upon himself and the United States Air Force.

REFERENCES

1. Charles B. DeBellevue, Photo Gallery, Official United States Air Force Website, https://www.af.mil/News/Photos/igphoto/2000593696/
2. DeBellevue, Charles Barbin, Col, Together We Served, https://airforce.togetherweserved.com/usaf/servlet/tws.webapp.WebApp?cmd=ShadowBoxProfile&type=Person&ID=108004
3. Charles B. DeBellevue, Hall of Valour Project, https://valor.militarytimes.com/hero/3621
4. DeBellevue, Charles B., Eagle Profile, https://goefoundation.org/eagles/debellevue-charles-b/
5. Charles B. DeBellevue, Veteran Tributes, http://veterantributes.org/TributeDetail.php?recordID=128
6. Bryan R. Swopes, Charles B. DeBellevue, This Day in Aviation, 9 September 1972, https://www.thisdayinaviation.com/tag/charles-b-debellevue/
7. Charles B. DeBellevue, Oklahoma History Center Hall of Fame, Inducted 2012, https://www.okhistory.org/historycenter/militaryhof/inductee.php?id=15
8. Charles Barbin DeBellevue, Wikipedia, https://en.wikipedia.org/wiki/Charles_B._DeBellevue

10

LIEUTENANT INDRA LAL ROY, DFC

The First and the Only Indian Air Ace

Lieutenant Indra Lal Roy.
Image Source: better2240.rssing.com

World War I had about 1.5 million Indians fighting in every theatre of the battle. Among the handful of Indians who fought in the air was a gifted combat pilot who would go on to become India's first and only fighter Ace, Indra Lal 'Laddie' Roy, the incredible air warrior from Bengal. Indra Lal Roy was the sole Indian World War I flying Ace. He is also designated as first Indian fighter aircraft pilot. While serving in the Royal Flying Corps (RFC) and its successor, the Royal Air Force (RAF), he claimed ten aerial victories; five aircraft destroyed (one shared), and five 'down out of control' (one shared) in just over 170 hours flying time.

Lolita Roy and her six children, listed in the 1911 census at their home at 77 Brook Green, W.
Image Source: greatwarlondon.wordpress.com

Initial Years and Family

He was the second son of Piera Lal Roy and Lolita Roy. Born on 02 Dec 1898 at Calcutta (now Kolkata), where his father was a barrister and Director of Public Prosecutions. He was nicknamed "Laddie." Theirs was a highly qualified and distinguished family, originally from the Barisal district in present-day Bangladesh. They were prominent Zamindars of East Bengal. Their family estate is called the Lakhutia Zamindar estate and was founded by Roop Chandra Roy in the late 17th century. His older brother, Paresh Lal Roy (1893–1979), served in the 1st Battalion, Honourable Artillery Company (HAC) was incorporated by Royal charter in 1537 by King Henry VIII and is considered the second-oldest military corps in the world. He was an Indian amateur boxer, credited with popularising the sport among Indians. Later became known as the "father of Indian boxing." His maternal grandfather, Dr. Surya Kumar Goodeve Chakraborty, was one of the first Indian doctors to be trained in Western medicine. His nephew, Subroto Mukerjee (1911–1960), served as a fighter pilot in World War II and later became the first Indian Chief of Air Staff of the Indian Air Force. Interestingly Indra Lal's elder brother Paresh was NDTV founder Prannoy Roy's paternal grandfather.

Roy Joins RFC. Rear Row right.
Picture Source: indiatimes.com

Roy Joins RFC

In 1901, Roy's family moved to London. When the First World War broke out, Roy was attending school, in London, England. Determined to do his bit, he signed up for the cadet force at his school, the 400-year-old St Paul's School for Boys at Hammersmith. The bright teenager also designed a trench mortar and sent the design to the War office along with notes on its advantages. Impressed by his academic record and innovative designs, Roy was awarded a scholarship by Oxford University.

The Royal Flying Corps (RFC) was the air arm of the British Army.
Picture Source: thebetterindia.com

Initially, he was rejected by the Royal Flying Corps on the grounds of poor eyesight. But Roy was not someone who would give up so easily. He sold his motorbike to pay for a second opinion from one of Britain's leading eye specialists. This time, he cleared the eye test. Five months after turning 18, on 4 April 1917, he joined the RFC and was commissioned as a Second Lieutenant on 5 July. After training and gunnery practice, he joined No.56 Squadron on 30 October. Roy was part of "A" Flight, commanded by Captain Richard Maybery, who himself became a flying Ace with 21 aerial victories, and was awarded bar to Military Cross.

Two months later, Roy was injured after he had crash-landed his S.E.5a fighter on 6 December. There are some unconfirmed version that he was actually shot by a German aircraft over France. An unconscious Roy was moved to the local hospital where he was taken for dead and laid out in the morgue. On regaining consciousness, he banged loudly on the morgue's door while shouting out in his school-boy French! After the door was finally opened by

Sketch by Roy.
Image Source: Indian Air Force Museum

the terrified hospital staff, the boy 'who had come back from the dead' was promptly sent back to England for further treatment. While recovering from the crash, Roy made numerous sketches of aircraft, many of which survive, and some are now displayed at the IAF museum in Delhi.

In May he returned to duty as an equipment officer and within a few weeks, the very determined Roy passed as medically fit to fly. More focused than ever before, Roy trained hard to become a good fighter pilot. He was transferred to Captain George McElory's flight in No. 40 Squadron in June 1918. Interestingly, Captain George Edward Henry McElroy MC and Two Bars, DFC & Bar (14 May 1893 – 31 July 1918) was a leading Irish fighter pilot and Air Ace of the RFC and RAF during World War I. He was credited with 47 aerial victories.

S.E.5a fighter.
Image Source: Pinterest

Captain George Edward Henry McElroy MC and Two Bars, DFC and Bar.
Picture Source: theaerodrome.com

On his return to active service, Roy achieved ten victories (two shared) in just thirteen days with just over 170 hours total flying experience. His first was a Hannover over Drocourt, Northern France, on 6 July. This was followed by three victories in the space of four hours on 8 July (two Hannover Cs and a Fokker D.VII); two on 13 July (a Hannover C and a Pfalz D.III); two on 15 July (two Fokker D.VIIs); and one on 18 July (a DFW C.V). Roy's final victory came the following day when he shot down a Hannover C over Cagnicourt. Unfortunately, the talented pilot didn't survive the war.

The Fokker D.VII: WWI German aerial superiority aircraft.
Picture Source: YouTube

Killed in Action and Final Resting Place

Roy was shot down on July 22, 1918, when his plane was attacked by four Fokker DVIIs. Roy fought back, however, was hit by the gunfire from one of the aircraft. His plane burst into flames before crashing into German-controlled territory. He was just 19 years old. As per some sources, Roy was so well known as a fighter pilot that Manfred von Richthofen "Red Baron" the top German WWI flyer with 80 aerial victories, paid him a tribute by dropping a wreath from the skies at the spot where his plane had crashed. This story pops up on social media now and then, but truth is that in fact, the Baron was himself killed in action on 21 April 1918, three months before Lt Roy's death. So that story is not based on facts. He was the first and only Indian flying Ace of the First World War. Indra Lal Roy was buried at the Estevelles Communal Cemetery, about 100 km north of the grave of another Indian combat pilot of World War I, Lieutenant Srikrishna Welingkar. His grave bears a simple

Indra Lal Roy's grave at Estevelles Communal Cemetery, Pas-de-Calais, France.
Picture Source: Wikiwand

inscription in Bengali that reads *"Maha birer samadhi; sambhram dekhao, sparsha koro na"* (The grave of the courageous warrior; respect it, do not touch it).

List of Aerial Victories

S.No	Date/Time	Aircraft/Serial No.	Opponent	Location
1	6 July 1918@ 0545	S.E.5a (B180)	Hannover C	Drocourt
2	8 July 1918@ 0645	S.E.5a (B180)	Hannover C	Drocourt
3	8 July 1918@ 0925	S.E.5a (B180)	Hannover C	East of Monchy
4	8 July 1918@ 1025	S.E.5a (B180)	Fokker D.VII	South-east of Douai
5	13 July 1918@ 0645	S.E.5a (B180)	Hannover C	West of Estaires
6	13 July 1918@ 2005	S.E.5a (B180)	Pfalz D.III	Vitry - Brebières
7	15 July 1918@ 2005	S.E.5a (B180)	Fokker D.VII	Hulloch
8	15 July 1918@ 2005	S.E.5a (B180)	Fokker D.VII	Hulloch
9	18 July 1918@ 2040	S.E.5a (B180)	DFW C.V	South-east of Arras
10	19 July 1918@ 1025	S.E.5a (B180)	Hannover C	Cagnicourt

Information Source: Wikipedia

Distinguished Flying Cross (DFC).
Image Source: Amazon.com

Honours and Awards

Roy was posthumously awarded the Distinguished Flying Cross (DFC) in September 1918 for his actions during the period of 6–19 July 1918. He was the first Indian to receive the DFC. His citation read:

Lieutenant Indra Lal Roy

"A very gallant and determined officer, who in thirteen days accounted for nine enemy machines. In these several engagements he has displayed remarkable skill and daring, on more than one occasion accounting for two machines in one patrol." (20 September 1918).

Stamp Release

Image Sources: Twitter Mintage World @MintageWorld

In December 1998, to mark the 100th anniversary of his birth, the Indian postal service issued a commemorative stamp in his honour. Another Stamp was released in 2019.

Stamp was released in 2019 to Honour Indians of World War I.
Image Source: Wikipedia

List of World War I Flying Aces from British India

1. Captain Lawrence Percival Coombes (15 aerial victories flying Sopwith Camel for the Royal Naval Air Service and the Royal Air Force).
2. Captain Maurice Douglas Guest Scott (12 aerial victories flying as both observer and pilot).
3. Captain (later Squadron Leader) Edward Dawson Atkinson (10 aerial victories in two combat tours, one while flying a Nieuport, and another flying the S.E.5a.
4. Lieutenant Indra Lal Roy (10 aerial victories while piloting a S.E.5a).
5. Captain (later Group Captain) Arthur Peck (8 aerial victories).
6. Captain Douglas Carbery (6 confirmed aerial victories while serving as an aerial observer in various squadrons).
7. Lieutenant Thomas Cecil Silwood Tuffield (6 aerial victories confirmed while flying as an observer).
8. Captain George M.Cox (5 aerial victories flying Sopwith Camel).

REFERENCES

1. K.S. Nair, Remembering Indra Lal Roy, India's 'Ace' Over Flanders, *The Wire*, July 22, 2017, https://thewire.in/history/indra-lal-roy-ace-flanders-india-failed-celebrate
2. Samyak Pandey, This forgotten pilot from India was just 19 when he shot down 9 German planes during WWI, *The Print*, July 22, 2019, https://theprint.in/theprint-profile/this-forgotten-pilot-from-india-was-just-19-when-he-shot-down-9-german-planes-during-wwi/266203/
3. Indra Lal Roy, Wikipedia, https://en.wikipedia.org/wiki/Indra_Lal_Roy
4. Sanchari Pal, This Forgotten Pilot Was Just 19 When He Became India's First and Only Flying Ace, *The Better India*, September 14, 2017, https://www.thebetterindia.com/115388/indra-lal-roy-indian-flying-ace-world-war-2/
5. Samyak Pandey, This forgotten pilot from India was just 19 when he shot down 9 German planes during WWI, *The Print*, July 22, 2019, https://theprint.in/theprint-profile/this-forgotten-pilot-from-india-was-just-19-when-he-shot-down-9-german-planes-during-wwi/266203/
6. Samyukhtha Sunil, India's First 'Ace' Fighter Pilot, *Amar Chitra Katha*, July 21, 2020, https://www.amarchitrakatha.com/history_details/the-life-of-indra-lal-roy/
7. Officer's record: Indra Lal Roy (RFC and RAF), *National Archives UK Government*, http://www.nationalarchives.gov.uk/pathways/firstworldwar/people/lalroy.htm
8. Joseph Noronha, INDRA LAL ROY (1898-1918), *SP Aviation*, 09/2017, http://www.sps-aviation.com/story/?id=2126&h=Indra-Lal-Roy-1898-1918
9. The Roy brothers: fighting for King and Emperor, *The Great War London*, https://greatwarlondon.wordpress.com/2013/10/09/the-roy-brothers-fighting-for-king-and-emperor/

11

Gabby Gabreski

America's Two-War Ace: WW II and Korea

Gabreski in the cockpit of his P-47 after his 28th victory.
Picture Source: avgeekery.com

Col Francis "Gabby" Gabreski. (Imperial War Museum FRE 13934).
Picture Source: thisdayinaviation.com

Francis Stanley "Gabby" Gabreski was a Polish-American career pilot in the United States Air Force (USAF). He was the top American and United States Army Air Force (USAAF) fighter Ace over Europe during World War II and a jet fighter Ace in the Korean War. Although best known for his credited destruction of 34½ aircraft in aerial combat and being one of only seven U.S. combat pilots to become an Ace in two wars, Gabreski was also one of the Air Force's most accomplished leaders. In addition to commanding two fighter squadrons, he had six command tours at Group or Wing level, including one in combat in Korea, totalling over 11 years of command and 15 overall in operational fighter assignments. After his Air Force career, Gabreski headed the Long Island Rail Road, a commuter railroad owned by the State of New York, and struggled in his attempts to improve its service and financial condition. After two and a half years, he resigned under pressure and went into full retirement.

Early Years

Gabreski's parents had emigrated from Poland to the Oil City, Pennsylvania, in the early 1900s. His father (Stanisław "Stanley" Gabryszewski) owned and operated a market, putting in 12-hour days. As in many other immigrant-owned businesses in those days, the whole family worked at the market. But Gabreski's parents had dreams for him, including attending the University of Notre Dame, Indiana. He did so in 1938, but, unprepared for real academic work, almost failed during his freshman year.

Taylor Cub Aircraft. Photo credit: Louis C. McGowan.
Picture Source: newenglandaviationhistory.com

Interest in Flying: Could Not Do Solo

During his first year at Notre Dame, Gabreski developed an interest in flying. He took lessons in a Taylor Cub and accumulated six hours of flight time. However, his autobiography indicates, he struggled to fly smoothly and did not fly solo, having been advised by his instructor Homer Stockert that he did not "have the touch to be a pilot".

Boeing Stearman PT-17 Kaydet Trainer.
Picture Source: Smithsonian timeandnavigation.si.edu

Joins U.S. Army Air Forces (USAAF)

At the start of his second year at Notre Dame, Gabreski enlisted in the U.S. Army Air Corps, volunteering as an aviation cadet. After his induction into the U.S. Army at Pittsburgh, he undertook primary flight training flying the Stearman PT-17, Gabreski was a mediocre trainee and was forced to pass an elimination check ride during primary to continue training. He advanced to basic flight training on the Vultee BT-13 and completed advanced training at Maxwell Field, Alabama, on the North American AT-6 Texan. Gabreski earned his Wings and his commission as a second lieutenant in the Air Corps in March 1941. He then sailed for Hawaii for his first assignment.

2nd Lt. Francis "Gabby" Gabreski (left), and 1st Lt. Emmett "Cyclone" Davis, second from the left, at Wheeler Field Officer's Club 1941.
Picture Source: Wikipedia

First Fighter Squadron: Meets His Future Wife

Assigned as a fighter pilot with the 45th Pursuit Squadron of the 15th Pursuit Group at Wheeler Army Airfield, Hawaii, he trained on both the Curtiss P-36 Hawk and the newer Curtiss P-40 Warhawk. He met his future wife, Catherine "Kay" Cochran, in Hawaii and became engaged shortly after the Japanese attack on Pearl Harbour. During that action, Gabreski joined several members of his squadron in flying P-36 fighters in an attempt to intercept the attackers, but the Japanese had withdrawn. During the spring and summer of 1942, Gabreski remained with the 45th

World War II: Curtiss P-40 Warhawk.
Picture Source: thoughtco.com

(renamed as 45th Fighter Squadron in May 1942), training in newer model P-40s and the Bell P-39 Airacobras that the unit began to receive.

Offers to Serve as Liaison Officer to Polish Squadrons

He closely followed reports on the Battle of Britain and the role played by the Polish RAF squadrons, especially by the legendary No. 303 Polish Fighter Squadron. Gabby felt strongly about what the Nazis had done to Poland and was anxious to get into the war. Hearing about Polish fliers who were helping the British fight the Luftwaffe in the Battle of Britain, he got an idea: "I was a fighter pilot, of Polish origin, and I could speak Polish. Why not see if I could get myself assigned to Europe so I could learn from the Poles and pass the information along to my own people." He became concerned that the US did not have many experienced fighter pilots. Polish squadrons had proved to be capable within the RAF. The idea was approved, and he left Hawaii for Washington D.C. in September 1942, where he was promoted to Captain.

RAF Duty With Polish Squadron

In October, Gabreski reported to the Eighth Air Force's VIII Fighter Command in England, at that time still a rudimentary new headquarters. After a lengthy period of inactivity, he tried to arrange duty with 303 Squadron, but that unit had been taken out of action for a period of rest. Instead, he was posted to No. 315 (Dublin) Squadron in January 1943. The well-seasoned Poles, who had been fighting the Germans since 1939, accepted Gabreski, if reluctantly at first. Squadron Leader Tadeusz Andersz helped him transition to the Spitfire and fly the squadron's "finger-four" formation as well as "rodeo" and "circus" manoeuvres designed to entice Luftwaffe pilots to come up and fight. Andersz also taught Gabby how to hold his fire until he got close behind an enemy aircraft, as well as to resist the tendency to overshoot by going too fast. And there was that basic lesson every fighter pilot should learn: Always be alert for enemy fighters attacking out of the sun. Gabby remembered an early combat mission with the Poles that reinforced this imperative. "One moment I had looked back into an empty sky above me," he recalled, "and the next moment it had been full of Focke Wulf 190s that seemed to come out of nowhere. I was lucky to have survived the lesson; a lot of inexperienced pilots didn't."

Captain Francis S. Gabreski, U.S. Army Air Corps, in the cockpit of his Supermarine Spitfire Mk.IX, PK E, BS410, with No. 315 Squadron, Royal Air Force, at RAF Northolt, England, 1943. This airplane was shot down 13 May 1943. It is currently under restoration. (Royal Air Force).
Image Source: thisdayinaviation.com

Gabreski flew the new Supermarine Spitfire Mark IX, flying patrol sweeps over the Channel. He first encountered *Luftwaffe* opposition on February 3, when a group of Focke-Wulf Fw 190s attacked his squadron. He was too excited about wanting to make a "kill". Later Gabreski realised and learned that he had to keep calm during a mission, a lesson that served him well later in the war. This squadron was part of the group that Winston Churchill praised during the Battle of Britain when in his famous quotation: "Never in the field of human conflict have so many, owed so much, to so few." Gabreski began

Gabreski and S/Sgt Ralph Safford, his crew chief. The assistant crew chief Felix Schacki is in the background.
Picture Source: Wikipedia

missions escorting bombers to the French Coast. But when he finally met the enemy in the air he did not get an opportunity to fire his guns. "Get in close so you cannot miss," the Poles advised, words that he never forgot. After flying 20 missions, Gabreski gained skill and a swagger in the air. He later spoke with great esteem about the Polish pilots and the lessons they taught him. In all, Gabreski flew 20 missions with the Poles, engaging in combat once. Gabby's experience with No. 315 Squadron gave him confidence in his abilities as a fighter pilot, which he might not have gained had he been sent immediately to an American unit upon his arrival in Britain. Although he didn't score any victories, he was awarded the Polish Cross of Valour.

CO 61st Fighter Squadron: Ill Will in the Unit

In February 1943, Gabreski joined the 61st Fighter Squadron of 56th Fighter Group (FG), flying the Republic P-47 Thunderbolt. Lieutenant Les Smith recalled 40 years later that the group's pilots at first did not accept Gabreski. "That was unfortunate," he said, "but I think not unexpected, since he had not trained with us in the States, had not shipped with us, and had no personal relationships within the group. There was another unfortunate factor over which he had no control—his rank as Captain. This put him in direct competition with the old Captains already assigned and in indirect competition with our older first Lieutenants who hoped to become Captains. We knew Second Lieutenants were not really involved in this rivalry, but we held the older pilots in great esteem, and if they didn't like the new stranger, we weren't going to be too friendly either. We eventually recognized Gabby's superior ability as a pilot and his very aggressive fighting spirit, and we respected him for them."

Republic P-47N Thunderbolt.
Picture Source: Wikipedia

Focke-Wulf Fw 190.
Picture source: Wikipedia

Learning to fly its P-47 Thunderbolt, best known as the Jug, he earned the admiration of his colleagues for his skill and the affectionate nickname of "Gabby." Before long Gabreski taught his men every tactic learned from the Poles. Then, after completing ten more combat missions, he received the Air Medal and was presented the Polish Cross of Valour by General Sikorsky, the Polish Premier in exile. He was quickly cleared as a flight leader. He was resented by many of his fellow pilots, and the fact that he was opinionated and verbose made it worse. In May, Gabreski was promoted to Major. On June 9, he took command of the Squadron when its commanding officer was moved up to Group deputy commander. This also stirred ill feelings toward him since he had jumped over two more senior pilots.

First Victory

Gabby scored his first confirmed kill, an Fw-190, on August 24. He described his feelings afterwards: "That evening before I went to sleep I thought about the implications of what I had done that day. I had killed a man, I was sure of it. Yet I felt no remorse. It wasn't that I particularly wanted to kill people, Germans or otherwise. But this was war, and for three years I had been preparing myself mentally and physically for the day when I would begin shooting down enemy aircraft. Yes, there was a man inside of the Fw-190 I'd destroyed today, but I never saw him, never heard him, and never knew his name or what he looked like."

Piper L-4 Grasshopper.
Picture Source: www.dday-overlord.com

The general ill will in the unit against him was soon exacerbated when both of his much more respected pilot colleagues were lost in combat on June 26 and did not subside until he recorded his first credited kill, an Fw 190 on August 24, 1943. Gabreski escorted the Allied bombers on their historic raid on ball-bearing factories at Regensburg and Schweinfurt. After his first air victory, began an incredible string. His first kill presaged criticism that followed him throughout his combat career, when his wingmen complained that his attack had been too hastily conducted to allow them to also engage.

Although he gained confidence on every flight, a simple mishap almost ended his career as a fighter pilot. He was hand-propping a Piper L-4 for a short flight when the engine backfired, causing the prop to kick back, barely missing his head and striking his right hand. One finger bled profusely, attached only by a piece of skin and a tendon. At first when Gabby was whisked to a nearby hospital it seemed the finger couldn't be saved, but a surgeon managed to set it in a permanently curved position. Gabby was grounded for the next three months.

Becomes an "Ace": Has a Close Shave

The top two American aces in the European theatre, Robert S. Johnson congratulates Gabby on a job well done.
Image Source: acesofww2.com

Gabreski's first mission following his return to duty was disappointing since he was forced to break off when his belly tank malfunctioned, while the remaining pilots downed five Messerschmitt Me-110 twin-engine fighters with no losses. That day, August 17, 1943, was called "Black Thursday" after the bomber force they were escorting lost 60 planes. On November 26, 1943, the 56th FG was assigned to cover the withdrawal of Boeing B-17 Flying Fortress bombers that had bombed Bremen, Germany. The P-47s arrived to find the bombers under heavy attack near Oldenburg and immediately dived into the fray. Gabreski recorded his fourth and fifth kills to become an "Ace". He was leading a bomber escort flight when they were attacked by a group of Me-110s. Gabby was making a stern attack on one of them when it suddenly exploded. Large pieces of the Messerschmitt skimmed off his canopy and smashed into the P-47's right wing. His plane was still flyable, though, so he climbed back up to continue the fight and downed a second 110. The 56th set a record that day with 23 confirmed victories, and Gabby was awarded the Distinguished Service Cross for his leadership. This was his 75th mission. He had a close brush with death on December 11, when a 20 mm cannon shell lodged in his engine without exploding, destroying its turbocharger. Low on fuel and ammunition, Gabreski out manoeuvred the Bf-109 until it succeeded in placing a burst of fire into his P-47, disabling the engine. Gabreski stayed in the airplane, however, until it restarted at a lower altitude, where the turbocharger was not needed.

Responsibilities at 56th FG

In November 1943, the Group Commander of the 56th, Colonel Hubert Zemke, was replaced in command for two months by Colonel Robert Landry, a staff officer at VIII FC. Because of Landry's inexperience, combat missions of the 56th were alternately led by deputy commander Lt Col Schilling and Gabreski, who acted as deputy group operations officer. When Zemke resumed command on January 19, 1944, Gabreski relinquished command of the 61st Squadron. In February

Lieutenant Colonel Gabreski's Republic P-47D-25-RE Thunderbolt, 42-26418, RAF Boxted, Essex, England, 1944. (U.S. Air Force 68268 A.C./American Air Museum in Britain UPL 33594)
Image Source: thisdayinaviation.com

1944, Gabreski brought two Polish pilots into the 56th, who had flown with him in 1943 while serving with the RAF, including future USAAF ace Squadron Leader Boleslaw "Mike" Gladych. With Gabreski's support and to ease a shortage of experienced pilots caused by many veterans reaching the completion of their tours, the 61st FS in April accepted five other Polish Air Force pilots into the squadron as the "Polish Flight".

Becomes Leading American Ace in European Theatre

Gabreski's victory total steadily climbed through the winter of 1943–44. Meanwhile, larger fuel tanks were installed on the P-47Ds, giving them greater range, and in early 1944 Eighth Air Force Commander Maj. Gen. James H. "Jimmy" Doolittle authorized fighters to leave the bombers and seek out enemy aircraft wherever they could be found. As a result, the fighters were able to destroy scores of German planes in the air and on the ground. In January 1944, when the Allies set out to

Lieutenant Colonel Gabreski (standing, just left of centre) with the pilots of the 61st Fighter Squadron, July 1944.
Picture Source: American Air Museum in Britain

cripple the Luftwaffe, Gabreski became Deputy Flying Executive Officer of the 56th Fighter Group, and before the month ended he was a double Ace. After "Operation Big Week" was launched to destroy enemy aircraft production, he scored his first triple victory. By March 27, he had 18 victory credits and had six multiple-kill missions to rank third in the "Ace race" that had developed within VIII Fighter Command. He downed only one more aircraft in the next two months, during which time the two pilots ahead of him (Majors Robert S. Johnson and Walker M. Mahurin also of the 56th FG) were sent home. In April 1944, the 56th FG moved to RAF Boxted and Gabreski was promoted to Lieutenant Colonel. During his most productive mission, on May 22, he downed three Fw-190s and scored a "probable." He resumed command of the 61st FS when its Commander was transferred out.

On May 22, Gabreski shot down three Fw 190s over a *Luftwaffe* airfield in northwest Germany. June 6th, 1944, D-Day, Gabreski led his squadron in long

The "Impregnable Quadrilateral" – 'Hub' Zemke (56FG C/O), Dave Schilling, Gabby Gabreski & Fred Christensen.
Image Source: acesofww2.com

sweeps over the beaches of Normandy. He tied with Johnson as the leading ace in the European Theatre of Operations on June 27, passing Eddie Rickenbacker's record from World War I in the process. On July 5, 1944, he became America's leading ace in the European Theatre of Operations (ETO), with his 28th victory, destroyed while leading the group on an escort mission over a German airbase in France, thus matching the total at that time of confirmed victories of the Pacific Theatre's top American

Lieutenant Colonel Gabreski, at right, with the ground crew of his Republic P-47D Thunderbolt, July 1944 (American Air Museum in Britain).
Picture Source: thisdayinaviation.com

Ace, Richard Bong. This total was never surpassed by any U.S. pilot fighting the *Luftwaffe*. His 61st Squadron's five kills that day brought its total to 230, the best record in the ETO.

The publicity resulting from Gabby's 28th victory was almost overwhelming for him. "I hardly had a moment to spare doing what I was in England to do," he said, "which was to fly airplanes. I felt I had an obligation in the war, just like everybody else. But I was being taken out of that environment and put on a little pedestal. It was an awkward position for me, and I never did fit into it very well."

All Set for Marriage, Becomes a Prisoner of War

Since Gabreski became America's leading Ace in Europe and the country's newest celebrity. The War Department, eager to take advantage of his newfound notoriety, immediately made arrangements to have the pilot shipped home so he could help sell war bonds. On July 20, 1944, Gabreski had reached the 300-hour combat time limit for Eighth Air Force fighter pilots and was awaiting an aircraft to return him to the United States on leave and reassignment. He had already advised "Kay" Cochran to proceed with wedding plans, and his hometown of Oil City, Pennsylvania, had raised $2,000 for a wedding present in anticipation of his return. Gabreski found, however, that a bomber escort mission to Russelheim, Germany, was scheduled for that morning, and, instead of boarding the transport, he requested to "fly just one more." Returning from the mission, Gabreski observed Heinkel He 111s parked on the airfield at Bassenheim, Germany, and took his airplane down to attack. He was dissatisfied with his first strafing run on a He 111, and he reversed for a second pass. When his tracers went over the parked bomber, he dropped the nose of his Thunderbolt to adjust, and its propeller clipped the runway, bending the tips. The damage caused his engine to vibrate violently and he was forced to crash land. Gabreski ran into nearby woods and eluded capture for five days. Gabreski soon realized that the Americans weren't the only ones anticipating his arrival. Dazed but unhurt, Gabby knew he was hundreds of miles inside Germany and needed to avoid capture. For the next five days, he managed to elude searchers, but was finally apprehended by a policeman who turned him over to the military at Oberursel on July 25. Hanns Scharff, an affable English-speaking German intelligence officer, interrogated him. "Hello, Gabby," he greeted him. "We have been waiting for you for a long time." He handed Gabreski a copy of a military newspaper that documented the pilot's historic 28th kill. After being captured and interrogated, he was sent to Stalag Luft I.

Two German officers stand on the wing of Lieutenant Colonel Gabreski's P-47D-25-RE Thunderbolt, 42-26418, near Bassenheim, Germany. Picture from *Luftwaffe*. Source: thisdayinaviation.com

POW Days

Gabby was transferred to Stalag Luft I at Barth on the Baltic Sea, north of Berlin.

Stalag Luft I at Barth on the Baltic Sea. Germany held nearly 9,000 USAAF and RAF airmen at Stalag Luft I.
Picture Source: americanairmuseum.com

By then his former Group Commander, Colonel "Hub" Zemke, who had been shot down in October 1944, was the camp's ranking officer. Stalag Luft I was a permanent prisoner-of-war camp holding Allied air officers. Gabreski received quarters in one of the 20-man shacks surrounded by two rows of barbed wire fence. There he shared the bad food, hunger and punishments of the prisoners. But he was proud of the men's spirits under such miserable circumstances. The prisoners had their own clandestine radios to listen to war news, a newspaper printed under the very noses of their guards, and supervision of the simultaneous digging of as many as 100 escape tunnels, few of which led to freedom.

Gabby's most enduring memory of the next nine months as a POW besides boredom was increasing hunger. Red Cross parcels kept the prisoners from starving, but as the war got worse for Germany, the parcels stopped coming. The winter and early spring of 1945 was a horrific period marked by sub-zero temperatures and increasingly inadequate sustenance? By March 1945, after Gabreski received command of a newly completed prisoner compound, food quality was at rock bottom. But he did not lose faith. Soon he began to hear artillery from the east. On April 30, however, Russian troops liberated the camp. Many of the captives wanted to open the gates and take off, but Zemke ordered them to stay put and wait for aircraft to fly them out to Camp Lucky Strike, near Le Havre, France. The war and its privations had ended.

Lieutenant Colonel and Kay Cochran Mrs. Francis S. Gabreski, 11 June 1945.
Picture Source: andrezejburlewicz.blog

Summary of WW II Flying

Gabreski flew 166 combat sorties and was officially credited by the USAAF with 28 aircraft destroyed in air combat and 3 on the ground. He was assigned five P-47s during his time with the 56th FG, none of which he named, but all of which bore the fuselage identification codes HV: A.

Repatriation and Fighter Command

Following his repatriation, Gabby managed a detour to visit his old unit in England and then persuaded authorities to allow him to take a flight directly to New York. Gabreski returned to the United States a hero, as America's top Ace in Europe. Gabreski married beautiful Kay Cochran on June 11, 1945. Like so many veterans after World War II, Gabby didn't know what to do next. He wanted to complete his college degree as well as continue flying. After a 90-day recuperative leave, he became Chief of Fighter Test Section at Wright Field, Ohio, and at the same time completed test pilot training at its Engineering Flight Test School. In April 1946, he left the service. He meanwhile received a job offer from Douglas Aircraft Co. as a foreign sales representative. He accepted the offer in May 1946 and toured Mexico, Argentina, Brazil, and Chile attempting to sell the Douglas DC-6, a pressurized version of the DC-4. The trip was not very successful, though, and Gabby found the traveling life and especially being away from his wife, now with one child and another on the way, uncomfortable.

Gabby with His Wife.
Picture Source: YouTube

He decided to see if he could re-join the Army Air Forces, and was pleased to be accepted in April 1947 as a regular Lieutenant Colonel assigned to command the 55th Fighter Squadron, flying North American P-51s at Shaw Air Force Base, S.C. "It was great to get back in the cockpit again," he said, "and it was great to be a squadron commander in peacetime conditions. The P-51 was a beautiful airplane with a lot of range. It was a joy to fly."

North American P-51 Mustang.
Picture Source: Wikipedia

The Air Force sent him to Columbia University in September 1947 to complete his degree and study Russian. In June 1949, he graduated with a Bachelor of Arts degree in Political science. He returned immediately to flying, becoming commander of his former unit, the 56th Fighter Group, now flying Lockheed F-80 Shooting Star, the Air Force's first operational jet fighter. It was a great leap forward for the prop plane pilots, but the early jets had their share of difficulties. The jet engines had a voracious appetite for fuel at low altitudes, and Gabby admitted to nearly running out of gas several times. The jets also suffered from slow engine acceleration, and in addition, there was the possibility of a compressor stall if the throttle was advanced too quickly. While in command of the 56th, Gabreski oversaw the conversion of the unit to North American F-86 Sabres and was promoted to Colonel on March 11, 1950.

In June 1950, the peacetime routine changed drastically for American fighter pilots when Communist North Korean forces invaded South Korea. The 56th exchanged its P-80s for North American F-86 Sabre jets, and Gabby was transferred to the 4th Fighter-Interceptor Wing (FIW), based in Japan with units in Korea. He embarked on his second combat career as the Deputy Wing Commander, and could fly with any squadron he chose to learn about tactics and techniques in a new kind of war.

North American F-86F Sabre.
Picture Source: flyingmag.com

Initial Days in Korean War

In June 1951, he and a group of selected pilots of the 56th Fighter Interceptor Wing (FIW) accompanied the delivery of F-86Es of the 62nd Fighter Interceptor Squadron (FIS) to Korea aboard the escort carrier USS Cape Esperance. The planes and pilots joined the 4th Fighter Interceptor Group at K-14 (Kimpo) Air Base, where most engaged in combat. Taking command of the 51st Fighter Interceptor Group, Gabreski converted it to F-86 Sabre jets and introduced the concepts of a flight of four and hot take-offs to increase combat effectiveness over targets. These innovations were highly successful and enabled his Wing to attain a 14 to 1 kill ratio.

Gabreski stepping out of his Sabre, April 1950.
Picture Source: Wikipedia

He was engaged in aerial combat again during the Korean War. On June 17, 1951, Gabby took off with some trepidation on his first Korean War mission. "I searched the deep blue sky for signs of enemy fighters and began to wonder if I still had what it took to fly combat," he recalled. "I was thirty-two years old now, and my eyesight might not be as sharp as it was in Europe. Had my reflexes slowed? Would I still have the old fire in my belly that made me want to climb up their tails before opening fire? Only time would tell."

The MiG Alley

When Communist MiG-15s appeared, he experienced one of the frustrations that dogged all F-86 pilots at that time. The higher the Sabres flew, the more unstable they became, while the MiGs could fly higher and loiter longer because they were lighter and were close to their bases on the Chinese side of the Yalu River.

Target MiG Alley.
Picture Source: reddit.com

When he flew his first mission against a target in "MiG Alley," south of the Yalu River, for the first time in his life he could see the enemy but was not permitted to attack them in their privileged sanctuary in Manchuria. The international border with North Korea was theoretically a line over which American interceptors could not cross since China was not officially involved in the war. It was a difficult way to fight a war.

Leading a flight of four F-86s on July 8, 1951, Gabby saw some F-80s and MiGs scrambling at 10,000 feet and barrelled down behind a MiG that was breaking away from the fight. As he had in Europe, he got on its tail, held his fire until close enough, and then blasted it to pieces with his six .50-caliber machine guns. He knew now he was still able to attack an enemy aircraft and make a kill, shooting down a MiG 15. Gabby got his second victory, over Pyongyang, on 02 September, a stray MiG-15 was heading home when he took it out with a deflection shot. A month later he downed another one on 02 October. Meanwhile, Communist air tactics changed. Taking advantage of their numerical superiority, they formed two long lines of 50 to 60 fighters—"MiG trains" flying down both sides of the peninsula. More F-86s and pilots were clearly needed, so two new Wings were formed in November, with Gabby commanding the 51st Fighter Interceptor Wing (FIW). Its first mission was on 01 December.

MiG Alley.
Image Source: forces.net

F-86 Sabres of the 51st Fighter-Interceptor Wing.
Picture Source: flyingfiendsinkoreanwar.com

Around this time, the Americans noted that the MiG pilots seemed more skilled, and suspected they were facing experienced Russian pilots, though they couldn't be sure. That suspicion was officially confirmed, much later, after the fall of Soviet Union. There was always the temptation to cross the Yalu River after them, and Gabby and others admitted they sometimes did so if they thought they could shoot down a MiG without going too far (reported in "MiG Madness," March 2008 issue).

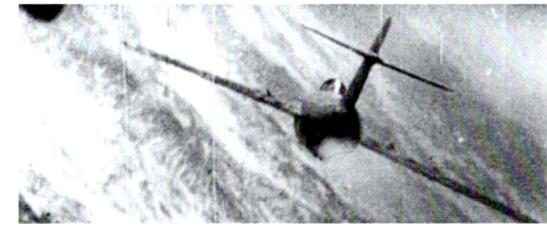

A MiG-15 sits square in the sights of F-86 Sabre.
Picture Source: historynet.com

Becomes a "Jet Ace"

Gabby flew all of the 51st's early missions and devised a new tactic known as the "fluid four," a more flexible version of WWII's finger-four that was better suited to jets. The growing MiG threat against Boeing B-29 Superfortress bomber attacks along the Yalu River caused the Fifth Air Force to create a second Sabre

127

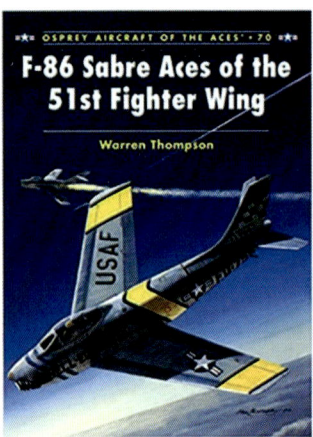

Gabreski 51st Fighter Wing Chronicled in this Book.
Image Source: amazon.in

Wing by converting the 51st FIW from F-80s to F-86s in a 10-day period. Gabreski was transferred to K-13 (Suwna Air Base, accompanied by most of the former 56th FIW pilots who had come with him to Korea, and took command on November 6, 1951. By January 1952, the Wing had destroyed 26 MiGs with only seven losses. Gabreski got one on January 11, and shared credit for another on February 20. During its first seven months as an F-86 Wing, the 51st, with only two operational squadrons, scored 96 MiG kills, comparing favourably to the 125 of the veteran 4th FIW, which operated three.

Combat operations picked up with the clearing spring weather. Gabby was leading a flight along the Yalu River on 07 April 1951 when he sighted the contrails of 30 MiGs climbing up from Antung in the safety of Chinese airspace. He was very conscious of the imaginary line he should not cross, but he couldn't pass up the opportunity. The sun was behind his formation as 15 MiGs came out of the contrails, and Gabby pounced on a straggler returning to his Chinese base. He fired until the enemy pilot blew off his canopy and bailed out. That brought his jet tally to 5½ victories and made him the first American to become an Ace in two wars. General Ridgway flew in to congratulate him. Gabby scored again on 12th April, and got his next, the last kill, taking his total to 6½. As per policy, American fighter Aces were to return back after crossing five kills. His combat days effectively ended.

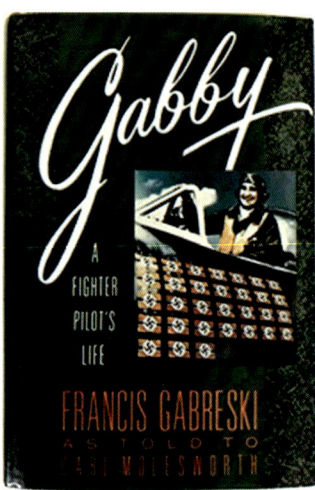

Image Source: abebooks.com

Combined with his 28 kills in Europe, that made him the third-highest-scoring American Ace of all time, after Lockheed P-38 pilots Dick Bong and Thomas McGuire. He was grounded after completing his 100th mission, returning Stateside to a ticker-tape parade in San Francisco and a visit with President Harry Truman at the White House. The airman commented in his memoirs that "it was quite a thrill for a Polish kid from Oil City who had almost flunked out of flight school."

Aggressive Commander: Overflies China

Colonel Gabreski in the cockpit of a North American Aviation F-86E Sabre, Korea, 1952.

He was an aggressive commander and fostered a fierce rivalry between the two F-86 wings, fuelled in part by the fact that the 4th had also been the keenest rival of the 56th FG during World War II. While this aggressiveness paid off in the destruction of MiGs and air superiority over all of Korea, it also led Gabreski to make the first intentional violation of rules of engagement that prohibited combat with MiGs over China. The MiG force was based in this ostensible sanctuary during the entire war. Gabreski and a fellow former 56th pilot, Colonel Mahurin, planned and executed a mission in early 1952 in which the F-86s turned off their IFF equipment and overflew two Chinese bases.

The Opinion of His Wingmen

Gabreski was also criticized for having a poor attitude towards wingmen. One historian, citing five interviews with pilots and an unpublished manuscript by a sixth, observed that Gabreski flew the fastest aircraft available and failed to notice when his slower wingmen could not keep up. These pilots, reportedly afraid to fly with him, commented that he was more interested in personal achievement than in his wingmen. He was also criticized for a lack of discipline among his off-duty pilots and for allegedly encouraging exaggerated kill claims. Nonetheless, at least three wingmen had different views. 1st Lieutenant Joe L. Cannon of the 51st FIW flew over 40 missions with him and described Gabreski as a mentor and "my kind of fighter pilot". 1st Lt. Harry Shumate, another 51st FIW pilot, stated that while flying wingman in Gabreski's flight, Shumate was the first to spot a MiG-15 heading for its base and Gabreski told him to "go get him" while the leader covered.

Gabby at left and Eugen Barnum watch as Frank Klibbe describes an aerial encounter.
Picture Source: starduststudios.com

A 4th FIW pilot, 1st Lt. Anthony Kulengosky, observed, "I moved up in the world of wingmen by flying Col. Francis Gabreski's wing on a mission. I was absolutely thrilled to fly on this legend's wing...He was a tiger and went on to become an ace again. When asked who I looked up to the most as a pilot and a gentleman in all my flying, I still have to say it was "Gabby" Gabreski. When he took over the 51st Wing, he asked me to move over as a flight leader in his outfit." Capt. Robert W. "Smitty" Smith, a 4th FIW pilot in Korea, recalled, "Shortly after my arrival, Gabby flew the first F-86E to arrive on base in simulated combat over the field against an F-86A and whipped the other guy badly, with every Sabre jock on the base as a witness. After he landed he briefed all pilots and announced that the limited number of E's would be reserved for flight leaders. I never forgot his response, when someone asked about the problem of wingmen staying with leaders. He replied "Wingmen are to absorb firepower" and I never knew him well enough to judge whether he had a dry sense of humour, but he made the right choice. One thing I know for sure, Gabby proved himself the greatest at our skills and talents, when he added 6 ½ MIG kills to his 28 victories in WW II and become the all-time American Fighter Ace, he did it in the P-47, not the better air-to-air P-51. And he didn't have a chance to fly the much more powerful F-86F, which arrived after us."

Francis Gabreski and a couple of his ground crewmen.
Picture Source: starduststudios.com

Combat Grace and Magnanimity

A noted pilot also rebuts some of the criticism. Major Whisner had been a P-51 double-ace with the 352nd FG in World War II and was one of the pilots Gabreski brought with him from the 56th FIW in June 1951. Before the mission of February 20, 1952, Gabreski and Whisner, each had four MiGs credited as destroyed. During the mission, Gabreski attacked and severely damaged a MiG 15 that fled across the Yalu River into China. He broke off the engagement and returned to base after his own airplane was damaged, where he claimed the MiG as a "probable kill". Whisner trailed the MiG deep into Manchuria trying to confirm Gabreski's kill, but his Sabre ran low on fuel. He completed the shoot-down and returned to K-14 where he confirmed the kill for Gabreski but did not claim it himself. Gabreski confronted him and angrily ordered him to change his mission report, confirming Whisner's own role in the kill. Whisner refused. Soon after, Gabreski recanted his anger and the two shared the claim, as a consequence of which three days later Whisner and not Gabreski became the first pilot of the 51st FW to reach jet Ace status.

Gabreski (left) congratulating Whisner.
Picture Source: Wikipedia

Stops Logging Sorties to Avoid Return to the USA

Gabreski's Korean tour was due to end in June. As he approached his mission limit in early April, he quit logging sorties to avoid being transferred from his command. He was, however, grounded by Fifth Air Force from further combat in mid-May when his deputy commander, Colonel Mahurin, was shot down. Gabreski was subsequently replaced by Colonel John W, Mitchell, the leader of Operation Vengeance, the mission to shoot down and kill Admiral Isoroku Yamamoto of Imperial Japanese Navy in World War II.

Humane Side

Air Ace Col. John William Mitchell.
Picture Source: Wikipedia

It wasn't all war all the time for Gabby, especially when winter weather prohibited operations. Although he didn't make it a practice to flaunt his faith, he was a deeply religious and compassionate man despite his "killer" military reputation. The Wing chaplain had located an orphanage in Suwon that was crowded with 300 poorly clothed, sick and starving children. When Gabby heard about it, he had the Wing sponsor the orphanage and appealed to the citizens of Oil City for help. They responded with mountains of donated clothing, medicine, school supplies and building materials. Soon it became a haven of happiness.

Mission Summary after Korea

On his return to the United States, Gabreski received the key to the city from San Francisco Mayor Elmer E. Robinson and was given a ticker-tape parade up Market Street on June 17. Gabreski's 6½ MiG-15 kill credits make him one of seven U.S. pilots to become an Ace in more than one war (the others being Whisner, Colonel Harrison Thyng, Colonel Hagerstrom, Colonel V. Garrison, Lt Col George A. Davis, and U.S. Marine Corps Lieutenant Colonel John F. Bolt). Gabreski was officially credited with 123 combat missions in Korea, totalling 289 for his career. Although he flew many F-86s in combat, his assigned aircraft was F-86E-10-NA 51-2740, nicknamed "Gabby".

Lt Col George Andrew Davis Jr. the Only American Ace to be shot in Korea.
Picture Source: nationalmuseum.af.mil

Image Source: YouTube

Aerial Victories Summary

As a member of Hubert Zemke's Wolfpack the 56th Fighter Group, and commanding the 61st Fighter Squadron, he became the leading American Ace in Europe with 28 victories in 17 months, before becoming a POW at Stalag Luft I in WW II. During the Korean Conflict, he scored 6.5 more combat victories over Korea, bringing his total to 34.5 and making him America's top living ace.

Post-Korea

Gabreski's Air Force career continued for another 15 years, during which time he held three Wing commands totalling nearly nine years of duty. His assignments were: Chief of Combat Operations Section,

Office of the Inspector General (July 1952 – June 1954), Student, Air War College, Maxwell (1954–1955), Deputy Chief of Staff, Headquarters 9th Air Force (July 1955 – August 1956), Commander, 342d Fighter-Day Wing (September 10, 1956 – November 19, 1956). Commander, 354th TFW (F-100 Super Sabre) (November 19, 1956 – July 13, 1960), Commander, 18th Tactical Fighter Wing (F-100) Okinawa (August 8, 1960 – June 19, 1962).

F-100Ds, Mid-Air Refuelling.
Picture Source: flickr.com (Photo by hondagl 1800)

The aerial refuelling capability of the F-100 and fighters like it enabled the Air Force to quickly send reinforcements to trouble spots around the globe. Gabby found that in-flight refuelling required special training, and after flying a number of such missions, described them as "a dramatic experience." He said later, "I would rather attack a squadron of Fw-190s alone in a P-47 than face one of those drogues again in an F-100. That was nightmare fodder."

Gabreski.
Picture Source: acesofww2.com

Director of the Secretariat, Headquarters Pacific Air Forces, Hawaii (July 1962 – July 1963), Inspector General, Pacific Air Forces – Hickam Air Force Base, Hawaii (July 1963 – August 1964), and Commander 52nd Fighter Wing (F 101 Voodoo) (August 17, 1964 – October 31, 1967). Gabreski retired on November 1, 1967. He left the Air Force bringing to an end a brilliant military career with 34.5 air victories and nearly every military air honour. As per his USAF official biography, he retired with more than 5,000 flying hours, 4,000 of them in jets.

Military Honours and Awards

Gabreski's military decorations and awards included Distinguished Service Cross (DSO), Distinguished Flying Cross (DFC) with two silver and two bronze oak leaf clusters, Distinguished Flying Cross (DFC) (United Kingdom), Legion of Honour (France), Polish Cross of Valour (Poland), Republic of Korea Presidential Unit Citation, among many others.

Post-Retirement Assignments

Immediately after retirement, in 1967 nearby Grumman Aerospace Co. offered Gabby a job as a marketing vice president, looking after Public Relations and Customer Relations before becoming assistant to the corporation's president. Gabreski worked for Grumman Aerospace until August 1978. He was asked by New York Governor to serve as President of the financially stressed and state-owned Long Island Rail Road in an attempt to improve the commuter line. After what he described as an 18-month struggle with the board of the Metropolitan Transport Authority, Gabreski resigned on February 26, 1981. With the exception of a 2½-year stint in 1978-81 as President of the

Picture Source: Poles in America Foundation poles.org

Col. Francis Gabby Gabreski during one of the Planes of Fames annual "air show" events.
Picture source: flickriver.com

Long Island Railroad, he stayed with Grumman until his retirement in 1987.

Personal Life and Death

Francis and Kay Gabreski had nine children (three sons and six daughters) in 48 years of marriage. Two of their three sons graduated from the USAF Academy and became career Air Force pilots. His daughter-in-law Terry Lee (Walter) Gabreski was promoted to Lt. Gen in August 2005, the highest-ranking woman in the USAF until her retirement in 2010. His wife died as the result of an automobile accident as they both were returning from the Oshkosh Air Show on August 6, 1993. Gabreski died of an apparent heart attack on January 31, 2002, and is buried in Calverton National Cemetery. Gabreski's funeral on February 6 was with full military honours and included a missing man formation flyover by F-15E Strike Eagles.

Daughter-in-law Lt Gen. Terry Gabreski.
Picture Source: Wikipedia

Summarised

The man who would become the top American fighter ace in Europe during World War II and a jet ace in Korea almost washed out of flight training. After six hours of civilian instruction in a Taylor Cub, he was deemed too tense at the controls, and the owner of Stockert Flying Services said he "didn't have the touch to be a pilot." Later, during Army Air Corps primary training, he barely survived a last-chance elimination flight in a Boeing-Stearman PT-17. But assigned a new instructor, he managed to complete his flight training. Francis Gabreski never looked back, embarking on a storied 27-year Air Force career that led in his twilight years to his designation as "America's greatest living ace."

Four F-15E Strike Eagles Flying the "Missing Man" Formation in Gabby's Honour.
Picture Source: USAF

Legacy

Francis S. Gabreski flew 289 combat missions in two wars and destroyed 34½ enemy aircraft. He was enshrined in the National Aviation Hall of Fame in 1978 "for outstanding contributions to aviation by his displaying unusual valour and new combat tactics in becoming a leading ace in two wars and by devotion to duty in peace." Their motto is "Honouring Aerospace Legends to Inspire Future Leaders". The Suffolk County Air Force Base in Westhampton Beach, New York, which became Suffolk County Airport in 1969, was renamed Francis S. Gabreski Airport in 1991. The collocated New York Air National Guard installation at the airport was also renamed Francis S. Gabreski Air National Guard Base. Gabreski Road at

Gabby Gabreski 1/6th Statue WWII on Sale.
Image Source: aikensairplanes.com

Shaw AFB, SC, is named in his honour. The Colonel Francis S. Gabreski Squadron of the Civil Air Patrol located in Bellport, New York is named in his honour. He wrote his autobiography, "Gabby: A Fighter Pilot's Life" in 1998 along with Carl Molesworth.

REFERENCES

1. Francis Stanley Gabreski (28 January 1919–31 January 2002), This Day in Aviation, https://www.thisdayinaviation.com/tag/gabby-gabreski/
2. Gabreski, Francis "Gabby", Military Combat, The National Aviation Hall of Fame, Enshrined 1978, https://www.nationalaviation.org/our-enshrinees/gabreski-francis/
3. Francis Stanley "Gabby" Gabreski, Aces of WW2, https://acesofww2.com/USA/aces/gabreski/
4. C.V. Glines, Gabby Gabreski: America's Two-War Ace, History Net, https://www.historynet.com/gabby-gabreski-americas-two-war-ace.htm
5. Walter J. Boyne, Gabreski, Air Force Magazine. November 1, 2005 https://www.airforcemag.com/article/1105gabreski/
6. Francis Stanley "Gabby" Gabreski, Wikipedia https://en.wikipedia.org/wiki/Gabby_Gabreski
7. Francis Gabreski, Ace Fighter Pilot, Pennsylvania Military Museum, March 9, 2020, https://www.pamilmuseum.org/news/2020/3/9/francis-gabreski-ace-fighter-pilot
8. Francis Stanley Gabreski, American Air Museum in Britain, https://www.americanairmuseum.com/person/65786
9. Jeremy Reynolds, Oil City works to keep legacy of WWII 'ace' in the public eye, Pittsburgh Post-Gazette, May 24, 2020, https://www.post-gazette.com/local/north/2020/05/24/Francis-Gabby-Gabreski-ace-pilot-PA-Memorial-Day-Oil-City-history-air-force/stories/202005220125
10. Legends of Airpower Season 1 Episode 6 Opening: "Gabby" Gabreski, YouTube, https://www.youtube.com/watch?v=jwbACRFoLqM
11. P-47 Ace Francis "Gabby" Gabreski Interview, YouTube, https://www.youtube.com/watch?v=ZF3YR-1ucIU

12

Hans-Ulrich Rudel

German Ace: Eagle of the Eastern Front

Hans-Ulrich Rudel.
Picture Source: warhistoryonline.com

Hans-Ulrich Rudel was a German ground-attack pilot, and the most decorated German serviceman of World War II, being the sole recipient of the Knight's Cross with Golden Oak Leaves, Swords, and Diamonds. Post-war, he was a prominent neo-Nazi activist in Latin America and West Germany. During the war, Rudel was credited with the destruction of 519 tanks, as well as one battleship, one cruiser, 70 landing craft, 150 artillery emplacements, and more than 800 vehicles of all types. He also claimed 11 aerial victories, earning flying ace status. Rudel flew 2,530 ground-attack missions exclusively on the Eastern Front, mostly flying the Junker Ju 87 "Stuka" dive bomber.

Hans-Ulrich Rudel.
Picture Source: military.wikia.org

Rudel surrendered to US forces on 8 May 1945 and immigrated to Argentina in 1948. A committed and unrepentant National Socialist, he founded the "*Kameradenwerk*", a relief organization for Nazi refugees that helped fugitives escape to Latin America and the Middle East. Together with Willem Sassen, Rudel helped shelter Josef Mengele, the notorious former SS doctor at Auschwitz. He worked as an arms dealer and a military advisor to the regimes of Juan Perón in Argentina, of Augusto Pinochet in Chile, and of Alfredo Stroessner in Paraguay. Due to these activities, he was placed under observation by the US Central

Intelligence Agency (CIA). In the West German federal election of 1953, Rudel, who had returned to West Germany, was the top candidate for the far-right German Reich Party but was not elected to the Bundestag. Following the Revolution in 1955, the uprising that ended the second presidential term of Perón, Rudel moved to Paraguay, where he acted as a foreign representative for several German companies. In 1977, he became a spokesman for the German People's Union, a neo-Nazi political party founded by the extremist politician Gerhard Frey. Rudel died in West Germany in 1982.

Early Years: Joins Luftwaffe at 20

Rudel was born on 2 July 1916, in Konradswaldau, in Prussia. His father was a Lutheran minister. As a boy, Rudel was poor in studies, but a very keen sportsman. He joined the Hitler Youth in 1933. After graduating in 1936, he participated in the compulsory Reich Labour Service (RAD). Later the same year Rudel joined the Luftwaffe and began his military career as an air reconnaissance pilot.

Early model Ju-87's with very unusual looking wheel fairings.
Image Source: acesflyinghigh

World War II Begins: Becomes a Stuka Pilot

German forces invaded Poland in 1939 starting the WW II in Europe. As an air observer, Rudel flew on long-range reconnaissance missions over Poland. In early 1941, he underwent training as a Stuka (Junkers Ju 87) bomber pilot. He was posted to 1 *Staffel* (flying unit) *Sturzkampfgeschwader* 2 (Dive Bombing Wing StG 2), which was moved to occupied Poland in preparation for "Operation Barbarossa", the invasion of the Soviet Union, in June 1941. In his first day of battle, Rudel flew four missions. In a little over a month he flew 100, receiving the Iron Cross First Class and new respect from his flying mates. "He is the best man in my squadron!" claimed Captain Ernst-Siegfried Steen. "But this crazy fellow will have a short life...." Rudel later wrote in his memoirs, "He knows that I generally dive too low a level, in order to make sure of hitting the target and not waste ammunition."

On 21 September 1941, Rudel took part in an attack on the Soviet battleship "Marat" of the "Baltic Fleet". Marat was sunk at her moorings on 23 September 1941 after being hit by a 1,000-kilogram (2,200 lb.) bomb

Soviet battleship *Marat* in 1939.
Picture Source: warhistoryonline.com

near the forward superstructure. It caused the explosion of the forward magazine which demolished the superstructure and the forward part of the hull. 326 men were killed and the ship gradually settled to the bottom. Her sinking was credited to Rudel. His backseat gunner, Sergeant Alfred Scharnowski, was with him all the way. The young East Prussian, the 13th child in his family, was accustomed to having the odds against him. "He seldom speaks," commented Rudel, "...nothing ruffles him." Rudel's unit then took part in "Operation Typhoon", Army Group Centre's attempt to capture the Soviet capital.

The Soviet battleship *Marat*, undergoes repairs in Kronstadt Harbour. Although *Marat* would contribute to the defence of Leningrad as a floating battery, it never steamed out of Kronstadt again. (National Archives).
Picture Source: historynet.com

Luftwaffe aerial photograph of the damaged Soviet battleship *Marat* in Kronstadt, leaking oil.
Picture Source: commons.wikimedia.org

On target approach, the flak was so intense that the Stukas, bobbing, weaving, and dodging, broke formation. Rudel held station on Steen's wing, and together they bored in. From miles away they could see *Marat* tied up with the heavy cruiser *Kirov* at its stern. Wing-mounted dive brakes were extended for greater stability and accuracy. Steen pitched over into the attack, with Rudel right behind him. The airspeed indicator wound up as the altimeter wound down. "I have already picked up *Marat* in my sights," Rudel recounted. "We race down towards her; slowly she grows to a gigantic size. Now all their A.A. guns are directed at us" quoted Don Hollway for historynet.com.

Steen closed his brakes, trying to get down through the flak before it blew him out of the sky. Rudel cut his brakes as well, "going all out. I am right on his tail, traveling much too fast and unable to check my speed." He passed so close to the lead plane that he could see Steen's rear gunner, Sergeant Helmut Lehmann, looking terrified that Rudel would ram them. *Marat* loomed up below, "large as life in front of me. Sailors are running across the deck….Now I press the bomb release switch on my stick and pull with all my strength." Already too low to use the Stuka's automatic dive-recovery system, Rudel was also well below his bomb's 3,000-foot safe release height. "My acceleration is too great," he wrote. "My sight is blurred in momentary blackout…when I hear Scharnowski's voice: 'She is blowing up, sir!'" They had pulled out a dozen feet or so above the water. Behind them, Rudel saw a 1,200-foot pillar of smoke and fire billowing from the battleship. His bomb had exploded in an ammunition magazine. *Marat*'s bow had blown off.

Misses a Mission: Saves His Life

The Stukas regrouped at their airfield to next target the cruiser "*Kirov*". Taxiing for take-off, his CO Steen's plane got mired in soft ground, so he switched to Rudel's Stuka. Rudel had to watch as the CO took off. In the midst of their attack dive, they took a hit on the tail. Unable to pull out, Steen aimed the Stuka at "*Kirov*", but hit the sea alongside. Even Rudel's ardent Nazism seemed shaken by the incident.

Hans Ulrich Rudel and Gunner Erwin Hentschel.
Picture Source: asisbiz.com

Rudel's Gunner – Erwin Hentschel

Rudel's gunner from October 1941 was Erwin Hentschel, who served with Rudel for the next two and a half years, both men earning the Knight's Cross of the Iron Cross. Hentschel completed 1,400 sorties with Rudel and drowned on 21 March 1944 when they were making their way to the German lines following a forced landing.

Battle of Stalingrad

Rudel took part in the Battle of Stalingrad. From May 1941 to January 1942, Rudel flew 500 missions. In early 1942, Rudel got married to his fiancée Ursula while home on leave. Having survived his first Russian winter and a summer commanding a Stuka training unit. By the time he re-joined StG.2, flying the new Ju-87D over Stalingrad, the German Sixth

Oberleutnant (First Lieutenant) Hans-Ulrich Rudel drinks from his Goblet after the presentation of the Luftwaffe Honour Goblet on 20 October 1941.
Image Source: ww2images.blogspot.com

Army had already cornered the Russians in 1,000 yards of the Volga River's west bank. Stuka bombing precision was essential in this situation. "We have to drop our bombs with painstaking accuracy," explained Rudel, "because our own soldiers are only a few yards away in another cellar behind the debris of another wall."

In February 1943, Rudel flew his 1,000th combat mission, which made him into a national hero. Within days the Sixth Army was encircled. The Stukas flew 10 to 15 sorties a day, dawn to dusk around Stalingrad's shrinking the *Kessel* (cauldron), where Soviets and Nazis fought to the death over the wreckage, rubble, and their dictators' prestige. "Because of the uninterrupted sorties and the stiff fighting," Rudel said, "...the whole squadron has at the moment scarcely more than enough aircraft to form one strong flight." StG.2 pulled out to a base 100 miles west of the city, only to find Soviet armour bearing down on the airfield. Rudel flew 17 sorties, stopping the last tank himself just a few yards short of his own runway. "We know the strength of the opposition," he wrote. "It is too late to free the Sixth Army."

Rudel demonstrates his preferred method of attack on a Russian Tank.
Picture Source: donhollway.com/hans-ulrichrudel

Stukas in action, 1943.
Image Source: German Federal Archives

Experimental Ju 87G: More Success – More Awards

He then participated in the experiments using the Ju 87 G in the anti-tank role. It had occurred to the German high command that the most efficient way to kill a tank wasn't by trying to hit it on the roof with a bomb. Armed with two 600-pound cannon pods, the Stuka became slow and unwieldy, unable to dive or carry bombs, but its 6-foot gun barrels could put 37mm tungsten-core shells through square-foot targets from the air at more than 150 yards. This Ju-87G – the *Kanonenvogel* (Cannon bird) or *Panzerknacker* (Tank cracker) – would become one of the war's supreme tank busters, largely in Rudel's hands. The anti-tank unit took part in operations against the Soviet Kerch–Eltigen Operation. The footage from one of his on-board gun camera was used in a Reich Ministry of Propaganda newsreel. In April 1943, Rudel was awarded the Knight's Cross of the Iron Cross with Oak Leaves, receiving the Oak Leaves from Hitler personally in Berlin.

A stricken Soviet T-34/76 burns, a fate shared by 519 enemy tanks targeted by Rudel. (Bundesarchiv Bild F016221-0014).
Picture Source: historynet.com

July 1943 Battle of Kursk

On 12 July 1943 as thousands of German and Russian tanks wheeled and fired at the point-blank range

Rudel – The Greatest tank Hunter Pilot.
Image Source: extraordinarybiographies.blogspot.com

below him, Rudel circled behind the enemy armour formations to attack from the rear. In his first attack he disabled four tanks, and by the end of the first day, he had bagged 12, the equivalent of a Soviet armour company. He came in so low he risked being caught in the target's explosion. "This happens to me twice in the first few days when I suddenly fly through a curtain of fire," he reported. "I come out, however, safe and sound on the other side, even though…my aircraft is scorched and splinters from the exploding tank have riddled it with holes."

Loyalty to Stuka

The Luftwaffe already intended to replace Ju-87G with the faster Focke-Wulf Fw-190F in the ground attack role (and Rudel would sometimes fly it), but his name would always be linked to the Stuka. "The evil spell is broken," he raved about the *Panzerknacker*. "In this aircraft, we possess a weapon which can speedily be employed everywhere and is capable of dealing successfully with the formidable numbers of Soviet tanks."

Stuka Ju-87G. Battle of Kursk.
Picture Source: flightlineweekly.com

Gets His Gunner a Knight's Cross

Soon appointed Wing commander, Rudel formed an elite tank-hunter squadron, a Stuka "fire brigade" tossed into the line wherever the latest Russian breakthrough threatened. His mission tally and score rose dramatically; by November he had racked up 1,500 missions and more than 100 tank kills. His back-seater, Sergeant Erwin Hentschel, became the most successful gunner in the Luftwaffe, with more than 1,200 missions and several enemy aircraft to his credit. Rudel recommended him for the Knight's Cross, but the paperwork had not gone through when he was called to Hitler's Wolf's Lair headquarters to receive his own Swords to his own Knight's Cross. Rudel took Hentschel along with him, and by sheer force of his personality and respect that he commanded even from Hitler, he arranged for the gunner to receive his medal directly from the Führer.

Forced Landing on a Highway

In the winter of 1943, on a reconnaissance mission, lost in thick fog and running low on fuel, Rudel made a forced landing on a

Stuka legion Hans Ulrich Rudel and Gunner Erwin Hentschel.
Picture Source: asisbiz.com

highway. "We taxi along the very broad highway as if we were driving a car," Rudel recounted, "obeying the usual traffic regulations and allowing heavy Lorries to pass. Many of them thought they were seeing a ghost plane." They taxied for nearly 25 miles along, surely some sort of taxi record, till an overpass blocked the way. Leaving Hentschel to guard the plane, Rudel caught a ride to base and returned to take off when the weather lifted.

Aerial Victory

He also proved that the *Panzerknacker* was effective against Soviet "flying tanks", the heavily armoured *Shturmoviks*. He dropped down alone through the fighter cover. Within three hundred feet of the IL-2 *"Shturmovik"*, he let loose a round of anti-tank ammunition, his target exploded.

Ilyushin IL-2 *Shturmovik*.
Picture source: en.wikipedia.org

Marking his 1,300th mission, Rudel, who was a teetotaller, would have left the champagne to his backseater Erwin Hentschel, then celebrating his 1,000th (National Archives).
Picture Source: historynet.com

Daring Rescue of a Downed Pilot behind Enemy Lines

Rudel was appointed *Gruppenkommandeur* of III. *Gruppe* on 22 February 1944. On 20th March 1944, when Luftwaffe thought that the Stuka's glory days as a dive bomber were over, Rudel joined the effort to cut the Soviet bridgeheads over the Dniester River near Nikolayev, Ukraine. On his eighth sortie that day, now-Major Rudel saw one of his crews forced down on the wrong side of the river. He landed there to pick them up. He had performed such rescues a half-dozen times before, and had been so rescued himself. But with two extra passengers, his Stuka bogged down in the mud. With Soviet troops closing in, Rudel, Hentschel, and the rest ran several miles in full gear. Dropping flight suits and boots, they slid down riverbank cliffs into the water. The 600-yard-wide Dniester was in full flood, a few degrees above freezing and full of ice. "Gradually one becomes dead to all sensation save the instinct of self-preservation," Rudel recalled. His athletic training

Dniester River.
Image Source: kids.britannica.com

saved him. Last into the water, he was second to reach the far bank. Eighty yards short, gunner Hentschel threw up his arms and went under. Rudel dived back in, but couldn't find his flying mate. The rest were soon captured. Rudel had been shot in the shoulder, and was wet, barefoot, and freezing. Although deep in enemy territory, he continued his escape. Soviet dictator Josef Stalin had announced a 100,000-ruble reward for the capture of the "Eagle of the Eastern Front," dead or alive. Rudel sheltered among refugees and locals who had no love for Stalinist Russians, and barely survived his trek across some 30 miles of enemy territory to reach German lines.

Rudel with Hitler.
Picture Source: carolynyeager.net

Refuses Hitler to Be Ground for Injury

His feet were so badly injured that when he next flew he had to be helped into his plane. Yet within the week Rudel chalked up his 1,800th mission, destroyed 17 enemy tanks in one day. On 29 March 1944, he went to Hitler's retreat to receive the Diamonds to his Knight's Cross, Germany's highest decoration at the time (one of only 27 awarded), and the tenth member of the Wehrmacht to receive this award. The Führer permitted him to wear padded flight boots for the ceremony. Reluctant to risk his hero again, Hitler grounded him, but relented when Rudel said he would refuse the medal if forbidden to fly.

Gets a Troop Doctor as His New Gunner

Upon his return, Ernst Gadermann, previously the troop doctor of III. *Gruppe*, had joined Rudel as his new radio operator and air gunner. He had put him up for the Iron Cross, but the same did not finally get cleared. He insisted and used his influence in Hitler's office and not only ensured he got one, but that the presentation was made by Hitler personally.

Rudel with Ernst Gadermann.
Picture Source: cieldegloire.com

Rudel's Repeat Injuries but Flies on Regard Less

Germany needed heroes in the summer of 1944, and Rudel was a great candidate. He completed 2,000 missions, and 300 enemy tanks destroyed. Shot down over Latvia, he crash-landed with his gunner, Ernst Gadermann. Both men were wounded, and both were immediately back in the air. Rudel's tally now stood at 320 tanks destroyed. Shot down again and wounded in the upper right leg, he "escaped" from the hospital to fly, with the leg in a cast.

Knight's Cross of the Iron Cross with Golden Oak Leaves, Swords, and Diamonds.
Image Source: flying-tigers.co.uk

Appointed Leader of SG 2: Receives the Highest German Award

Rudel was promoted to *Oberstleutnant* (Lt Col) on 1 September 1944, and appointed leader of SG 2, replacing Stepp, on 1 October 1944. On 22 December 1944, Rudel completed his 2,400th combat mission, and the next day, he reported his 463rd tank destroyed, approximately equal to three Soviet tank Corps. Colonel General Ferdinand Schörner claimed, "Rudel alone is worth an entire division!" On 29 December 1944, Rudel was promoted to *Oberst* (Colonel), and was awarded the Knight's Cross of the Iron Cross with Golden Oak Leaves, Swords, and Diamonds, the only person to receive this decoration. No other German soldier has ever received it. This award, intended as one of 12 to be given as a post-war victory award for Nazi Germany, was presented to him by Hitler on 1 January 1945, four months before Nazi Germany was defeated. Present at the Eagle's Nest, the Nazi western

Rudel with Hitler Receiving his medal.
Picture Source: agenda21radio.news

headquarters, were Col. Gen. Alfred Jodl, Grand Adm. Karl Dönitz, Field Marshal Wilhelm Keitel, and Imperial Marshal Hermann Göring. Again Hitler ordered him grounded; again Rudel refused. The Führer smiled and said: "Very well then, fly. But be careful, the German people need you."

Leg is Amputated: Hits Many More Targets – Surrenders to Americans

On 8 February 1945, with his leg still in a cast, Rudel shot up a dozen tanks that had breached the Oder River. He used his last cannon round to score an unlucky 13th, a Stalin, but his Stuka was hit by Soviet 40mm anti-aircraft fire. On the verge of passing out, he called back to Gadermann, "Ernst, my right leg is gone." Gadermann (who would survive the war with 850 missions and earn his own Knight's Cross) talked his half-conscious pilot down to a crash-landing, pulled him from the wreckage, and stopped the bleeding. Rudel woke up in a hospital with his leg amputated below the knee. He returned to flying on 25 March 1945. He claimed 26 more tanks destroyed by the end of the war.

Rudel with a leg amputated below the knee.
Image Source: ww2gravestone.com

Offers to Fly Out Hitler to Safety

On 19 April 1945, the day before Hitler's final birthday, Rudel met with Hitler in the *Führerbunker* at the Reich Chancellery in Berlin. By April 26, it was barely possible to fly into the embattled capital. Rudel phoned Hitler's adjutant, Colonel Nicolaus von Below, offering to land a Stuka on a Berlin road the following morning, and implying he could evacuate the Führer. Hitler refused, and within the week Hitler was dead. On 8 May 1945, Rudel fled westward from an airfield near Prague, landing in US-controlled territory, and surrendered. The Americans refused to hand him over to the Soviet Union.

Post-War: Clandestine Move to Argentina

The unforgettable friendship between Juan Domingo Perón and the Nazi Hans Ulrich Rudel.
Picture Source: agenda21radio.news

While Rudel had been interned, his family fled from the advancing Red Army and had found refuge with Gadermann's (Gunner) parents in Wuppertal. Rudel was released in April 1946 and went into private business. In 1948, he immigrated to Argentina, travelling via Austria to Italy first. In Rome, with the help of smugglers, and aided by an Austrian bishop, he bought himself a fake Red Cross passport with the cover name "Emilio Meier", and took a flight from Rome to Buenos Aires, where he arrived on 8 June 1948. Rudel authored books on the war, supporting the Hitler regime and attacking the High Command of the German Armed Forces for "failing Hitler".

Years in South America: Reignites Nazi Past

After Rudel moved to Argentina, he became a close friend and confidant of the President of Argentina Juan Perón, and Paraguay's dictator Alfredo Stroessner. In Argentina, he founded the "*Kameradenwerk*" (comrades work), a relief organization for Nazi war criminals. Prominent members included many Nazi SS officers, including many who were declared war criminals and whose extradition had been demanded by the Soviet Union on war crime charges. In addition to these war criminals that fled to Argentina, the

Rudel with Juan and Isabel Peron.
Picture Source: donhollway.com

"*Kameradenwerk*" also assisted Nazi criminals imprisoned in Europe, including Rudolf Hess and Karl Dönitz, with food parcels from Argentina and sometimes by paying their legal fees. In Argentina, Rudel became acquainted with notorious Nazi concentration camp doctor and war criminal Josef Mengele. Rudel, together with Willem Sassen, a former Waffen-SS and war correspondent for the Wehrmacht, who initially worked as Rudel's driver, helped to relocate Mengele to Brazil by introducing him to Nazi supporter Wolfgang Gerhard. In 1957, Rudel and Mengele together travelled to Chile to meet with Walter Rauff, the inventor of the mobile gas chamber.

Writes Wartime Memoirs: Stuka Pilot

In Argentina, Rudel lived in Villa Carlos Paz, roughly 36 kilometres from the populous Córdoba City, where he rented a house and operated a brickworks. There, Rudel wrote his wartime memoirs "Trotzdem" ("Nevertheless" or "In Spite of Everything"). In the book, he supported Nazi policies. The book was published in November 1949 by the Dürer-Verlag in Buenos Aires. Dürer-Verlag became very unpopular for publishing books for many former Nazis, and later went bankrupt. Rudel's book was later re-edited and published in the United States, as the Cold War intensified, under the title, "Stuka Pilot", which supported the German invasion of the Soviet Union. Pierre Clostermann, a French fighter pilot, had befriended Rudel and wrote the foreword to the French edition of "Stuka Pilot".

Image Source: amazon.com

Becomes Arms Dealer and Military Adviser

In the 1950s, Rudel befriended Savitri Devi, a writer, and proponent of Hinduism and Nazism, and introduced her to a number of Nazi fugitives in Spain and the Middle East. With the help of Perón, Rudel secured lucrative contracts with the Brazilian military. He was also active as a military adviser and arms dealer for the Bolivian regime, Augusto Pinochet in Chile and Stroessner in Paraguay. He was in contact with Werner Naumann, formerly a State Secretary in Goebbels' Ministry of Public Enlightenment and Propaganda in Nazi Germany. Following the revolution, in 1955, a military and civilian uprising that ended the second presidential term of Perón, Rudel was forced to leave Argentina and move to Paraguay. During the following years in South America, Rudel frequently acted as a foreign representative for several German companies, including Salzgitter AG, Dornier Flugzeugwerke, Focke-Wulf, Messerschmitt, Siemens and Lahmeyer International, a German consulting engineering firm. Rudel's input was used during the development of the A-10 Thunderbolt II, a United States Air Force aircraft designed solely for close air support, including attacking ground targets such as tanks and armoured vehicles.

Left to right: Werner Baumbach, Hans-Ulrich Rudel & Adolf Galland. Photo taken in 1951.
Picture Source: Twitter @DownedWarbirds

Returns to Germany 1953 Joins Politics

Rudel ran a huge network of Nazi sympathisers and SS officers, and helped German companies sell discarded equipment to South American countries. Rudel returned to West Germany in 1953 and became a leading member of the Neo-Nazi nationalist political party, the German Reich Party (DRP). In the West German federal election of 1953, Rudel was the top candidate for the DRP, but was not elected to the Bundestag. Rudel had an egocentric character. In his political speeches, Rudel heavily criticized the Western Allies during World War II for not having supported Germany in its war against the Soviet Union. Rudel's political demeanour subsequently alienated him from his former comrades.

Multiple Marriages

Rudel was married three times. His 1942 marriage to Ursula, nicknamed "Hanne", had two sons. They divorced in 1950. It was reported that one reason for the divorce was that his wife had sold some of his decorations, including the Oak Leaves with Diamonds, to an American collector, but she denied it. Also she refused to move to Argentina. Rudel married his second wife, Ursula née Daemisch in 1965. The marriage gave his third son. Rudel survived a stroke on 26 April 1970. Following his divorce in 1977, he married Ursula née Bassfeld.

Death and Funeral

Rudel died after suffering another stroke in Rosenheim on 18 December 1982 and was buried in Dornhausen on 22 December 1982. During Rudel's burial ceremony, two Bundeswehr F-4 Phantoms appeared to make a low altitude fly-past over his grave. Dornhausen was situated in the middle of a flight path regularly flown by military aircraft, and Bundeswehr officers denied deliberately flying aircraft over the funeral. Four mourners were photographed giving Nazi salutes at the funeral, and were investigated under a law banning the display of Nazi symbols. The Federal Minister of Defence Manfred Wörner declared that the flight of the aircraft had been a normal training exercise.

Rudel's Grave.
Picture Source: YouTube

Summary of Military Career

In all, Hans-Ulrich Rudel was credited with 2,530 missions. The majority of these were undertaken while flying the Junkers Ju 87, although 430 were flown in ground-attack variants of the Focke-Wulf Fw 190. He was credited with one battleship "Marat", sinking one Cruiser, and heavily damaged a destroyer "Minsk". Also 70 landing craft, some 800 vehicles, 150 gun positions, numerous armoured trains and bridges, and 519 tanks. Rudel was also credited with 11 aerial victories, including 7 Ilyushin IL-2s. He had been shot down more than 30 times by anti-aircraft guns, but never by an enemy pilot. He was wounded five times and rescued six stranded aircrew from the enemy-held territory.

Germany's Most Successful Stuka Pilot. Image Source: hubpages.com

Honours and Awards

Rudel was awarded Iron Cross (1939) 2nd Class (10 November 1939) & 1st Class (15 July 1941). Knight's Cross of the Iron Cross with Golden Oak Leaves, Swords, and Diamonds. He was the first and only one to get the Golden Oak Leaves on 29 December 1944. He was the first and only foreign to get the Hungarian Gold Medal of Bravery (14 January 1945).

Rudel's Philosophy

Rudel's first "flight," was at age 8, when he jumped off a roof with an umbrella, in 1924, which earned him his first broken leg. Young Rudel, an avid skier, and athlete, came of age in early 1930s Germany at the same time as Nazism and the dive bomber. Plummeting from on high with sirens wailing and bombs whistling, Stukas struck terror long before they struck targets. Yet withstanding rapid changes in air pressure as he plunged thousands of feet, not to mention a near-blackout on pull-up, proved difficult for Rudel. And as the son of a Lutheran minister, he didn't exactly fit into the Stuka fraternity. "He doesn't smoke, drinks only milk, has no stories to tell about women, and spends all his free time playing sports," wrote one of his instructors. "Senior Officer Cadet Rudel is a strange bird!" Rudel's fighting philosophy, and that of his life, came to him not in the air, but when he was on foot, on a Ukrainian hillside the same afternoon he swam the half-frozen Dniester River. With a bullet through his shoulder, his comrades gone and enemy troops closing in fast, he remained defiant: "Only he is lost who gives himself up for lost."

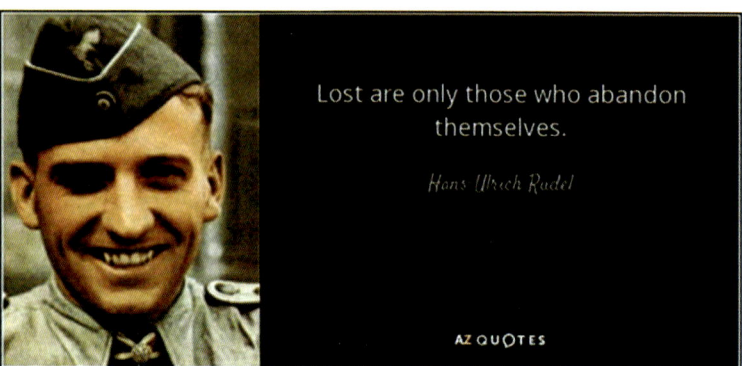

Image Credit; AZ Quotes

REFERENCES

1. Hans-Ulrich Rudel – The Surrender of Germany's Most Decorated Ace 1945, YouTube, Mark Felton Productions, https://www.youtube.com/watch?v=zZHCvPim0eA
2. Don Hollway, Hans-Ulrich Rudel: Eagle of the Eastern Front, July 2011 issue of Aviation History, History Net, https://www.historynet.com/hans-ulrich-rudel-eagle-eastern-front.htm
3. CIA Library Document, https://www.cia.gov/library/readingroom/docs/RUDEL,%20HANS_0118.pdf
4. The book "Stuka Pilot," Autobiography of WW2 Nazi pilot Hans Rudel.
5. Rudel, Hans Ulrich "Eagle of the East Front", World War II Graves, https://ww2gravestone.com/people/rudel-hans-ulrich/
6. Hans-Ulrich Rudel, Wikipedia, https://en.wikipedia.org/wiki/Hans-Ulrich_Rudel
7. The Sirens of Death – 11 Amazing Facts About the Ju 87 Stuka, Military History Now, June 03, 2015 https://militaryhistorynow.com/2015/06/04/screaming-death-10-amazing-facts-about-the-ju-87-stuka/

13

VIETNAMESE FIGHTER ACE NGUYỄN VĂN CỐC

The Highest Scoring Pilot in Vietnam War

Image Source: tintuc24.host

Nguyễn Văn Cốc is a former North Vietnamese MiG-21 fighter ace of the Vietnamese People's Air Force's (VPAF) 921st Fighter Regiment. Văn Cốc was shot down before scoring his first kill on January 2, 1967 in Operation Bolo, a U.S. aerial ambush. He ejected and survived, and then went on to destroy an F-105 in an attack out of the sun on April 30, and then scored eight more kills through December 1969. Of nine victories, two were drones, and for the aircraft, six of the seven can be confirmed in U.S. records, making him the top-scoring pilot of the war no matter how you count it. All of them flying the MiG-21, and in all cases using the heat-seeking R-3S Atoll missile. Undoubtedly he was the "Tiger on the prowl over Vietnam's Jungle".

Young Days: Early Family Tragedies

Nguyễn Văn Cốc was born in 1943, in the Việt Yên District of the province of Bac Giang in French Indochina, north of Hanoi. He had his first birthday the same year Lt. Robin Olds scored his first aerial victory over the Luftwaffe. When he was 5 years old, his father, Nguyen Van Bay, Chairman of the Viet Minh in the district, and his uncle (also a member of the Viet Minh), were executed by the French. Fearing further trouble with the French, his mother moved the family.

Capt. Nguyễn Văn Cốc, 921st Fighter Regiment, Vietnam People's Air Force.
Image Source: lyonairmuseum.org

145

Joins Air Force: MiG 17 and MiG 21 Conversion

Nguyễn spent the rest of his childhood near Chu air base, which kindled an interest in aircraft. He attended Ngō Sĩ Liên School in Bắc Giang, and upon completion of his schooling, enlisted in the VPAF in 1961 at the age of 18. He underwent his initial training at Cat Bi Airbase in Haiphong. At this point in his life he had never driven a car, graduating directly from bicycle to aircraft as a mode of transportation. Nguyễn subsequently spent four years undergoing pilot training in the Soviet Union at Bataysk and Krasnodar Soviet Air Force bases. Of the 120 trainees in Nguyễn's batch, he was one of the seven who graduated as a MiG-17 pilot. After a brief spell back in North Vietnam serving with the 921st "Red Star" Fighter Regiment, Nguyễn was among the group of 13 Vietnamese MiG-17 pilots who were selected to convert to the best fighter in the Soviet inventory, the MiG-21. He returned to the Soviet Union and underwent conversion training on the MiG-21 before returning to the 921st Fighter Regiment in June 1965. He began operational flying in December 1965. Being 26 years old, Nguyễn was older than the average Vietnamese prentice.

Bac Giang in French Indochina, north of Hanoi.
Image Source: Wikipedia

MiG-21PFs of the 921st Fighter Regiment, Vietnam People's Air Force.
Image Source: lyonairmuseum.org

Daunting Odds

The pilots of the 921st faced daunting odds as they attempted to blunt the massive American air offensive. They were outnumbered and outclassed in nearly every regard, including aircraft, weapons, training and combat experience. Because their numbers were so few, the VPAF had no "rotation" program like their American counterparts, who were able to go home after 100 combat missions. They were already home and they "flew till they died". He was assigned to fly his first combat sorties in December 1966, without scoring victories at that time, but acquiring valuable experience.

Discussing Combat.
Image Source: congnghe.vn

Aerial Trap and an Early Ejection

On 2 January 1967, he was among a group of pilots who fell into an aerial tarp set up by the United States Air Force's (USAF) F-4s of 8th Tactical Fighter Wing the "Wolf Pack". The American air defence fighters flew a mission to Hanoi using the same flight patterns and radio call-signs as the fighter-bomber F-105 formations. VPAF MiG-21 who came to intercept the so-called bombers, actually faced interceptors armed with air-to-air missiles. In a matter of few seconds, the five MiG-21s were shot. Nguyễn and four other Vietnamese pilots, all managed to eject safely. He learned that one should be always alert and expect the unexpected. Two more MiG-21s were downed by American Phantoms four days later. VPAF couldn't sustain such high losses. The MiG-21 operations were reduced for several months (January-March) until new tactics were evolved.

Senior Lieutenant Nguyễn Văn Cốc, at right, with two other pilots. A MiG 21PF is in the background. VPAF.
Image Source: thisdayinaviation.com

Aerial Victories

VPAF MiG-21PFM.
Image Source: airliners.net

Nine air-to-air combat kills of United States aircraft and two AQM-34 Firebee UAV kills were credited to him during the Vietnam War. Of these, seven have currently been acknowledged by the United States Air Force. While sometimes U.S. forces may have attributed aircraft losses to surface-to-air missiles, since it was considered "less embarrassing", there was often doubt about the cause of the loss. Cốc also claimed an F-4 Phantom and F-105 Thunderchief in November and 17 December 1967 but there are no corresponding American losses. Flying the MiG-21PF, Nguyễn Văn Cốc normally flew as a wingman. He scored all his victories using the heat-seeking R-3S Atoll missile. All the kills, credited to Nguyễn Văn Cốc by the VPAF are described below.

30 April 1967: First Air Victory

USAF F-105D piloted by Robert A. Abbott of the USAF 355th TFW was his first air victory and occurred while he was acting as a wingman to Nguyễn Ngọc Độ who also downed an aircraft. They took off from Noi Bai and due to the superb controlling by the VPAF ground controllers they could place in an excellent attack position. The events started to happen very fast. 1st Lt. Robert A. Abbott, pilot of one out of 40 F-105D Thunderchiefs of the 354th TFS/355th TFW while on course to attack a power station in Hanoi, heard on the radio that the crew of one of the F-105 F Wild Weasels of the formation

Senior Lieutenant Nguyễn Văn Cốc with other air and ground crew.
Image Source: zingnews.vn

F-105 D Thunderchief.
Image Source: boxartden.com

(responsible of the SAM suppression) reported that they had been hit by a missile fired by a North Vietnamese MiG-21 that had appeared out of nowhere, and that they were ejecting. He tried to see where the MiGs were in an attempt to avoid the attack, but it was too late. So sudden was the attack that his airplane was violently shaken by an explosion and Abbott couldn't control it anymore. The MiG-21 had come from the sun, and fired an R-3 infrared missile which struck the fuselage of his F-105D Bu No 59–1726. Abbott could eject, but was captured by North Vietnamese soldiers as soon as he touched the ground. It was the evening of April 30, 1967, and even when Abbott didn't know at that time, that he became the first victim of the pilot who would be the leading ace of the Vietnam War, Senior Lieutenant Nguyễn Văn Cốc.

Nguyễn recalled the battle: "I was scrambled as the wingman of Nguyen Ngoc Do. I noticed F-105s flying beneath us at an altitude of 2500 m, at 30 degrees to our course. My leader also saw the Thunderchiefs. We both increased our speed and dived at the US fighter-bombers, which were unaware of our presence. My leader shot down the second airplane of a group of four F-105s. Until now, I had been protecting my leader, but with an enemy fighter filling my sights, I also opened fire, downing another Thunderchief. We received an order to return to base and made a successful landing, while the eight F-105s dropped their bombs and started a search for the lost pilots".

The victim of Nguyễn's leader Nguyen Ngoc Do was the F-105 °F Bu No 62–4447 flown by Leonard K. Thorness and Harold E. Johnson (357th TFS/355th TFW, both of them were captured). That was followed seconds later by Nguyễn's kill. That was an outstanding day for the VPAF. A third Thunderchief fell destroyed by the MiG-21 pilot Le Trong Huyen a few minutes later, the American pilot, Joseph S. Abbott (333rd TFS), perished. There was no MiG loss that day.

VPAF Increased Losses Again

Despite the fact that during May 1967 the VPAF could engage the US combat planes with a high degree of success, their losses also began to increase again, forcing it to reduce its operations during June and the first half of July. Since that point the MiG-21s of the 921st Fighter Regiment became pretty active again, and so did Senior Lieutenant Cốc.

23 August 1967: The "Double Attack"

On August 23 1967 the USAF launched a mid-day raid against Hanoi with about 40 aircraft, and in response, four MiG-17s of the 923rd Fighter Regiment were scrambled. They were followed at 1345 h by two MiG-21PF of 921st Fighter Regiment piloted by Nguyen Nhat Chieu (leader) and Nguyễn Văn Cốc (wingman). At that time the favourite Vietnamese tactic was the "double attack". It was essentially a classic coordinated attack from different directions. The MiG-17s acted as bait performing a head-on pass and distracting the escort, while the MiG-21s attacked from the rear, catching the fighter-bombers and escorts by surprise. Cốc shot down one USAF's F-4D over Nghia Lo. The "double attack" had been perfectly executed, shooting down no less than three USAF jets.

McDonnell F-4D Phantom II.
Image Source: aircraftcompare.com

Nguyễn Văn Cốc recollects "My leader Nguyen Nhat Chieu and I went the long way round to get into a better attacking position behind the enemy formation. He fired an AAM, bringing down a Thunderchief, while I also successfully attacked a Phantom with an R-3S AAM. In the meantime, my leader began another attack with his second missile but it missed. He went into a cloud overhead, only to reappear moments later firing with his cannon. I also attacked the Phantom, using a missile, but I was too close, and I strayed into Nguyen Nhat Chieu's line of fire as he dived from above. My airplane was damaged, but all the controls were working normally so I asked permission to carry on the engagement with the damaged aircraft. However, ground control ordered to return to base. Because of the damage, my MiG-21 was only able to do a maximum speed of 600 km/h".

Cốc's prey was the F-4D BuNo 66–0238 flown by Major Charles R. Tyler and Captain R.N. Sittner (555th TFS/8th TFW, Tyler was captured and Sittner was killed), Chieu shot down the F-105D of Elmo Baker (who was taken POW), and one of the MiG-17 F fliers, Nguyen Van Tho, bagged the F-4D BuNo 66–0247. Captain Larry E. Carrigan (pilot) was captured and 1st Lt. Charles Lane (radar specialist) perished.

Fourth Victory

At 1348h on October 3 1967 Nguyễn Văn Cốc and another pilot were scrambled in their MiG-21s. Their targets were two radar contacts over Hai Duong towards Hanoi at an altitude of 7,000 meters, most likely reconnaissance aircraft. The actual number of enemy planes were three. An RF-4 and an escort of two F-4Ds, one of them had been hit by the Vietnamese AAA and lost one engine. The task of finding the US planes was not easy. American EB-66 Destroyers were effectively jamming of the Vietnamese radars, so sufficient intercept information was not coming. Cốc climbed to 7,500 meters to perform a visual search for the targets, and about 1354 h (only six minutes after take-off) he spotted enemy planes on a south-westerly course. He stealthily approached the trailing F-4, which was the one damaged by the AAA, and fired an R-3 Atoll. The Phantom crew were forced to eject. Both could be rescued. This victory was considered only

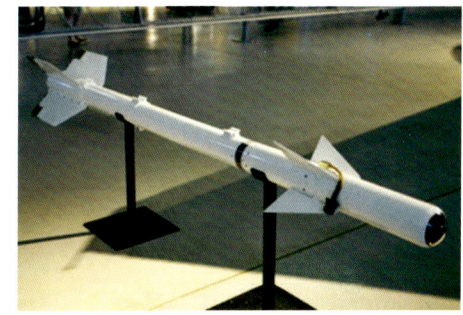

R-3 Atoll (K-13).
Image Source: Wikipedia

"probable" by the VPAF at that time, as credit was also given to the AAA battery. This was followed a few days later (October 7) by Nguyễn Văn Cốc's fourth kill the F-105 Bu No 63–8330 (13th TFS/388th TFW) whose pilots were captured after they bailed out.

Becomes An Ace: Great month for VPAF

Next month was very busy for the MiG-21 pilots of the 921st Fighter Regiment. They shot down one F-4D on November 8, two F-105s on November 18, two F-4Bs downed by MiG-17 pilots of 923rd Fighter Regiment the next day, and one more Thunderchief on the 20th. One of the Thunderchiefs destroyed on November 18 1967 was the fifth victory of Nguyễn Văn Cốc, making him an ace. His prey was the F-105 BuNo 63–8295 of the 34th TFS/388th TFW, the crew, Oscar Dardeau and Edward Leinhoff crashed to their death. The unknown Cốc's wingman shot down the second US jet (F-105D Bu No 60–0497 of 469th TFS/388th TFW). November was thus an excellent month for the Vietnamese, because with these six confirmed victories, the VPAF lost only one MiG-17 to a Phantom on November 6.

VPAF MiG-17.
Image Source: jetphotos.com

The Sixth Victory

Almost a month later (December 19) Cốc claimed a sixth kill (another F-105) not confirmed by USAF loss records. The actual victory six of this skilful flier occurred on 3 February 1968, when he shot the F-102A Bu No 56–1166, killing its pilot 1st Lt. Wallace L. Wiggins (509th FIS/405th FIW).

The Great Engagement of 7 May 1968: Last Fighter Victory

Nguyễn Văn Cốc would wait three months for his next victory. Three flights of MiG-21 fighters from the VPAF 921st Regiment were flown to Tho Xuan Air Base, as part of redeployment in response to the U.S. bombing halt above the 19th Parallel. The flights were led by Dang Ngoc Ngu, Nguyen Van Minh, and Cốc. On May 7 1968 he took off in his MiG-21PF from Tho Xuan airfield (at that time in southern North Vietnam) as the wingman of Dang Ngoc Ngu, followed by other two MiG-21s. The original target of Ngu and Cốc was an EKA-3B, but both MiGs were detected by an AEW plane E-1, and five F-4Bs of VF-92 were sent to the area. Cốc later recounted the mission, "My leader Dang Ngoc Ngu and I took off from Tho Xuan. A second pair of MiGs, flown by Nguyen Dang Kinh and Nguyen Van Lung, acted as our escorts. Because of poor coordination with local air defence forces, our MiGs were mistaken for American fighters, and the AAA opened upon us. This was not the only mistake—even Dang Ngoc Ngu initially mistook the escorting MiGs for Americans and dropped his fuel tanks in preparation for an attack, but he soon recognised them as North Vietnamese. We flew three more orbits over Do Luong before being told of fighters approaching from the sea, this time they were real Americans. The U.S. flight detected were a formation of five F-4B Phantom II from Fighter Squadron 92 (VF-92), USS *Enterprise*, led by Lieutenant Commander Ejnar S. Christensen. Over North Vietnamese airspace, a U.S. Navy EKA-3A electronic warfare aircraft tried to jam North Vietnamese communications but failed, and Nhu's flight of MiG-21 fighters was guided towards their target by ground controllers. Dang Ngoc Ngu noticed two F-4 Phantoms five kilometres to starboard. There were a lot of clouds, and he was unable to get into a firing position. I wanted to follow him, but I noticed I was running low on fuel. I was planning to land back at Tho Xuan when suddenly I spotted a Phantom ahead of me at an altitude of 2500 m. I went after him and launched two missiles from 1500 m. The Phantom crashed in flames into the sea."

EKA-3B Skywarrior.
Image Source: commons.wikimedia.org

The action gave the VPAF their first aerial victory over the airspace above the Military Zone IV of North Vietnam and gave Cốc his seventh aerial victory. The U.S. Navy confirmed that the downed F-4B had been Bu No 151485, call-sign Silver Kite 210, of VF-92 launched from "Enterprise". The pilot of the ill-fated plane was Lieutenant Commander Ejnar S. Christenson, and his Radar Intercept Officer, Lieutenant Worth A. Kramer who ejected safely from their aircraft before impact and were recovered a short time later. All the MiG pilots – Ngu, Cốc, Kinh, and Lung landed safely at Tho Xuan airbase. That was the last kill of Cốc that year. The end of the "Rolling Thunder" bombing campaign on October 31, as ordered by the then US President Lyndon B. Johnson, meant the end of Cốc's chances of shooting down more American combat planes.

RS AQM-34 Drone.
Image Source: commons.wikimedia.org

Last Victories UAVs

However, the USAF kept on sending UAVs to perform recce flights over North Vietnam, and two of these drones became the last victories of Cốc in December 1969. First a USAF AQM-34 Firebee unmanned aerial vehicle. The second was initially presumed a USAF AQM-34 Firebee. This could have even been an OV-10 Bronco whose two crew died when it was shot down in the same area on December 20, 1969. The USN reported that the OV-10 Bronco Bu No 155503 of the VAL-4 "Black Ponies" downed by a MiG near the DMZ. The Vietnamese pilots were not familiar with these types of airplanes. One thing was sure: with 8 enemy planes and 2 drones downed he was certainly the leading Ace of the Vietnam War. Furthermore, 7 airplanes are fully confirmed by US sources, so he is admitted by both sides as the Top Ace of the Vietnam War, and that is something commendable. Two out of his nine kills were UAV Firebees (not counted by USAF as losses in air combat). Among his remaining seven claims, an amazing amount of six had been fully confirmed by USAF loss records.

North American Rockwell OV-10 Bronco.
Picture Source: Wikipedia

Operation "Bolo": Clash of Generations over North Vietnam

The exploits of the legendary Col. Robin Olds and the "Wolf Pack" of the 8th Tactical Fighter Wing during the Vietnam War are well known among military aviation enthusiasts. Col. Olds' skill as a fighter pilot and reputation as an effective combat leader, earned during three decades of distinguished military service, are held in high regard by peers and historians alike. The story of his principal adversaries in Southeast Asia, the pilots of the fledgling VPAF, is not well known in the West. Preeminent among these pilots was Capt. Nguyễn Văn Cốc of the 921st Fighter Regiment, the top-scoring fighter Ace of the Vietnam War. On 2 January 1967, the destinies of these two formidable air warriors came together during a brief and decisive air battle over North Vietnam. This encounter ended in a lopsided victory for the "Wolf Pack," but that was not the whole story. The VPAF proved to be resilient and resourceful in recovering from this defeat, and lessons learned from this epic air battle shaped strategy and tactics employed by both opposing forces throughout the remainder of the war. This story has been beautifully documented by Lyon Air Museum's Docent Jeff Erickson.

Operation "Rolling Thunder"

In March 1965, the United States commenced large-scale offensive air operations against North Vietnam. Operation "Rolling Thunder" was an intensive and sustained bombing campaign designed to halt the flow of men and materiel into South Vietnam and persuade the North Vietnamese regime to cease support for the communist insurgency in the south. Missions targeted industry, storage facilities, transshipment points, lines of communication, and air defences. U.S. Air Force and Navy combat aircraft struck at the heart of the Hanoi regime's infrastructure with ever increasing frequency and devastating impact on its capacity to make war. A large proportion of USAF bombing missions over North Vietnam were flown by the F-105D Thunderchief fighter/bomber (nicknamed "Thud") operating from bases in South Vietnam and Thailand.

MiG 21s attacking F-4 Phantoms from Sun and Clouds.
Image Source: YouTube

"Hit and Run" VPAF

North Vietnam took steps to establish a force of fighter/interceptor aircraft, using equipment and pilot training provided by their Soviet and Chinese allies. The VPAF received its first jet aircraft, the Soviet designed MiG-17 (Chinese J-5), in February 1964. Initially based in China, these aircraft equipped the first operational jet fighter unit, known as the 921st "Sao Do" (Red Star) Fighter Regiment. During the next two years, the VPAF capability was upgraded substantially to include supersonic MiG-19 (Chinese J-6) and MiG-21 interceptors. By 1966, the 921st Fighter Regiment was operating the second generation MiG-21PF all-weather variant, equipped with short-range air-to-air guided missiles.

Operations conducted by fighter/interceptor aircraft of the VPAF were entirely defensive, and remained so throughout the war. MiG pilots engaged U.S. bombers and strike aircraft in "hit and run" attacks over friendly territory and actively avoided contact with American escort fighters. Despite the vast numerical and qualitative superiority of U.S. airpower, this "air insurgency" strategy met with some success, taking a toll on U.S. aircraft, reducing bombing effectiveness, and diverting US combat aircraft resources from strike missions to defend against the MiG threat.

MiG 21s "Hit and Run".
Image Source: Pinterest (Illustration by Dave Seeley)

U.S. Armed Forces Overwhelming Superiority in Assets

U.S. armed forces went to war with a number of fighter/interceptor aircraft such as the F-100, F-102 and F-104, but only the F-4 Phantom and F-8 Crusader achieved substantial success in air combat against the VPAF. The combat effectiveness of U.S. fighters was hampered, to some degree, by rules of engagement that precluded firing on targets beyond visual range. Insufficient pilot training in air combat skills and lack of an internal gun in most versions of the F-4 had a negative impact on performance in air-to-air engagements.

Vietnamese Aggression

Most U.S. strike missions were flown by F-105 fighter/bombers. Heavy loads of air-to-ground ordnance severely limited F-105 flight performance and they were vulnerable to attack by faster and more manoeuvrable interceptors. MiGs flew ground-controlled intercepts (GCI) against U.S. bomber formations, with guidance provided by a network of Soviet-built radar stations and command centres. These installations were "off-limits" to U.S. strike aircraft due to concerns about killing Russian or Chinese advisors. As a consequence, ground controllers were able to position interceptors optimally for "hit-and-run" attacks against the bomber formations while

Image Source: Facebook, Real Air Power

minimizing exposure to threats from U.S. combat air patrols. MiG-17s often engaged in frontal attacks with guns, while MiG-21s generally attacked from the rear to take advantage of their greater speed and heat-seeking missile armament. The MiGs attacked aggressively, usually from multiple directions, and often with devastating effects. Bombers were frequently forced to jettison ordnance prematurely to evade destruction.

Colonel Olds and New Tactics

U.S. corrective action came in the form of "Operation Bolo." With the consent of Gen. William Momyer, 7th Air Force Commander, Col. Olds and the senior staff of the 8th Tactical Fighter Wing conceived and planned an elaborate deception, designed to lure North Vietnamese MiGs into an engagement with a superior force of U.S. fighters. The plan was assigned the code name "Bolo" in reference to the fearsome Filipino edged weapon that is readily concealed, but lethal at close range.

Colonel Olds.
Image Source: lyonairmuseum.org

The plan called for a "West Force," composed of seven flights of F-4Cs from the 8th TFW based at Ubon Air Base, Thailand to simulate an F-105 strike against a target in North Vietnam. The Phantoms of the West Force would employ ingress routes, altitudes, speeds, formations, call signs, and communications jargon typical of an F-105 strike package. The F-4s were also equipped with the QRC-160 jamming pod normally carried by the F-105s, enabling them to mimic the Thud's electronic signature. An "East Force" of seven additional flights of F-4Cs from the 366th TFW, based at Da Nang Air Base, South Vietnam would cover avenues of escape for the MiGs, including alternate VPAF airfields and routes to sanctuaries in China. Arrival times over target airfields were spaced at intervals to maintain continuous coverage, preventing surviving MiGs from landing and forcing them to exhaust fuel. The plan also called for radar surveillance by Lockheed EC-121 airborne early warning aircraft, six flights of F-105s to provide SAM suppression, and stand-off radar jamming by EB-66s, escorted by F-104s of the 435th TFS. The plan specified that the target area would be clear of other U.S. aircraft, enabling F-4 crews to engage hostile targets without the positive visual ID normally required by 7th Air Force rules of engagement.

Lockheed EC-121 airborne early warning aircraft.
Image Source: Wikipedia

Col. Robin Olds vs. Nguyễn Văn Cốc

Op Bolo commenced on 2 January 1967. F-4 flights were identified using radio call signs based on names of contemporary auto manufacturers, including Ford, Olds, and Rambler. Leading the Olds flight, Col. Robin Olds arrived first over the target area near Phúc Yên Air Base at 1500 local time. After some delay due to overcast, VPAF ground controllers took the bait and directed MiG-21s of the 921st Fighter Regiment to the intercept. Among the pilots dispatched on this mission was Nguyễn Văn Cốc, flying as wingman to a more senior pilot. Expecting to encounter a bomber formation, the MiGs employed familiar tactics, emerging from cloud cover sequentially to approach the formation from multiple directions. Much to their surprise and dismay, they were confronted by deadly Phantoms with a full complement of air-to-air weapons and ready for a stand-up fight. The number 2 aircraft of Olds flight, piloted by Lt. Ralph Wetterhahn, scored the first

AIM-7 Sparrow missile.
Image Source: military-today.com

victory with an AIM-7 Sparrow missile. After several missiles failed to launch or guide, Col. Olds and his WSO, 1st Lt. Charles Clifton, scored a second kill shortly thereafter with an AIM-9 Sidewinder. A third victory was scored with an AIM-9 fired by the flight's number 4 aircraft, piloted by Capt. Walter Radeker III.

The battle was over in a matter of minutes. Two other flights from the 8th TFW also scored aerial victories over MiG-21s, and the Wing claimed a total of seven aircraft destroyed and two probable kills during Operation Bolo with no losses. Among the North Vietnamese aircraft destroyed was the MiG-21 flown by Nguyễn Văn Cốc and another piloted by another future VPAF, Ace, Vu Ngọc Đỉnh. Both pilots escaped their damaged aircraft and survived.

North Vietnamese air force pilots confer in front of a MiG-21.
Image Source: historynet.com

With the loss of more than half of their operational MiG-21 force, the VPAF ceased intercept operations for several months to recover, re-equip and re-think strategy. In the aftermath of the January debacle, senior North Vietnamese leaders undertook an intensive after-action review, surfacing a number of deficiencies in tactics, pilot training and decision-making within the chain of command. The revamped VPAF strategy and tactics finally realized results on 30 April 1967, to the detriment of the American 355th TFW, when Nguyễn Văn Cốc and his flight leader both claimed victories over F-105s. Other pilots of their squadron downed three more F-105s.

As might be expected in any protracted conflict, both sides continued to evolve technology, strategy, and tactics in pursuit of advantage and in response to initiatives by their adversary. The North Vietnamese continued to exploit the "target-rich environment" presented by U.S. strike missions, claiming to have downed a total of 266 U.S. aircraft with no less than 17 VPAF pilots claiming status as "Ace". U.S. authorities acknowledge the loss of 89 aircraft in air-to-air engagements, while claiming 195 aerial victories, for a kill ratio of 2.2:1. Only two U.S. pilots qualified as Aces, with three additional weapon systems officers achieving five or more victories. The comparatively small number of U.S. Aces can be attributed to the relative scarcity of targets for U.S. fighter crews and their shorter terms of service in the theatre of operations.

As Commander of the 8th TFW, Col. Olds remained active in the air, flying a total of 152 combat missions, 105 of them over North Vietnam. He went on to destroy three more MiGs in combat, making him a "triple Ace" with a total of 16 aerial victories during his illustrious career. After ejection, Capt. Cốc returned to active duty with his regiment and went on to become the highest-scoring fighter Ace of the Vietnam War, with seven victories against US Air Force and Navy combat aircraft. Among his victims were three F-4s, three F-105s, and a single F-102. Two of the Phantoms downed by Capt. Cốc were F-4Ds of the 8th Tactical Fighter Wing, the same unit that inflicted the January 1967 disaster on the Sao Do Regiment. He was also credited with shooting down two AQM-34 unmanned drones. All of his air-to-air victories were achieved with the K-13 infrared-guided missile.

Nguyễn Văn Cốc Proudly Holding a MiG 21 model in 2015.
Image Source: soha.vn

In some respects, these two men stood in striking contrast to one another. At age 44, Col. Olds was a seasoned combat veteran, approaching the end of his career and seizing one last opportunity to apply his considerable talents as a fighter pilot. Capt. Nguyễn Văn Cốc went to war as an inexperienced but aggressive and highly motivated rookie, anxious to prove himself in combat. On further reflection, however, it appears that these men had a great deal in common, and that their differences are largely generational and cultural. Robin Olds was a member of America's "Greatest Generation," raised during the Great Depression. When his nation was threatened by axis aggression, he went to war as a volunteer and fulfilled his obligation with courage and skill. Nguyễn Văn Cốc was also raised in a time of adversity, during the French occupation of Indochina, and volunteered at an early age to risk his life in defence of his homeland. Both men acquired a compelling desire to fly at an early age and pursued their careers as military aviators with zeal and commitment. Their skills as pilots and combat leaders distinguished them among their peers, and they met the challenge from their adversaries with courage, tenacity and resilience. Both men were patriots, and repeatedly demonstrated a willingness to put their lives on the line in defence of the values they held dear. These two fine aviators were held in high regard by their peers and have earned the respect and admiration of the nations they served.

Why so many Vietnamese Aces?

Why did so many VPAF pilots score higher than their American adversaries? Mainly because of the numbers. In 1965 the VPAF had only 36 MiG-17s and a similar number of qualified pilots, which increased to 180 MiGs and 72 pilots by 1968. Those brave six dozen pilots confronted about 200 F-4s of the 8th, 35th, and 366th TFW, about 140 Thunderchiefs of the 355th and 388th TFW, and about 100 USN aircraft (F-8s, A-4s, and F-4s) which operated from the carriers on "Yankee Station" in the Gulf of Tonkin, plus scores of other support aircraft (EB-6Bs jamming, HH-53s rescuing downed pilots, Skyraiders covering them, etc.) Considering such odds, it is clear why some Vietnamese pilots scored more than the Americans; the VPAF pilots simply were busier than their US counterparts, and they "flew till they died." They had no rotation home after 100 combat sorties because they were already home. American pilots generally finished a tour of duty and rotated home for training, command, or flight-test assignments. Some requested for a second combat tour, but they were the exceptions.

Book Cover. Nguyễn Văn Cốc on the right with one Knee down.
Image Source: amazon.com

In mid-1960's the American pilots were focused on the use of air-to-air missiles (like the radar homing AIM-7 Sparrow and IR AIM-9) to win the air battles. However, they had forgotten that a skillful pilot in the cockpit was as important as the weapons he uses. The VPAF knew it, and trained its pilots to exploit the superb agility of the MiG-17, MiG-19 and MiG-21, getting into close combat, where the heavy Phantoms and "Thuds" were at a disadvantage. Only in 1972, when the "Top Gun" program improved the skills in aerial combat of USN Phantom pilots, and the F-4E appeared with a 20 mm built-in Vulcan cannon, could the Americans neutralize that Vietnamese edge.

Lastly, the overwhelming US numerical superiority meant that, from the point of view of the Vietnamese pilots, the aerial battlefield was a "target-rich environment." For the American airmen, Vietnam was a "target poor

Lt. Gen Nguyễn Văn Cốc.
Image Source: Facebook

environment." The VPAF never had more than 200 combat aircraft. Officially, there were 16 VPAF Aces during Vietnam War. 13 were MiG-21 pilots, and three were MiG-17. There were no MiG-19 aces. Americans had five Air Aces.

Cốc moved out from Combat Flying

The end of Operation Rolling Thunder on 31 October 1968 removed Cốc from the opportunity for further air combat. In that year, Nguyễn Văn Cốc was transferred from operational duties so that his valuable combat experience could be put to use in training new pilots. Despite the fact that he did not participate in the furious battles of 1972, many of his apprentices did, showing that his skills were not lost when he stopped to fly combat sorties. His best student, Nguyen Duc Soat, learned well from the master, and shot down five F-4s and one A-7 during 1972.

Flying Ace Nguyen Duc Soat. Image Source: vi.wikipedia.org

Captain Nguyễn Văn Cốc is congratulated by Hồ Chí Minh, President of the Democratic Republic of Vietnam.
Image Source: thisdayinaviation.com

Honours and Awards

In 1969, Nguyễn Văn Cốc was awarded the prestigious Huy Hiệu medal for each of his nine aerial victories and was recognized as a Hero of the Vietnamese People's Armed Forces.

Post War Years

After the war, he remained in the Vietnamese National Air Force, retiring in 2002 as Chief Inspector with the rank of Lieutenant General, after declining health. Details on Nguyễn's current whereabouts are scarce, but reports from 2015 indicated that the retired Ace was alive, but in poor health, suffering from an ailment involving paralysis.

Nguyễn Văn Cốc in 2015. Image Source: vietnamnet.vn/vn

If one looks past the controversy and different ideologies associated with the Vietnam War, Nguyễn was simply a man who loved to fly. That said, as the most successful Ace of the Vietnam War, he was certainly one of the 'best of the best.'

REFERENCES

1. Nguyễn Văn Cốc – MiG-21 Ace, YouTube, December 08, 2008, https://www.youtube.com/watch?v=9pk5RE9ypDc
2. Nguyễn Văn Cốc, This Day in Aviation, 20 December 1969, https://www.thisdayinaviation.com/tag/nguyen-van-coc/
3. Lyon Air Museum, Docent Jeff Erickson. http://www.lyonairmuseum.org/blog/operation-bolo-clash-generations-over-north-vietnam
4. Nguyễn Văn Cốc, Wikipedia, https://en.wikipedia.org/wiki/Nguy%E1%BB%85n_V%C4%83n_C%E1%BB%91c
5. Trung Hieu, The West writes about the tactics of the Vietnamese "Ace" pilots, December 18, 2016 https://soha.vn/phuong-tay-viet-ve-chien-thuat-cua-cac-phi-cong-ace-cua-viet-nam-20161218204836785rf20161218204836785.htm
6. Hoang Sang, "Bird cut number 2" of the Vietnam Air Force, December 02, 2015 https://soha.vn/quan-su/chim-cat-so-2-cua-khong-quan-viet-nam-20151202030526329rf20151202030526329.htm
7. Hoang Anh, Farewell to the legend of the sky: "Bird cut number 2" tells about "silver swallow wings" November 19, 2015, https://soha.vn/quan-su/chia-tay-huyen-thoai-bau-troi-chim-cat-so-2-ke-ve-canh-en-bac-20151119022445997rf20151119022445997.htm
8. General Nguyen Van Coc – legendary pilot shot down 9 American planes, December 21, 2019, https://giadinh.net.vn/xa-hoi/tuong-nguyen-van-coc-phi-cong-huyen-thoai-ban-roi-9-may-bay-my-20191221110826747.htm

14

IVAN KOZHEDUB

The Highest Scoring Allied and Soviet Air Ace of World War II

Ivan Kozhedub.
Picture Credit: sputniknews.com

Ivan Nikitovich Kozhedub was a Soviet World War II fighter ace. Credited with over 62 victories. He was the highest-scoring Soviet and Allied fighter pilot of World War II. Ivan is one of the few pilots to have shot down a Messerschmitt Me 262 jet. He was made a Hero of the Soviet Union on three occasions. After the war, he remained in the military and commanded the 324th Fighter Air Division during Soviet operations in the Korean War. He finally retired in 1985 as the Marshal of Aviation.

Early Life

Kozhedub was born into a poor rural family on 8 June 1920 in the village of Obrazhiyivka, near the city of Shostka (now Sumy Oblast, Ukraine) in the western part of the USSR. He was the youngest of five children in a Ukrainian family. He had a hard time when he was a child and never had enough to eat as a teenager. He had to work all the time back then. His only toys were handmade stilts, a rag ball, and skis made of barrel planks. His father was rather an unusual person for his social status; working at the factory and doing country work, he found time to read books and even to compose verses. He was religious, strict, and a persevering tutor. One time Ivan's father, despite the protests of Ivan's mother, sent his 5-year-old son to guard a garden at night. Much later Ivan asked his father about this situation

Chief Marshal of Aviation Ivan Nikitovich Kozhedub.
Image Source: ww2gravestone.com

pointing out that thieves were rare in that area and no matter what happened such a watchman would have been of little use. His father answered, "I accustomed you to the difficulties." At the age of six, Ivan learned to read and write and soon went to school. After finishing a seven-year program he graduated and was admitted into the Shostka Chemistry Technological College.

Inspired By Great Soviet Aviators

The nearby aero club fascinated him. Later, no matter what he might be doing, solving a difficult math problem or playing, he would forget instantly about everything, as soon as he heard the rumble of an aircraft motor. In the 1930s, the Komsomol (Young Communist League) was a patron of aviation

Young Kozhedub.
Image Source: ww2gravestone.com

and, naturally enough, all were crazy about flying. He remembers well the words of his school teacher, "Choose the life of an outstanding man as a model, and try to follow his example in everything." For him, a boy of 16, and for thousands of other Soviet teen-agers, the famous pilot Valery Chkalov was such a man. The whole world admired his bold long distance flights in the Tupolev ANT-25, such as his 1936 flight from Moscow to Udd Island, Kamchatka, a 9,374 kilometre trip in 56 hours, 20 minutes. Also a little shorter but more hazardous flight of 8,504 km in 63 hours, 16 minutes from Moscow to Vancouver, Wash., via the North Pole, on June 18-20, 1937. He was also a fearless test pilot, and it was during a test flight that he lost his life on December 15, 1938.

Valery Chkalov, the USSR's most popular and record-breaking pilot.
Image Source: rbth.com

Begins to Fly

In 1938 he became a member of the aeronautic club. In April 1939 Kozhedub made his first flight the Polikarpov U-2 (U stands for Russian uchebny meaning "learning"). The inquisitive young man was greatly impressed by the beauty of his native land that opened up to him from a height of 1500 meters.

Recalled of His Younger Days Later

Kozhedub recalled that his country was absolutely ready to rebuff any aggression. Any fighting on our own territory was considered unthinkable. Everything we read or heard over the radio about the war to the west seemed very remote to us. At that time we did not know that more than 40,000 of the most talented military leaders had been killed by Stalin's purges a few years earlier. We realized what had happened much later. Every report about the retreat of our troops made our hearts bleed.

Polikarpov U-2.
Image Source: crewdaily.com

Joins the Red Army

In 1940 Kozhedub was called up for military service in the Red Army. In 1941 he was admitted to the Military Pilots' Aviation School in Chuguev. He was one of the best students and graduated as an aviation instructor. At the start of the German invasion of the Soviet Union, he continued to be retained as an instructor. Kozhedub remained at the school for nearly two years where he trained many young Soviet pilots. In 1941, following the start of the Great Patriotic War, the aviation school together with Kozhedub was shifted to the Asian part of the country. Kozhedub was desperate to take part in military action and asked to be sent to the frontline. But he was not allowed to participate in the war until November 1942.

Transferred to Operational Unit

Feeling his talents would be better used in combat, Kozhedub requested a transfer to an operational unit. Kozhedub remembers, "I requested a transfer to the front more than once. But the front required well-trained fliers. While training them for future battles, I was also training myself. At the same time, it felt good to hear of their exploits at the front. In late 1942, I was sent to learn to fly a new plane, the Lavochkin La-5. After March 1943, I was finally in active service." In March 1943 he was posted, as a Senior Sergeant, to the 240th Fighter Aviation Regiment, one of the first units to receive the new Lavochkin La-5 aircraft.

The first five Marshals of the Soviet Union in November 1935. (left to right, front row first): Mikhail Tukhachevsky, Kliment Voroshilov, Aleksandr Yegorov, Semyon Budyonny, Vasily Blyukher. Only Voroshilov and Budyonny survived the Great Purge.
Image Source: Wikipedia

Lavochkin La-5.
Image Source: world-war-2.wikia.org

War Action Begins

His first combat mission was on 26 March 1943 on a La-5 fighter. His plane was badly damaged by a pair of Messerschmitt Bf 109s. He was able to land his fighter but the aircraft was badly damaged. He did not get a single scratch, though his plane was written-off. His leader told him later, "Make haste only when catching fleas." But he was in a hurry and thought he could down at least two or three enemy planes at one go. Carried away by the attack, he did not notice many Messerschmitt Bf-109s approaching from behind. Of course, that was a bitter experience and a serious lesson for him.

Messerschmitt Bf-109.
Image Source: Wikipedia

He operated on the Voronezh Front and, in July over the Kursk battlefields. After that first flight, he flew many missions. Then he was allotted a new La-5 with the cowl number 75. The plane was named after the famous Soviet aviator and Hero of the Soviet Union Valery Chkalov. The 23-year-old pilot opened his fighting account with first aerial kill of a Junkers Ju 87 Stuka bomber, shot down during the Battle of Kursk on 6 July 1943 while engaging a fight with 12 enemy planes.

Air Battle for Kursk.
Painting Image Source: weaponsandwarfare.com

The next day he gained a new victory by bringing down another Ju-87. On 9 July Kozhedub simultaneously destroyed two Bf-109 fighters.

Promoted Junior Lieutenant

By 16 August he had claimed eight air victories. He was promoted to Junior Lieutenant. Then his unit moved towards Kharkiv. At this time he usually flew escort for Petlyakov Pe-2 twin-engine bombers. By October 1943 Senior Lieutenant Kozhedub had made 146 combat missions and brought down 20 enemy planes. By that time he was fighting as an equal with German air experts. Kozhedub skilfully combined his piloting technique with firing skills.

In air fights over the Dnieper, pilots from Kozhedub's air regiment engaged Jagdgeschwader 51 (JG 51) Mölders squadron and won the air duels. Ivan Kozhedub increased his account. In 10 days of intense fighting, he shot down 11 enemy planes. He was awarded the order of the Hero of the Soviet Union on 4 February 1944.

Hero of the Soviet Union.
Image Source: Wikipedia

Be Kind to the Aircraft: He Will Be Kind to You

Aggressive, tireless, brave, and skillful, Kozhedub was the ideal fighter pilot. His aircraft was his religion. Kozhedub once said: "The motor works accurately. The plane is obedient to my every movement. I am not alone – my fighting friend is with me." For Kozhedub this was not a poetic exaggeration or a metaphor; approaching the cockpit before take-off he always found some kind words for his plane.

Gets a New Plane: A Gift from a Farmer

In May 1944 Kozhedub was promoted to Captain and became the commander of a squadron. With 38 air victories under his belt, he received a new La-5F, a gift from a farmer named Vasily Konev. Konev gave money to the Red Army and asked that a plane be constructed in the name of his nephew, Lieutenant Colonel Georgy Konev, a fighter pilot who died at the front. The request of the patriot was executed and the plane was transferred to Ivan Kozhedub.

Ivan Kozhedub's Famous La-7 "White 27".
Image Source: redbanner.co.uk

Lone-Wolf Operations

In July 1944, Kozhedub was posted to the 1st Belorussian Front as Vice Commander to the 176th Guards Fighter Regiment, and received La-7 No. 27, in which he would score his final 17 victories. Kozhedub was first upset by the new appointment but later found that he could fly with Aces who went on lone-wolf operations. The 176th Guards Fighter Regiment carried out 9,450 combat missions, of which 4,016 were lone-wolf operations. They conducted 750 air battles, in which 389 enemy aircraft were shot down.

By mid-1944 Guard Captain Ivan Kozhedub had flown 256 combat missions and shot down up to 48 enemy planes. On 19 August 1944, he was awarded a second medal, this time a Gold Star. Once, in air combat over enemy territory, Kozhedub's La-7 was hit. When the engine stalled Kozhedub didn't give up, but chose a target on the ground and began to dive. When he was close to the ground the engine suddenly began to function again and Kozhedub brought the plane out of the dive and returned safely to his base.

Messerschmitt Me 262.
Image Source: YouTube (Flugmuseum Messerschmitt)

Shoots Down a Me-262

On 19 February 1945 during an operation near Frankfurt (Oder), Kozhedub shot a Me-262 jet that was piloted by Kurt Lange. The Me-262 was a latest German jet plane. Kozhedub's La-7 was a turbo-prop. The German plane was flying at a much higher speed that was unreachable by the La-7 at a height of 3500 meters. Kozhedub later described this combat: "What is it? My formation member hurried opening fire at the enemy. But suddenly the German plane began to move to the left towards me. The distance was sharply reduced and I approached the enemy. With involuntary excitement, I opened fire. The jet plane collapsed and fell down."

Kozhedub.
Image Source: commons.wikimedia.org

He served as a fighter pilot in several combat zones, including Steppe Front, 2nd Ukrainian Front, and 1st Belorussian Front and at different ranks, starting from senior airman up to deputy commander of his air regiment. He claimed his 61st and 62nd victories – his final claims of the War– over Berlin on 16 April 1945.

On 22 April 1945 Kozhedub was reportedly attacked by a pair of American P-51 "Mustang" fighter planes. After just two minutes of engagement, one of the "Mustangs" has hit and shattered into pieces; the second pilot barely had time to jump with a parachute. These kills were not confirmed.

American P-51 "Mustang".
Image Source: Wikipedia

Image Source: Pinterest (Small Scale Art)

Greatest Soviet Pilot

Kozhedub was attributed with the highest number of air combat victories of any Soviet pilot during World War II. He is regarded as the best Soviet flying Ace of the war, and was associated with flying the Lavochkin La-7. He was reputed to have a natural gift for deflection shooting, i.e. aiming ahead of a moving target at the time of firing so that the projectile and target will collide.

Kozhedub's World War II Summary

He flew 330 combat missions, 120 aerial engagements, 62 enemy aircraft shot down, including one Me 262 jet fighter (possibly Uffz Kurt Lange of 1/KG(J)54.). On 18 August 1945, he was again awarded the title Hero of the Soviet Union for his military skills and personal courage.

Recalls About Tactics and Air Power

In an interview with Aviation History, Kozhedub recalled his experiences of hostilities in the early months of the war. He felt they required a change in the tactics and organizational structure of fighter aviation. The famous formula of air-to-air combat was: 'Altitude-speed-manoeuvre-fire.' A flight of two fighters became a permanent combat tactical unit in fighter aviation. Correspondingly, a flight of three planes was replaced with a flight of four planes. The

Kozhedub at the Museum many years later.
Image Source: russiapedia.rt.com

formations of squadrons came to include several groups, each of which had its own tactical mission (assault, protection, suppression, air defence, etc.). The massive use of aviation, its increasing influence on the course of combat and operations, required that its efforts be concentrated in those major specialties.

With His Squadron Boys.
Image Source: commons.wikimedia.org

Fighter air corps making up a part of air armies were set up for that purpose. Hundreds of fighters took part in crucial tactical and strategic operations. Quite often, air-to-air combat developed into a virtual air battle. The arsenal of combat methods used by Soviet fighter aces came to include vertical manoeuvres, multi-layered formations, among others. Out of the 44,000 aircraft lost by Germany on the Soviet-German front, 90 percent were downed by fighters.

Since the war was teaching its bitter lessons, the Soviets had to change tactics as they went along. The Air Force went over from 60-plane regiments, which appeared to be too heavy, to regiments consisting of 30 fighters (three squadrons). Practice showed that this structure was better, both because it made the Commander's job easier and because it ensured higher flexibility in repelling attacks.

Kozhedub's impression about the Nazi pilots was that the sinister colours of the German Messerschmitt Bf-109s and Focke-Wulf Fw-190s with the drawings of cats, aces, arrows, and skulls on their sides, were designed to scare Soviet pilots. But they didn't pay much attention to them, and concentrated on finding weak spots in their tactics. But he always respected the courage of the German Aces. Caution is all-important and one had to turn the head 360 degrees all the time. The victory belonged to those who knew their planes and weapons inside out and had the initiative. After August 1943, the supremacy in the air finally went over to the Soviet pilots. The onetime conceit of invincibility claimed by Göring's Aces had gone up in smoke.

Nose Art on Bf-109.
Image Source: Pinterest (enrique262.tumbler.com)

Recalls Importance of Battle for Kursk

Kozhedub felt the battle for Kursk was a landmark in the development of the forms and methods for operational and tactical use of Soviet aviation in the war years. In its first defensive stage, Soviet airmen flew 70,219 sorties. Tactical aviation accounted for 76 percent of the total, long-range aviation for 18 percent, and air defence fighters for six percent. During that period, they destroyed 1,500 enemy planes. Soviet losses were 1,000 aircraft. During the counter-offensive, they made 90,000 sorties, about 50 percent of which were designed to support attacking troops, and 31 percent to achieve supremacy in the air. The enemy lost up to 2,200 planes in that time.

Soviet MiG 15 in Korean War.
Image Source: migflug.com

The Korean War

After the war, Ivan went back to the Air Force Academy and graduated in 1949. In April 1951, promoted to *Polkovnik* (Colonel), he commanded the 324th IAD (Fighter Air Division) and was sent to Antung airfield on the China-North Korea border to fly the MiG 15 during the Korean War supporting the North Korean forces. He was not given permission to participate in combat missions. Under his leadership, the 324th IAD claimed 239 victories, including 12 Boeing B-29 Superfortress for the loss of 27 MiG-15s and 9 pilots in combat.

Ivan.
Image Source: Wikipedia

Higher Command

In 1956 he graduated from the High Command Academy, after which he was promoted to a General officer. From 1971 he served in the Central Office of the Soviet Air Force and from 1978 in the general inspection group of the Ministry of Defence of the USSR. He was made an Aviation Marshal in 1985 just before retirement.

Image Source: Pinterest (Scoffie Riffle on Warrior Badass)

List of Aerial Victories

The victories are well documented in the book Stalin's Falcons, by Mikhail Bykov. Kozhedub's all 64 air victories include the target aircraft and locations. There is a general consensus among historians. He mostly flew the La-5, and later part, its advanced variant La-7. The Lavochkin La-7 was a piston-engine single-seat Soviet fighter aircraft developed during World War II by the Lavochkin Design Bureau. It was a development and refinement of the Lavochkin La-5, and the last in a family of aircraft that had begun with the LaGG-1 in 1938.

Junkers Ju 87 "Stuka".
Image Source: militaryfactory.com

His first victory was on 6 July 1943 when he shot a Junkers Ju 87 "Stuka", west of Zavidovka, north of Kharkiv. There after he was achieving regular air victories. He shot a total of 18 Junkers Ju 87, 19 *Messerschmitt* Bf-109, 21 *Focke-Wulf* Fw 190, 2 *Heinkel* He 111, One Polish PZL P.24, three Henschel Hs 129, and one *Messerschmitt* Me 262. His last victory was on 17 April 1945, an Fw 190 at Kinitz. Until August 1944 Kozhedub was flying on Lavochkin La-5, after that Lavochkin La-7. He was never shot down, though his damaged fighter was often cited at various airfields.

Alleged shooting down of two USAAF P-51 fighters

Kozhedub allegedly shot down two USAAF P-51 Mustang fighters in a friendly fire incident 17 April 1945. He encountered a group of American B-17 Flying Fortresses under attack by *Luftwaffe* aircraft. His aircraft was apparently mistaken by American escort fighters for the enemy and attacked. Kozhedub, having no other option, defended himself by shooting down two of the P-51s. So far, this story is not confirmed completely. Film footage exists that had been touted as Kozhedub's actual gun camera film from the event; however, the footage was shot using Zeiss equipment, which was used primarily by the *Luftwaffe*.

Alexander Pokryshkin, Marshal Zhukov, Ivan Kozhedub at Kremlin.
Image Source: commons.wikimedia.org

Honours and Awards

Marshal Ivan Kozhedub was one of only two Soviet fighter pilots to be awarded the Gold Star of a Hero of the Soviet Union three times during World War II. The other, Aleksandr Pokryshkin, had flown from the German invasion in the summer of 1941 through the end of the war, during which time he scored 43 individual and 3 shared victories aerial victories in MiG3s, Bell Airacobras, Lavochkin La-5s and Yakovlev Yak-9Us.

Ivan Nikitovich Kozhedub photo at Central Armed Forces Museum of Russian Federation.
Image Source: tripadvisor.com

Alexander Pokryshkin, Marshal Zhukov, Ivan Kozhedub at Kremlin. Image Source: commons.wikimedia.org of the Soviet Union three times (4 February 1944, 19 August 1944 and 18 August 1945). Two Order of Lenin (4 February 1944 and 21 February 1978). Seven Order of the Red Banner (22 July 1943, 30 September 1943, 29 March 1945, 29 June 1945, 2 June 1951, 22 February 1958, and 26 June

1970). Order of Alexander Nevsky (31 July 1945). Order of the Patriotic War 1st class (11 March 1985). Two Order of the Red Star (4 June 1955 and 20 October 1955). Many campaign and jubilee medals.

Stamp Release and Legacy

Ivan Kozhedub is associated with a single fighter type, the series of radial engine, wooden aircraft designed by Semyen Lavochkin. The last of them, La-7 No. 27, like its pilot, survived graceful retirement, and in the airplane's case is at the Monino Air Museum.

Ivan Kozhedub died on 8 August 1991. A special stamp was released to honour him by Russia in 2020. A military university in

Kiev, Ukraine: Monument to the hero of World War II fighter pilot Ivan Kozhedub in the park. Inscription: Hero of the Soviet Union, Marshal of Aviation Ivan Nikitovich Kozhedub.
Image Source: dreamstime.com

Kharkiv is named in his honor, the Kozhedub University of the Air Force.

Image Source: Wikipedia

Ivan Kozhedub died 08-08-1991, aged 71, in Moscow, and is buried on the Novodevitchy Cemetery in Moscow.
Image Source: ww2gravestone.com

REFERENCES

1. Interview by Aviation History with the Ace pilot. https://www.historynet.com/aviation-history-interview-with-world-war-ii-soviet-ace-ivan-kozhedub.htm
2. Ivan Nikitovich Kozhedub, Wikipedia, https://en.wikipedia.org/wiki/Ivan_Kozhedub
3. Prominent Russians: Ivan Kozhedub, Russiapedia, https://russiapedia.rt.com/prominent-russians/military/ivan-kozhedub/
4. Kozhedub, Ivan Nikolayevich, Biography, Traces of War, https://www.tracesofwar.com/persons/34712/Kozhedub-Ivan-Nikolayevich.htm

15
"Zero" Fighter Ace Tetsuzō Iwamoto "Tiger Tetsu"

Highest Scoring Japanese

Iwamoto.
Picture Source: Wikipedia

Tetsuzō Iwamoto was one of the top-scoring aces among Imperial Japanese Navy Air Service (IJNAS) fighter pilots. He joined the Imperial Navy in 1934 and completed pilot training in December 1936. His first combat occurred over China in early 1938. He emerged as one of the top Aces of Imperial Japan during WWII, credited with at least 94 aerial victories including 14 victories in China. He flew Zero fighters from the aircraft carrier *Zuikaku* from December 1941 to May 1942, including at the Battle of the Coral Sea. He was Nicknamed Zero Fighter Ace Kotetsu "Tiger Tetsu".

In late 1943, Iwamoto's Air Group was sent to Rabaul, New Britain, resulting in three months of air combat against Allied air raids. Subsequent assignments were Truk Atoll in the Carolines and the Philippines. Following the evacuation of the Philippines, Iwamoto served in home defence and trained kamikaze pilots.

As a result of the Japanese use of the British naval practices, the IJNAS scoring system was based on the system the Royal Navy and the Royal Air Force (RAF) adopted from World War I until World War II. This system differed from the scoring system used by some other nations during World War II. Tetsuzo Iwamoto was one of the greatest Air Aces of the Imperial Japanese Navy. During his career, as per his personal records in his diary, he achieved 242 air victories, 14 in China, 202 confirmed in the Pacific, 26 shared victories and 27 unconfirmed victories. He is supposed to have damaged two planes and destroyed two on the ground. Research by academics surnamed Izawa and Hata in 1971 estimated his score at about 87. Irrespective of the scoring system, Tetsuzō Iwamoto was Japan's top Ace. Iwamoto was known as the Chûtai leader (leader of Flying Company, a squadron of 8 to 16 fighters). Iwamoto was one of few survivors of the IJNAS through the war. He fought over the Indian and the Pacific Ocean and also trained young pilots even in the last months of the war. Like many Japanese veterans, Iwamoto was reported to have fallen into depression after the war. His diary was found after his death, with records of his 202 Allied aircraft claimed as destroyed.

Battle of Rabaul (1942). Map depicting Papua New Guinea.
Image Source: nc.cdc.gov

Young Days

Tetsuzo was the third son of the Iwamoto family, born in a border town, the southern part of Karafuto (now Sakhalin, Russia) on 15 June 1916. His father was a chief police officer. Later he grew up in Sapporo, Hokkaidō, Japan. He enjoyed skiing in his elementary school days in Sapporo. When he was 13, his father retired and Tetsuzo moved with his family to his father's hometown, Masuda, Shimane Prefecture. He studied at

With father and brothers, second from the right.
Image Source: ameblo.jp.

the Prefectural Masuda Agricultural and Forestry High School. His favourite school subjects were mathematics and geometry, and he always scored high in them. Whilst he was a gifted student both academically and physically, his popularity with his teachers was poor due to his insubordinate, rebellious, and sometimes outright rude nature. He was regarded as the most opinionated student in his school, and often talked down his teachers in discussions, which was considered impolite. He was otherwise an active and nimble boy. He joined a school club brass band as a trumpeter.

Shimane Prefecture Japan.
Image Source: Wikipedia

Another hobby was growing plants and flowers. He helped local fishermen in the fishing season, going out to the sandy beach early in the morning, and driving fish into the nets.

Tetsuzo Iwamoto sunglass.
Image Source: snappygoat.com

Chooses a Military Career

Iwamoto started his military career in 1934 after he graduated the school at 18. Following the advice from his parents to study while young, Tetsuzo left for a large city where he was supposed to take a college entrance examination. He, however, secretly applied for and passed the examination for acceptance as an Imperial Japanese naval airman 4th class, and was promoted to 3rd class 5 months later. His parents were very disappointed, for they became reliant upon Tetsuzo rather than his eldest brother, who was already studying at some university in a large city and would not return to Masuda.

Gets Enrolled For Flight Training

In 1936, when he was a naval mechanic 2nd class and a crewman on the light aircraft carrier Ryûjō, he studied hard and passed the difficult IJNAS exam, taken by thousands of applicants. He was enrolled in the class 34th Sojyu-Renshusei flight trainee program for naval petty officers and sailors. He graduated as one of the select 26 young aviators of the class 34th in December of that year.

In April 1936, he was sent to Kasumigaura-Ku as a probationer. While his training going on November 1, 1936, he was promoted to naval mechanic 1st class. Finally, on December 26, he graduated 34th class of Sojyu-Renshusei, was promoted to airman 1st class (old rank name of pre-war Japan, equivalent to senior airman). At the flight training school at the Tomobe branch of Kasumigaura-Ku, his fighter course instructor was the famous Chitoshi Isozaki.

Imperial Japanese Navy (IJN) Air Ace Chitoshi Isozaki, 12 Victories.
Image Source: Wikipedia

Joins Naval Air Group

In December 1936. Iwamoto entered Saeki Kōkûtai Naval Air Group, which was based both at land and on-board a carrier, for 6 months of advanced training. After this, he joined Omura *Kôkûtai* on July 16, 1937 for operational training. He had tough training from senior pilots including Air Petty Officer 1st class Toshio Kuroiwa, who was the IJNAS legendary dogfight master pilot. Tetsuzo Iwamoto had to wait for his operational debut till February 10, 1938.

China Front

The Second Sino-Japanese war was already on. After combat training, on February 10, 1938, Tetsuzō Iwamoto was led by his leader APO 1/C Toshio Kuroiwa, flying for two and a quarter hours over the

Nanchang Dajiaochang Airport, China. 800 m Runway in 1938.
Image Source: Wikipedia

China Sea from Omura Airbase at Kyûshû Japan, ferrying to the airfield outside of Nanjing China. Tetsuzo's ability as a fighter pilot was recognized by all on his first air mission with his squadron, the 13th Flying Group on February 25, 1938, over Nanchang, China. This Flying Group was highly regarded and was famed as the *Nango Fighter Squadron*, named after its former squadron leader, Mochifumi Nango, who had shown considerable courage and conspicuous leadership.

First Combat Mission: Four Victories

On February 25, 1938, his Squadron's fighters escorted bombers Type 96, land-based attack aircraft. 16 Chinese fighters attacked the formation, and the squadron leader Lieutenant Takuma was lost on this mission. Iwamoto described his first combat in his notes. During the escort mission, the squadron was intercepted by sixteen I-15s and I-16s at an altitude of 5000 meters. Iwamoto claimed 4 victories (1 probable) in the combat. He secured his first victory by firing when within 50m of the enemy fighter. He first saw white smoke, then the enemy burned up and crashed. He was then at an altitude of 4000 m. When he looked back, there was an enemy fighter just behind him. He instantly made a Split S manoeuvre and narrowly escaped.

Mitsubishi G3M Type 96 Bomber Aircraft.
Image Source: alchetron.com

Polikarpov I-15 Soviet biplane fighter aircraft.
Image Source: commons.wikimedia.org

He got his second victory against an I-15. He saw it below him, turned, and attacked from its 6 o'clock high. When it was hit, it climbed sharply and went spinning downward out of control, and crashed into the ground. He kept his altitude of 4,000 m. He got an I-16 at the top of its roll in his gunsight and fired a burst, its engine burning and out of control; Tetsuzo lost sight of it before it crashed, and he reported this as probable. Another I-15 came down to him from 12 o'clock ahead. Both made a climb and were soon in a dogfight. The I-15 tried to break free of him and made a straight dive. That action made it easier for Tetsuzo to aim. He downed this I-15 on farmland near the airfield. He was flying at an altitude of 2000 m.

Above him, many enemy fighters were manoeuvring. He found one of them coming down with landing gear down. He chased it to an altitude of 200 m and fired a burst. The I-16 was surprised and made a split S manoeuvre, but crashed at a corner of the airfield. This was his 4th victory.

Anti-aircraft guns started firing heavily, and he found himself in an intense barrage of flak. Rushing to escape at full throttle with a number of enemy fighters behind him, he succeeded in returning safely from the battlefield. His section leader Kuroiwa with whom he had separated had already returned to the Wuhu

Iwamoto (one the left, back row).
Image Source: snappygoat.com 2

Soviet Polikarpov I-16 the world's first low-wing cantilever monoplane fighter with retractable landing gear.
Image Source: Wikipedia

airfield, Anhui China, waiting for his return. Kuroiwa scolded Tetsu severely for the rash attacks he made on the day.

13th and 12th Flying Groups Merge

On 22 Mar 1938, the 13th Flying Group Fighter Squadron was merged with the 12th Fighter Squadron on March 22, 1938, where Type 96 carrier fighters for 1st Chutai had landing gear painted in red and were called "Red legs squadron" while 2nd Chutai had gear painted in blue and were called "Blue legs squadron".

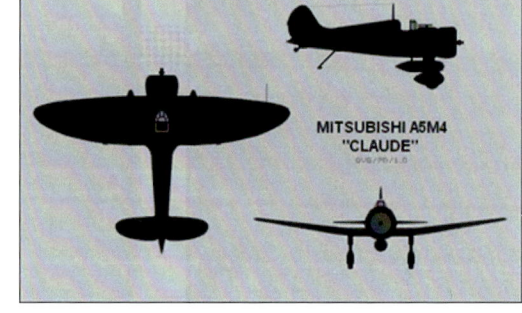

Mitsubishi A5M4 Type 96 Carrier fighter (Allies called it "CLAUDE").
Image Source: craymond.no-ip.info

On 29 Apr 1938, he fought Chinese Air Force fighters and scored several victories, and was later awarded a citation by Commander Tsukahara for his extreme courage and conspicuous gallantry in action above and beyond the call of duty as a fighter pilot against intense Chinese air force.

Iwamoto Becomes Top IJNAS Ace in China

By Sep 1938, he had completed 82 missions, and had 14 kills on his records and had become the top Japanese and IJNAS Ace. His activities subsequently earned him "Order of the Golden Kite – 5th class" recommendation in 1940. In September 1938, 22-year-old Iwamoto was ordered back to Japan, where he became a member of the Saiki Air Group and appointed to the training staff.

Pacific War: Pearl Harbour & Carol Sea

After a tour as an instructor, Petty Officer First Class Iwamoto returned to the front line on-board the carrier Zuikaku, now flying the legendary A6M 'Zero' fighter. Iwamoto was airborne for the day of infamy – the attack on Pearl Harbour – but flew air cover over the carrier group itself rather than escorting the actual raid. Iwamoto was heavily involved in the air war in the Pacific from the outset, regularly leading flights of A6Ms against their American, British and Australian adversaries. Iwamoto flew in the violent air engagements of the Battle of the Coral Sea in May 1942, during which Zuikaku's air group suffered significant losses. This necessitated a return to Japan

Japanese Order of the Golden Kite, 5th Class.
Image Source: warthunder.com

Legendary A6M 'Zero' fighter.
Image Source: warthunder.com

for resupply and to train replacement aircrew which resulted in Iwamoto and his comrades missing the Battle of Midway. This small team in Japan frantically tried to train new aircrew to stem the advancing allies.

The Caroline Islands.
Image Source: Wikipedia

Rabaul, New Britain

After a year of instructional duties, Chief Petty Officer Iwamoto returned to the front line and joined the 253rd Air Group, flying A6Ms from Rabaul, New Britain, in November 1943. Involved in daily air combat, the experienced fighter leader led his cadre of increasingly junior and less experienced pilots against the might of the US Navy and USAAF. Whilst operating from Rabaul, Iwamoto filed claims for a staggering 142 enemy aircraft shot down.

Caroline and Philippine Islands

After withdrawing from Rabaul, Iwamoto returned to Japan in June 1944 for a brief respite from the front line, before fighting in the skies over Formosa, then transferred to Truk Atoll in the Caroline Islands and then the Philippine Islands during autumn and the winter. In November 1944 Iwamoto's skill and leadership were further recognised when he was commissioned as an officer in the ranks of the Japanese Navy, holding the rank of Ensign. Iwamoto's last operational sorties were flown with the 203rd Air Group, defending Kyushu and Okinawa in the furthest Southwest reaches of Japan against the long-ranged B-29 attacks and the might of the US Navy's carrier-borne air power. The last few months of Iwamoto's war were spent training kamikaze pilots at Iwakuni airfield on Honshu Island.

Tactics

He was best at one-vs.-one dogfights. He often employed, quick roll tactic combining with skidding sideways for sudden deceleration within 1/2 quick roll to make the opponent on the tail to overshoot. Corkscrew loop (*Hineri-Komi Senpô*) was a short-cut or twist-in loop tactic, also called the skidding loop. He had also mastered the High and Low Yo-yo manoeuvres (*Suichoku-Senkai Kasoku Senpô*). His favourite formation tactic, was a two-group linked formation attack, where one section plays offence, zooming and diving formation attack, and another section plays defence, positioned on the higher altitude to cover and support the offensive section. Keeping his groups

Chiran high school girls wave farewell with cherry blossom branches to departing kamikaze pilot.
Image Source: Wikipedia

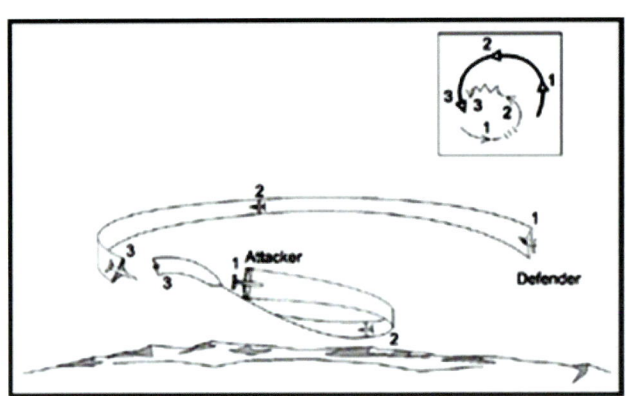

Low Yo-yo.
Image Source: Wikipedia

underneath thick clouds to hide his formation and waiting until the small number of opponent aircraft group coming down, then diving and zooming attack with all in a formation.

Attacking the opponent formations after their mission was over and they were on the way to the regroup point, to fly back across the distance range over the sea. This tactics was taken when his own group had much fewer aircraft. Aerial Bomb attack tactic, was a twelve o'clock high vertical dive attack from the front top in inverted flight (*Haimen Suichoku Kôka Senpô*). It was almost vertical diving (about 60 degree) attack because the 30 kg aerial bomb needed the releasing speed over 280 knots to work timer correctly for the explosion. Inverted flight at the starting point because Zero Fighter could not keep steep angle while diving due to its high flight stability.

Another tactic that the Japanese often employed was called "Send Wolf", which essentially meant first sending a strike against US targets, and then attacking those US aircraft that got airborne after the base had been attacked by surprise. He was very good at using the 30 kg No. 3 Aerial Bomb, which was difficult to handle, although it was used against aircraft, and caused great damage.

Flight leader Masao Sato with his pilots aboard Zuikaku, 6 Dec 1941, one day before Pearl Harbour Attack; Tetsuzo Iwamoto second row, right-most.
Image Source: ww2db.com

Aerial Victories Claimed in His Diary

His own diary claimed that he had downed 202 aircraft during the Second Sino-Japanese War and the Pacific War while sharing credit with others for another 26 kills; this translated to somewhere between 80 and 90 kills based on the scoring system used by American airmen during WW2. 94 victories is what has been assigned to him.

Japanese Type 99 30kg High-explosive bomb.
Image Source: commons.wikimedia.org

He had claimed seven victories against Grumman F4F Wildcat; four Lockheed P-38 Lightning; and 48 victories against Vought F4U Corsair, plus one unconfirmed. He also claimed two Bell P-39 Airacobra, one Curtiss P-40 Warhawk, 39 Grumman F6F Hellcat, one Republic P-47

Thunderbolt, and one North American Aviation P-51 Mustang. He later claimed four British Spitfire burned on the ground, and two shot over Indian Ocean. His other claims were 48 US Navy's Douglas SBD Dauntless with seven unconfirmed. He also reportedly shot down 30 Douglas SBD Dauntless using No.3 Aerial Bombs over Rabaul, in late 1943–1944. In addition, he claimed five Grumman TBF Avenger, five Curtiss SB2C Helldiver, five North American B-25 Mitchell medium bombers, and two Martin B-26 Marauder American twin-engine medium bomber at Rabaul. He claimed six Consolidated B-24 Liberator heavy bombers using No. 3 Aerial Bombs and damaged two at Truk, reportedly confirmed by the ground crew. He claimed one Boeing B-29 Superfortress four-engine propeller-driven heavy bomber at Kagoshima, Kyushu, Japan, in April 1945. He had also claimed a PBY5A flying boat over the Indian Ocean on April 5, 1942. He reportedly shot a Martin Mariner flying boat victory at Rabaul, on 19 February 1944. He strafed three Destroyers on the night of February 5, 1944. He strafed many Landing crafts at Kerama Islands, Okinawa, on night March 26, 1945. He strafed airfields Lae, Eastern New Guinea, on January 23, 1942, and Torokina, Bougainville, Solomons, at night in 1944.

Vought F4U Corsair.
Image Source: Wikipedia

US Navy's Douglas SBD Dauntless Dive Bomber.
Image Source: blenderartists.org

Air Kills Debate: Yet Greatest Japanese Fighter Ace of All Time

Iwamoto's final number of kills remains open to debate. A combination of the Japanese practice of crediting victories to a squadron rather than an individual, lost records, and discrepancies between the confirmation process in Japan and allied nations mean that the final tally will never be known. Most sources officially credit Iwamoto with 80 kills in WW II plus 14 in China, his war diary claims 202 individual victories, 26 shared and 22 unconfirmed. An outspoken, opinionated, and brash man on the ground, Iwamoto was conversely a tactically minded and cool-headed aviator who favoured hit and run tactics over dog-fighting. There is a very good chance that he was, and forever will be, the greatest Japanese fighter ace of all time.

The "Zero Fighting Tiger"

Zero Fighting Tiger.
Image Source: snappygoat.com

Tetsuzo had reportedly flown 8,000 hrs by March 1944. He was an operational leader, cleared for solo fighter reconnaissance and attack mission across night and over ocean. He was cleared for night carrier operations. It is believed that his aircraft in the Rabaul days was dyed in pink with more than 60 shot-down marks and was well known. He was called the "Zero Fighting Tiger".

The Kill Marks. Tetsuzo Iwamoto's 53-102. Single cherry = fighter, Double cherry = bomber. Art Work: Fuku.
Image Source: ww2aircraft.net

Imperial Japanese Navy Ensign Shoulder Boards

Promotions

- Sailor Fourth Class (Seaman Recruit) – June 1, 1934
- Sailor Third Class (Seaman) – November 15, 1934
- Sailor Second Class (Able Seaman) – November 2, 1935
- Sailor First Class (Leading Seaman) – December 26, 1935
- Petty Officer Third Class – May 1, 1938
- Petty Officer Second Class (Petty Officer) – November 1, 1939
- Petty Officer First Class (Chief Petty Officer) – May 1, 1941
- Chief Petty Officer (re-graded Petty Officer First Class) – November 1, 1942
- Commissioned an Ensign – November 1, 1944
- Promoted to Sub-Lieutenant upon retirement – September 5, 1945

Awards

- Order of the Golden Kite, Fifth Class – August 1, 1942
- Order of the Rising Sun, Green Paulownia Leaves Medal – Seventh Class – August 1, 1942

Rising Sun, Green Paulownia Leaves Medal – Seventh Class.
Image Source: worthpoint.com

Post-war Life

Iwamoto was promoted to Sub-Lieutenant on his enforced retirement from the Japanese Navy following his nation's surrender. The Allied Occupation Forces searched for war criminals in the Japanese Officer Corps. Iwamoto was summoned twice for questioning to Douglas MacArthur's Allied GHQ office in Tokyo. Although he managed to avoid being declared a war criminal, he was nevertheless blacklisted from public sector employment. Managers of nongovernmental businesses and local factories in his hometown also did not dare to employ him, in order to comply with the wishes of the new Allied GHQ. In general, anyone who had been an officer in the IJA or IJN was disliked by the Allied Occupation Forces.

253 Ku at Truk, 29 April 1944. Front row, extreme left, Tetsuzo Iwamoto.
Image Source: tumblr.com

His wife knew him as a "wartime ace pilot", and in fact was a fan who later married through a matchmaking agency. He very methodically kept all records. Japanese journalists who had promoted the Japanese militarism campaign during the war started a radio program of anti-militarism post-war called *Shin-Jitsu wa Kô Da* ("The Truth Is This"). The program considered people such as Iwamoto the cat's-paws of militarism. Iwamoto struggled to survive until the San Francisco Peace Conference was held, after which, in the spring of 1952, the Allied Occupation Forces finally left Japan. In 1952, Iwamoto finally obtained employment at the Masuda spinning mill of Daiwa Bōseki (since renamed to "Daiwabō" Co. Ltd).

Death due Medical Complications

In the summer of 1953, Iwamoto developed a stomach ache. A surgeon examined him and diagnosed enteritis. It was found later to be appendicitis. After a series of operations, he complained of a backache. Doctors decided to operate on him again. For reasons which are not entirely clear, the surgical team decided to remove three or four ribs without anaesthesia. This led to sepsis. Iwamoto died on 20 May 1955, at the age of just 38 years. His wife recalled his final words: "When I get well, I want to fly again."

Mitsubishi A6M Zero

A light and nimble fighter, the Mitsubishi A6M Zero was the first carrier-based fighter capable of besting its land-based opponents, and was Japan's main fighter of WWII. The Zero's design sacrificed protection for speed, manoeuvrability, and long-range, on the theory that superior speed and manoeuvrability were protections in their own right, with long-range an added bonus. The A6M came as a shock to Allied pilots when first encountered, because it could outmanoeuvre every airplane it faced at the time. A better dogfighter than anything the Allies had at the start of the Pacific War, the Zero's superior performance,

Mitsubishi A6M Zero.
Image Source: aviation-history.com

especially in the hands of Japan's elite naval aviators, exceeded anything the Allies had hitherto expected from the Japanese. In the war's early days, Japanese naval aviators flying Zeroes achieved a 12:1 kill ratio.

To counter the Zero's advantages, American pilots adopted teamwork tactics such as the "Thach Weave" which required pilot pairs to work in tandem, or the "Boom and Zoom", in which American pilots engaged the Zero only in diving attacks, as the acceleration of their heavier planes in a dive allowed them to flee if the diving attack failed. While holding considerable advantages in manoeuvrability and speed, the Zero's lack of protection for either the pilot or the fuel tanks proved a steadily mounting disadvantage as the war progressed, since the heavier and more rugged American fighters could absorb considerable punishment from Zeroes, while a single machine gun burst from the American plane could disintegrate a Zero.

Grumman F6F Hellcat.
Image Source: warhistoryonline.com

By 1943, attrition had thinned the ranks of Japan's elite aviators, and the Japanese Navy's training pipeline could not produce enough replacements of similar calibre. As a result, there were fewer and fewer

Image Source: cs.finescale.com

Japanese pilots capable of extracting the most out of the Zero's advantages while minimizing its disadvantages. Which was bad news for the Japanese, as the quality of American aviators was increasing, due to wartime experience as well as an extensive training program that produced capable aviators at a rate Japan could not match. That was exacerbated by the introduction of new American fighters, such as the F4U Corsair and the F6F Hellcat, which were a significant improvement over their predecessors, and proved more than a match for the Zero, with greater firepower, armour, speed, and similar manoeuvrability.

By 1944 the Zero was rapidly becoming obsolete, but it remained in front line service because the Japanese faced production difficulties in fielding a replacement. From its heyday at war's beginning when it ruled the skies of the Pacific while flown by elite pilots, A6Ms were reduced by war's end to flying kamikaze missions under the controls of barely trained novices.

List of Japanese Imperial Navy Air Aces on Mitsubishi A6M "Zero"

This is a list of Imperial Navy Air Aces flying the Mitsubishi Zero fighter during the Pacific War.

- Tetsuzo Iwamoto: 94 (including 14 in China/personal diary accounts for a total of 202 kills)
- Shoichi Sugita: 70 (some sources say 80)
- Saburō Sakai: 64 (2 in China)
- Takeo Okumura: 54 (4 in China)
- Hiroyoshi Nishizawa: 36 official (102 claimed)
- Toshio Ohta: 34
- Kazuo Sugino: 32
- Junichi Sasai: 27
- Sito Origami: 10 (9 in China, 1 disputed)
- Toshiyuki Sueda: 9

Shôichi Sugita. Second Highest Japanese Ace.
Image Source: wikidata.org

Famous Japanese Units

The air unit "Kōkûtai" (air group) with most Air Aces, was the Tainan Kōkûtai (in Formosa). It was the most famous group. They operated from Rabaul, New Britain, and acquired their legendary fame over Taihoku (The Philippines), the Dutch East Indies, Lae, and Buna

Fighter pilots of the Tainan Air Group pose at Lae in June 1942. Several of these aviators would be among the top Japanese Aces, including Saburô Sakai (middle row, second from left), and Hiroyoshi Nishizawa (standing, first on left).
Image Source: Wikipedia

(New Guinea) and, in the last stages of the war, in defence of mainland Japan. Saburō Sakai was another member in this unit after Dutch Indies operations, from the Denpasar base on Bali.

Other famous units with Air Aces were the 3rd Air Corps (including Yoshiro Hashiguchi), 253rd Air Corps (with Tetsuzo Iwamoto and Hiroyoshi Nishizawa among its members), Genzan Air Group, and other groups.

Kawanishi N1K2-J *343 A-15* of 301st Fighter Squadron/343rd Naval Air Group, Matsuyama airbase, 10 April 1945.
Image Source: Wikipedia

The Imperial Japanese Navy Air Service, similar to the German Luftwaffe idea of organizing an "all aces" select unit Jagdverband 44 equipped with Messerschmitt Me 262A-1a jet fighters, decided to create an all-ace unit (the 343 Kōkutai) with Kawanishi N1K2-J fighters towards the end of the conflict; this was commanded by Minoru Genda.

REFERENCES

1. C. Peter Chen, Tetsuzo Iwamoto, World War II Database, https://ww2db.com/person_bio.php?person_id=561
2. Tetsuzo "Tetsu" Iwamoto, Aces of WW2, https://acesofww2.com/japan/aces/iwamoto/
3. Tetsuzō Iwamoto, Biography, https://peoplepill.com/people/tetsuzo-iwamoto/
4. Tetsuzō Iwamoto, Wikipedia, https://en.wikipedia.org/wiki/Tetsuz%C5%8D_Iwamoto
5. Mark Barber, War Thunder Historical Consultant, a former pilot from the British Royal Navy's Fleet Air Arm, Ace of the Month – June – Lt JG Tetsuzo Iwamoto, June 01, 2015, War Thunder, https://warthunder.com/en/news/3142—en
6. Lt (jg) Tetsuzo Iwamoto, top gun pilot of the Imperial Japanese Navy, https://www.starduststudios.com/tetsuzo-iwamoto.html
7. Iwamoto, Tetsuzo, Biography, Traces of War, https://www.tracesofwar.com/persons/61886/Iwamoto-Tetsuzo.htm

16

GERMAN MAJOR HEINZ-WOLFGANG SCHNAUFER

Highest Scoring Night Ace

Heinz-Wolfgang Schnaufer and BF 110 Night Fighter Aircraft

Heinz-Wolfgang Schnaufer was a German Luftwaffe night-fighter pilot and the highest-scoring night fighter Ace in the history of aerial warfare. All Schnaufer's 121 victories were in World War II, and mostly against British four-engine bombers. For his excellent combat record, he was awarded the Knight's Cross of the Iron Cross with Oak Leaves, Swords and Diamonds, Germany's highest military decoration at the time, on 16 October 1944. He was nicknamed "The Spook of St. Trond", from the location of his unit's base in occupied Belgium. By the end of hostilities, Schnaufer's night-fighter crew held the unique distinction that every member – radio operator and air gunner – was decorated with the Knight's Cross of the Iron Cross. Schnaufer was taken prisoner of war by British forces in May 1945. After his release a year later, he returned to his home town and took over the family wine business. He sustained injuries in a road accident on 13 July 1950 during a wine-purchasing visit to France, and died in a Bordeaux hospital two days later.

Family of Wine Merchants and Winery

Heinz-Wolfgang Schnaufer was born on 16 February 1922 in Calw, located in the Free People's State of Württemberg of the German Reich, during the Weimar Republic era. He was the first of four children of a mechanical engineer and merchant. His father owned and operated the family business, the winery Schnaufer-

Heinz-Wolfgang Schnaufer.
Image Source: Wikipedia

Schlossbergkellerei ("Schnaufer's Castle Mountain Winery"), in the Lederstraße, Calw. The winery had been founded by both his father and his grandfather, in 1919, shortly after World War I. When his father unexpectedly died in 1940, his mother ran the business until the children took over the winery after World War II. The company then expanded the business and in addition to the winery offered wine imports, sparkling wines, and a distillery for wine and liqueur.

Four children of a mechanical engineer and merchant Alfred Schnaufer and his wife Martha.
Image Source: ww2gravestone.com

Early Years: Nazi Schooling

At an early age, he expressed his wish to join an organisation of military character and joined the "Deutsches Jungvolk" (German Youth) in 1933. After completing his sixth grade he cleared the entrance exam to join the National Political Institutes of Education, a secondary boarding school (Napola) founded under the recently established Nazi state. Napola schools were to raise a new generation for the political, military, and administrative leadership of the Third Reich. Schnaufer finished top of his class every year. At seventeen he graduated with his diploma in November 1939 with distinction. He also received the Reich Youth Sports Badge, the base-certificate of the German Life Saving Association, the bronze Hitler Youth-Performance Badge, and completed his B-license to fly glider aircraft.

Deutsches Jungvolk (German Youth).
Representative Image Source: ww2gravestone.com

Serious Interest in Flying

In 1939 Schnaufer was one of two students posted to the Napola in Potsdam, a Flying Platoon centralised for all the destined flyers from all the Napolas. Here he learned to fly glider aircraft, the DFS SG 38 Schulgleiter, and later on the two-seater Göppingen Gö 4 which was towed by a Klemm Kl 25.

Joins the Luftwaffe

Following his graduation from school, he cleared the entrance exam and joined the Luftwaffe on 15 November 1939

Göppingen Gö 4 Towed Glider.
Image Source: flugzeuginfo.net

and underwent his basic military training at the "Fliegerausbildungs regiment" 42 (42nd Flight Training Regiment). His further flight training was at the "Flugzeugführerschule" FFS A/B 3 flight school for the pilot license, which he completed on 20 August 1940. He was trained to fly the Focke-Wulf Fw 44, Fw 56 and Fw 58, and the Heinkel He 72, HD 41 and He 51, the Bücker Bü 131, the Klemm Kl 35, the Arado Ar 66 and Ar 96, the Gotha Go 145 and the Junkers W 34 and A 35.

Junkers W 34.
Image Source: armedconflicts.com

Friedrich "Fritz" Rumpelhardt, Schnaufer's Radio/Radar Operator.
Image Source: Wikipedia

Advanced Flying Training

Schnaufer then attended the advanced FFS C 3, advanced flight school, and later the blind flying school "BFS 2" from August 1940 to May 1941. This qualified him to fly multi-engine aircraft. During this assignment, he was promoted to Cadet Sergeant, and he became Second lieutenant on 1 April 1941. He was then posted for ten weeks to the "Zerstörerschule" (destroyer school). He was assigned radio operator Friedrich Rumpelhardt as an aircrew team on 3 July 1941. Schnaufer's previous radio operator had proved unable to cope with aerobatics, and Schnaufer thoroughly tested Rumpelhardt's ability to cope with aerobatics before they teamed up.

Volunteers to be a Night Fighter

The two decided to volunteer to fly night fighters to defend against the increasing Royal Air Force (RAF) Bomber Command offensive against Germany. After their training, the two were sent to the Nachtjagdschule 1 (1st night fighter school) near Munich, to learn the rudiments of night-fighting. The night fighter training was carried out on the Ar 96, the Fw 58 and the Messerschmitt Bf 110. Training at night focused on night take-offs and landings, cooperation with searchlights, radio-beacon direction finding, and cross country flights.

"Lichtenstein" SN-2 VHF band, and B/C UHF band night fighter radar antennas on the nose of a Bf 110 G-4.
Image Source: warhistoryonline.com

Joins Night Fighter Squadron

In November 1941, Schnaufer was posted to the II/NJG 1 (2nd group of the 1st Night Fighter Wing) at that time based at Stade near Hamburg. Here, Schnaufer was assigned to the 5th squadron of 1st Night Fighter Wing. The Bf 110's of II/NJG 1 at the time were not equipped with airborne radar such as the Lichtenstein radar. Night fighter intercept tactics had matured since their early beginnings in July 1940, and II. Gruppe had already been credited with 397 victories.

An early-model Bf 110G with Matratze UHF radar antennas.
Photo: Bundesarchiv.
Image Source: warhistoryonline.com

Initial Night Intercept Concept

Missions against enemy bombers at the time were usually flown by means of ground-controlled interception, although the Luftwaffe was already experimenting with airborne radar. German air defence system, consisted of a series of radar stations with overlapping coverage, layered three deep, was conceived by Lieutenant General (equivalent to British Major General) Josef Kammhuber and called "Kammhuber Line". Conceptually, the system was based on a combination of ground-based radar stations, searchlights and a fighter pilot controller officer. It consisted of a series of control sectors equipped with radars and searchlights, and an associated night fighter. Each sector would direct the night fighter into visual range with target bombers. The controller had to vector the airborne night fighter by means of radio communication to a point of visual interception of the illuminated bomber. These interception tactics were referred to as the "Himmelbett" (canopy bed) procedure.

Unit Moves to Belgium: Little Initial Action

On 15 January 1942, II/NJG 1 transferred to Saint-Truiden (Also called St. Trond) in Belgium. Schnaufer entered front-line service at a time when the RAF was reassessing the air offensive against Germany. The effectiveness of British Bomber Command to accurately hit German targets had been questioned by the British War Cabinet Secretary David Bensusan-Butt who published the Butt Report in August 1941. The report in parts concluded that the British crew failed to navigate to, identify, and bomb their targets. Although the report was not widely accepted by senior RAF commanders, Prime Minister Winston Churchill, instructed Commander-in-Chief Richard Peirse that during the winter months only limited operations were to be conducted. Flight operations were also hindered by bad weather in the first months of 1942, so II/NJG 1 only saw very limited action during that period.

Channel Dash Operation

On 8 February 1942, the II Group was transferred to Koksijde Air Base without having scored any victories while stationed at Saint-Truiden. The objective of this assignment was to give the German battleships "Scharnhorst" and "Gneisenau" and the heavy cruiser "Prinz Eugen" fighter protection in the breakout from Brest to Germany.

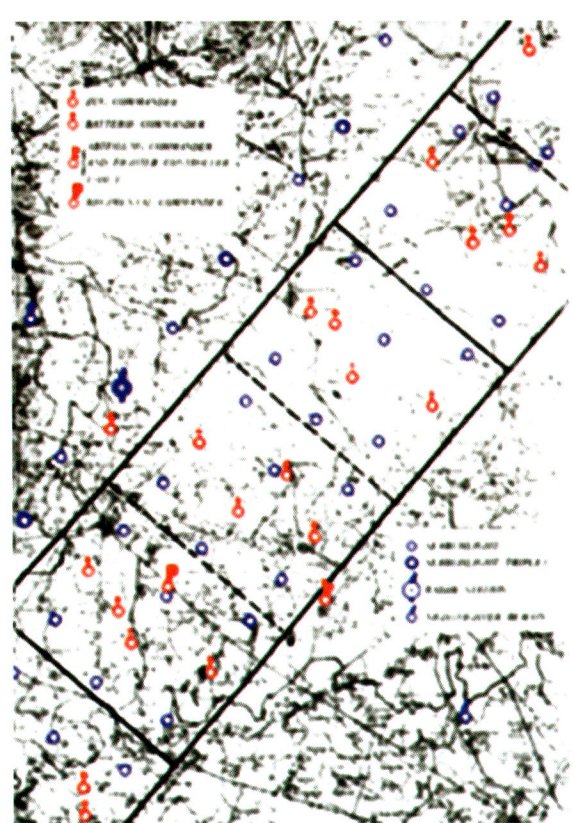

A map of part of the *Kammhuber Line* stolen by a Belgian agent and passed-on to the British in 1942. The 'belt' and night fighter 'boxes' are shown.
Image Source: Wikipedia

The Channel Dash operation (codenamed Operation Cerberus) of German Navy was between 11 and 13 February 1942. The operation was to support the return of the German ships to German bases as Britain planned invasion of Norway.

Operation Donnerkeil

In support of Operation Cerberus, the Luftwaffe under the leadership of General of the Fighter Force, Adolf Galland, formulated an air superiority plan dubbed Operation Donnerkeil for the protection of the three German capital ships. II/NJG 1 was briefed of these plans on the early morning on 12 February. The plan called for the protection of the German ships at all costs. The crew were told that if they ran out of ammunition they must ram the enemy aircraft. To the relief of the night fighters, they were assigned to the first-line reserves. The operation, which took the British by surprise, was successful and the night fighters were kept in their reserve role.

Channel Dash operation. A long line of German ships sailed through the Channel.
Image Source: warfarehistorynetwork.com

Repeat Unit Relocations: Yet No Contact with Enemy

On the evening of 12 February, II/NJG 1 was relocated to Amsterdam Airport Schiphol. On the afternoon of 13 February, Schnaufer flew a reconnaissance mission over the Ijsselmeer and the North Sea and then relocated to Westerland on the island of Sylt near the Denmark border. They further relocated to Aalborg-West in north Denmark from where they made a low-level flight in close formation over the Skagerrak,

Sir Arthur Harris, 1st Baronet, named "Bomber" Harris by the press and often within the RAF as "Butcher" Harris.
Image Source: Wikipedia

landing at Stavanger-Sola in Norway. Over the following days, they operated from the airfield at Forus in Norway, making a short-term landing at Bergen-Herdla in Denmark. In total, Schnaufer made two operational flights without contact with the enemy. Following this assignment, they relocated to the 5th squadron's new base in Germany at Bonn-Hangelar.

British Plan Area Bombing

Following the detailed analysis of the Butt Report, the British High Command made a number of decisions in February 1942 that changed the nature of the bomber war against Germany. On 14 February, Air Chief Marshal Norman Bottomley issued the "Area Bombing Directive", which lifted the restrictions placed on the bombers in 1941. Air Chief Marshal Arthur Harris, commonly known as "Bomber" Harris, was appointed commander-in-chief of Bomber Command. Around the same period was introduced "Gee", a radio navigation system that enabled better target-finding and bombing accuracy. It essentially measured the time delay between two radio signals to produce a fix, with accuracy on the order of a few hundred meters at ranges up to about 350 miles (560 km).

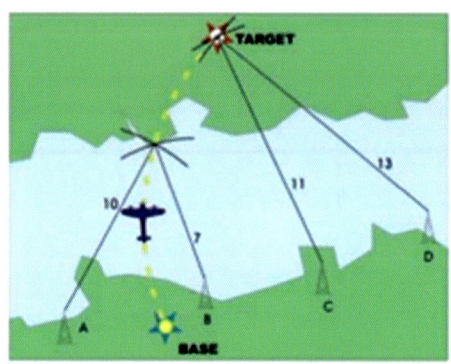

Image Source: RAF Training Notes

First 1,000 Bomber Allied Raid: Operation Millennium

This led to the first Allied 1,000 bomber raid. In Operation Millennium, the RAF targeted and bombed Cologne on the night of 30/31 May 1942. Schnaufer did not participate in the missions in defence of Cologne. The "Himmelbett" procedure had limitations in the number of aircraft which can be controlled. Therefore, only the most experienced crew were deployed, and Rumpelhardt and Schnaufer, who had yet to achieve their first aerial victory, were left out. Prior to Operation Millennium, Schnaufer had been appointed TO (Technical Officer) on 10 April 1942 and was located at Saint-Truiden again. As a Technical Officer, Schnaufer was responsible for the supervision of all technical aspects such as routine maintenance, servicing, and modifications of the *Gruppe*. In this role, he was no longer a member of the 5. *Staffel* but was then a member of the staff of II/NJG 1.

RAF Bombing of Cologne.
Image Source: placenote.info

First Aerial Victory: Ground Controlled Intercept

Schnaufer claimed his first aerial victory on their thirteenth combat mission flown one day after the attack on Cologne on the night 1/2 June 1942. Nominally this was the RAF's second 1,000 bomber raid against Germany, although the attacking force actually numbered 956 aircraft. Schnaufer shot down a Handley Page Halifax south of Louvain in Belgium. The aircraft probably was Halifax W1064 from No. 76 Squadron piloted by Sergeant Thomas Robert Augustus West, which was shot down at 01:55 on 2 June 1942 and crashed at Grez-Doiceau, 15 kilometres south of Louvain. West and another member of the crew were killed. This victory was achieved by ground-controlled interception through the Kammhuber Line. Once near to the target, Rumpelhardt had visually found the bomber and directed Schnaufer into attack position from below and astern. The Halifax caught fire after two firing passes. During this mission, the Himmelbett flight officer vectored them to a second bomber, a Bristol Blenheim. The attack had to be aborted after *Hauptmann* (Captain) Walter Ehle shot down the bomber from a more favourable attack position.

Handley Page Halifax.
Image Source: Wikipedia

Schnaufer Gets Hit: Lands Back Safely – Hospitalised

Heinz-Wolfgang Schnaufer in his Bf 110.
Image Source: Twitter @DownedWarbirds

On 2 June 1942 itself, a little later, around 3 PM, they spotted another target. Schnaufer made two unsuccessful attacks. During their third attack, which closed the distance to 20 metres (66 ft), they were hit by the defensive gunfire. Schnaufer was hit in his left calf, the port engine was burning, the rudder control cables were severed, and an electrical short circuit caused the landing lights to be permanently on. Rumpelhardt and Schnaufer considered bailing out but decided to make an attempt for their home airfield after they managed to put out the flames and restart the engine. While Rumpelhardt made radio contact with the Saint-Truiden airbase, Schnaufer landed the aircraft without rudder control and on ailerons and engine-power alone. This was the only time that their aircraft sustained damage in combat or any member of the crew was wounded. Both Rumpelhardt and Schnaufer were awarded the Iron Cross 2nd Class for their first aerial victory. Schnaufer had hoped that he could stay on active duty and that the bullet lodged in his calf would isolate itself. However, he had to be admitted to a hospital in Brussels from 8–25 June for surgery. Rumpelhardt was given home leave until 26 June while Schnaufer was in the hospital.

Back in Action: More Aerial Victories

Schnaufer had to wait two months to achieve another victory, claiming the destruction of two Vickers Wellingtons and one Armstrong Whitworth Whitley within the space of 62 minutes in the early hours of 01 August. The first Wellington, originally identified by the crew as a Halifax, was severely damaged 3,000 metres (9,800 ft) above the Netherlands and forced to crash land, killing the air gunner at 02:47 hours. The second Wellington was shot down 3,800 metres (12,500 ft) over Brussels, killing everyone on board at 3:17 hours. Rumpelhardt and Schnaufer flew their

Messerschmitt Bf 110G Zerstorer NJG Lichtenstein radar 02.
Image Source: asisbiz.com

first combat mission with the Lichtenstein radar on the night 5/6 August 1942. Though they managed to make contact with an enemy aircraft they failed to shoot it down.

Becomes an Air Ace

On the night of 24/25 August 1942, Schnaufer became an Ace (his fifth aerial victory), when he filed a claim for another Wellington, probably *BJ651*, which was shot down with the loss of Sergeant Eric Bound and crew. This was the first time Rumpelhardt had guided him into contact using the FuG 202 Lichtenstein B/C UHF-band airborne radar. His next claim was made on the night of 28/29 August. This was probably No. 78 Squadron Halifax II W7809, piloted by Sergeant John A. B. Marshall of the Royal Australian Air Force. All crew died in the crash. On the night of the 21/22 December 1942, Schnaufer shot down Avro Lancaster R5914; his first victory against this type. The aircraft crashed at Poelcapelle, killing three on board. It was Schnaufer's seventh victory. Schnaufer may also have been responsible for the destruction of another Lancaster that night. Rumpelhardt and Schnaufer had attacked a Lancaster and observed it catching fire followed by the aircraft plunging earthwards. German Captain Wilhelm Herget of the 4th Night Fighter Wing had also attacked a four-engine bomber in the same vicinity. The draw decided in favour of Herget who was given credit for the destruction of the Lancaster.

Rumpelhardt Temporary Grounded

By the end of 1942, Schnaufer's total stood at seven, with three victories recorded on the night of 1 August, which had earned him the Iron Cross 1st Class in early September 1942. From 29 November to 16 December 1942, Rumpelhardt was confined to the hospital bed with a high fever. Rumpelhardt then attended various officer training courses from February to October 1943. Between 14 May and 3 October 1943, Schnaufer claimed 21 further aerial victories in Rumpelhardt's absence; 12 with four different radio operators.

Early 1943: Lean Period

II/NJG 1 saw little action in the first few months of 1943, and Schnaufer did not claim his next aerial victory until 14 May 1943. II/NJG 1 Himmelbett control areas were located to catch the bombers heading for the Ruhr Area. RAF Bomber Command had made only ten major attacks in that region from January to April 1943. Consequently, II/NJG 1 claimed no victories in January, two in February, one in March, and three in April.

Short Stirling was a British four-engine heavy bomber.
Image Source: Wikipedia

Battle of Ruhr: Action Begins

Schnaufer's number of aerial victories increased again during the Battle of the Ruhr. Schnaufer, with Baro as his radio operator, shot down a No. 214 Squadron Short Stirling R9242 at 02:14 hours on 14 May 1943 on an attack mission against Bochum. Four members of the crew, including pilot Sergeant Raymond Gibney, lost their lives. His next victory on the same mission at 03:07 hours, his 9th overall, a No. 98 Squadron Halifax JB873 returning from Bochum. The captain, Sergeant G. Dane, and co-pilot Sergeant J. H. Body were killed in the crash. On the night of 29/30 May, Bomber Command attacked Wuppertal. Schnaufer and Baro took off on the first wave at 23:51 on 29 May and returned at 02:31 on the 30 May. They shot down two Stirlings, one at 00:48 and the other at 02:22, and one Halifax at 01:43.

Promoted First Lieutenant: Awarded the Honour Goblet of the Luftwaffe

In June 1943, Schnaufer filed claims for a further five aerial victories. Schnaufer and Baro shot down a Stirling from No. 218 Squadron on 22 June 1943 at 01:33. With Baro on the radio and radar, they managed another victory over a Wellington on 25 June 1943 at 02:58. On 29 June 1943, the two shot down three bombers in another attack on Cologne, a Lancaster and two Halifax bombers at 01:25, 01:45, and 01:55 respectively. This brought the number of aerial the victories he was credited with up to seventeen. Schnaufer was promoted to First lieutenant on 1 July 1943. Schnaufer claimed his last two aerial victories with Baro operating the radio on the night of 3/4 July, when RAF Bomber Command had again targeted Cologne. Their victims were a No. 196 Squadron Wellington shot down at 00:48 and a No. 149 Squadron Stirling at 02:33, bringing his total to 19 victories. His next radio operator was Oberleutnant Freymann. They shot down a No. 49 Squadron Lancaster, on another Cologne bombing mission, on 9 July 1943 at 02:33. He was awarded the Honour Goblet of the Luftwaffe on 26 July 1943.

Honour Goblet of the Luftwaffe.
Image Source: Wikipedia

Operation Gomorrah: British Introduce Chaff – German Counters

An Avro Lancaster dropping Window (the crescent-shaped white cloud on the left of the picture) from within the accompanying bomber stream.
Image Source: warhistoryonline.com

In mid-July, the Battle of the Ruhr was coming to an end and Bomber Command refocused its efforts on the port city of Hamburg in northern Germany. The codename for the attack was Operation Gomorrah, and the objective was the destruction of Hamburg. The operations began on 24 July 1943 and during four major night-attacks by the RAF and two minor day-attacks by United States Army Air Forces (USAAF) between 40,000 and 50,000 civilians were killed. To counter the mounting success of the German night fighter force, which was directly attributed to the introduction of the Lichtenstein radar, the RAF introduced Window (Chaff).

Window was a radar countermeasure in which aircraft spread a cloud of small, thin pieces of aluminium which effectively made it impossible for the German radar operator to identify the genuine target. Saturation of the Himmelbett control areas by a bomber stream and the introduction of Window practically made the previous Himmelbett procedure obsolete. This was also evident to the German high command.

To counter these British measures two new strategies were pursued, Wilde Sau (Wild Boar) and Zahme Sau (Tame Boar). Wilde Sau, conceived by Hans-Joachim Herrmann, was a technique by which the RAF bombers were mainly engaged by single-seat fighter planes, illuminated by searchlights, over the target area. The Zahme

The radio control centre for night fighters, with Jägerleitoffiziere (fighter controllers) and assistants plotting courses for directing the airborne fighters.
Photo: Bundesarchiv. Image Source: warhistoryonline.com

Sau procedure, proposed by Viktor von Loßberg, called for a night fighter to infiltrate the bomber stream. The position, altitude, and general directions were then broadcast. The information was received by other night fighters, who navigated to the bomber stream by themselves. In Zahme Sau, the German night fighters were tracked and radio-controlled by means of Y-Control. Schnaufer did not make any claims during Operation Gomorrah. Their next success came when he and Freymann shot down a Lancaster on 10/11 August 1943 at 00:32. The target that night was Nuremberg and it was the first aerial victory of the entire German night fighter force achieved by Y-Control. This was also the last victory with Freymann and his last as a member of II Group.

Y-Control Fighter Guidance System

Allied jamming of existing VHF voice radio links and MF navigation beacons was becoming extremely effective. The Y-Control System was based on the work done for the Y-Gerät system of bomber guidance. There were differences in frequency and a dedicated transponder was not required. The system worked by using sites known as Y-Stations. Each station had 5 radio operation systems. Each consisted of an omnidirectional transmitter an omnidirectional receiver and a direction finder. This allowed each site to control 5 fighters. The transmitter would send out a signal which was picked up by the "FuG16ZY" system in the fighter (known as the Y Fighter), this repeated the signal back on a frequency 1.9 MHz lower than the transmitted frequency. Using the direction finder (DF) the angle was measured. The range was calculated by a timing system connected to both the transmitter and the omnidirectional receiver. Height could not be measured. However, voice could be transposed onto the ranging signal allowing the ground controller to talk and listen to the pilot of the plane he was controlling, hence the pilot could report altitude when requested to do so. A Y-Fighter was part of a group of fighters intercepting an allied bomber stream. The Y-Fighter was painted a distinctive colour and the rest of the flight simply followed him. The flight leader could also listen to the ground controller channel and hear what the ground controller was saying while also being able to talk to the flight. The Y-Fighter was never the group controller. The system was susceptible to jamming on the FuG16ZY wavelength but was at least partially usable for the rest of the war. The controllable range was approx. 250 km depending on aircraft height. Relay stations could be used to extend the radio range. The radio system used was a modified FuG16 radio. The Y-Site also contained other systems such as radio communications for talking to area control.

Squadron Leader of 12.Staffel/NJG 1

Bf 110 of Nachtjagdgeschwader 4 (Night Fighter Wing 4).
Image Source: warhistoryonline.com

Schnaufer was transferred to IV/NJG 1(4th group of the 1st Night Fighter Wing), based in the Netherlands at Leeuwarden Air Base, where he was appointed Squadron Leader of the 12/NJG 1(12th squadron of 1st Night Fighter Wing) on 13 August 1943. At the time, IV/NJG 1 was under the leadership of Group Commander Hans-Joachim Jabs (50 night victories). Jabs' first impression of Schnaufer was not entirely favourable. Shortly after Schnaufer's arrival, on one of his first missions in Leeuwarden, Schnaufer had taken right of way during taxiing. This forced Jabs into second place in the order of take-off, an act of insubordination and perceived as arrogant by Jabs.

Initial Y Control Missions

Schnaufer, who had received the German Cross in Gold on 16 August 1943, flew his first operational mission with 12/NJG 1 on the night of 17/18 August 1943. RAF Bomber Command had targeted Peenemünde and the V-weapons test centre that night. Schnaufer, who had been tasked with leading one of the first Zahme Sau missions under Y-Control, had to abort the mission early due to engine trouble. Around mid-September 1943, the two-man Bf 110 crew was augmented by a third member, sometimes referred to as air mechanic or air gunner. The reason for this was that the decline of the Himmelbett procedure, the introduction of the broadcast procedure Zahme Sau, and the growing threat of RAF intruder night fighter operations, had necessitated the need for another pair of watchful eyes to the rear. Under Officer Wilhelm Gänsler, who had

The wreckage of a Handley Page Halifax of No. 77 Squadron, shot down on the night of August 24, 1943, (Schnaufer's 25th victim).
Image Source: historynet.com

already contributed to 17 claims made by Captain Ludwig Becker (night fighter Ace with 44 victories), was Schnaufer's new lookout man. With Handtke and Gänsler as his crew, Schnaufer claimed his 26th aerial victory on 23 September 1943 over a No. 218 Squadron Stirling during a Wilde Sau intercept mission.

Upward-firing Auto-cannon "Schräge Musik"

Following its May 1943 debut in action, during the second half of 1943, Schnaufer and his crew began experimenting with upward-firing auto-cannons, in German called "Schräge Musik". This allowed the night fighter to approach and attack the bombers from below – outside the enemy crew's usual field of view. An attack by a Schräge Musik equipped night fighter typically came as a complete surprise to the bomber crew, who realised a night fighter was close by only when they came under fire.

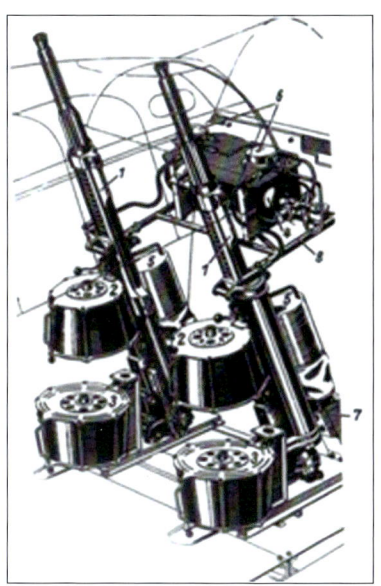

"Schräge Musik" Upward Facing Gun.
Image Source: acesflyinghighthesurvivors.wordpress.com

Rear view of a Bf 110G's rear cockpit glazing with Upward Firing MG FF/M Schräge Musik Gun.
Image Source: Wikipedia

It is not exactly known when Schnaufer's Bf 110 was equipped with Schräge Musik. Rumpelhardt stated that the weapons system was installed prior to his return from officer training. Reportedly Schnaufer's 20 to 30 victories were claimed using the upwards firing cannons. Later the Japanese also fitted such canons on their night fighters.

Knight's Cross of the Iron Cross for 42 victories

Rumpelhardt had returned from his officer training courses in early October 1943 and re-joined Schnaufer's crew. Gänsler, Rumpelhardt and Schnaufer claimed aerial victories 29 and 30 on 9 October. Schnaufer was awarded the Knight's Cross of the Iron Cross

Oberleutnant Heinz Wolfgang Schnaufer with his colleagues after 47th victory bar on the rudder of his Messerschmitt Me-110G at St. Trond in February 1944.
Image Source: albumwar2.com

German Night experts at St. Trond, early spring 1944. From left to right: Major Helmut Lent, bandmaster, Oberleutnant Heinz-Wolfgang Schnaufer, and Hauptmann Hans-Joachim Jabs. The picture is part of the collection of Martin Drewes, another Night Fighter.
Image Source: ritterkreuztraeger.blogspot.com

for 42 victories on 31 December 1943. On the night before his 22nd birthday, on 15 February 1944, Schnaufer and his crew claimed aerial victories 45 to 47. Bomber Command had sent 561 Lancaster and 314 Halifax four-engine bombers, supported by de Havilland Mosquito night-fighters and bombers, destined for Berlin. Schnaufer, who had been suffering from stomach pains all day, and his crew returned to Leeuwarden at 00:14. Rumpelhardt had been the first to congratulate him on his birthday over the intercom. Their fellow airmen had prepared a birthday celebration.

Appendicitis Operation: Break from Flying Operations

The stomach pains had become unbearable and Schnaufer was taken to a hospital with appendicitis. He stayed in the hospital for about two weeks before, together with Rumpelhardt, he went on vacation back home. Carelessly lifting his suitcase, he burst his stitches, resulting in further hospitalisation. He flew his first mission after these events on 19 March 1944.

Group Commander of the IV/NJG 1: Ace in a Day

Schnaufer was appointed Gruppenkommandeur IV/NJG 1 on 1 March 1944, taking over command from Jabs who was given command of NJG 1. He was promoted to Hauptmann (Captain) on 1 May 1944. Schnaufer became an Ace-in-a-day for the first time on 25 May 1944 when he claimed five RAF bombers shot down between 01:15 and 01:29 for victories 70 to 74. The bomber raid had targeted the railway marshalling yard at Aachen.

Luftwaffe Aces Heinz Wolfgang Schnaufer and Helmut Lent (110 aircraft) in 1944.
Image Source: asisbiz.com

Operation Overlord: Normandy Landing

On 6 June 1944, the Western Allied forces landed in Normandy, during Operation Overlord. In support of the invasion of Normandy General Dwight D. Eisenhower, the Supreme Allied Commander, assigned Bomber Command to support the ground forces. On the night of 12/13 June, Schnaufer claimed his first victory following the invasion when 671 bombers attacked various railway targets in France. Schnaufer claimed three bombers shot down that night, the first was a Lancaster and the second and third were a Lancaster or Halifax, between 00:27 and 00:34.

Knights Cross of the Iron Cross with Oak Leaves and Swords.
Image Source: kelsmilitary.com

Awarded Knight's Cross of the Iron Cross with Oak Leaves and Swords

Schnaufer was awarded the Knight's Cross of the Iron Cross with Oak Leaves on 24 June following four aerial victories claimed on 22 June, which took his total to 84 victories. For Schnaufer, July 1944 was less successful than the previous three months. He claimed two bombers on the night of 20/21 July and three on 28/29 July, taking his total to 89 aerial victories. One day later, on 30 July, he received a letter from Göring telling him that he had been awarded the Knight's Cross of the Iron Cross with Oak Leaves and Swords. Hitler himself made the presentation. It is said that when Hitler came to the presentation his first words were, "Where is the night fighter?" Shortly following the presentation, both Rumpelhardt and Gänsler received the Knight's Cross of the Iron Cross on 8 August. His crew was the only night fighter crew in the entire Luftwaffe of which all crew members wore this decoration.

Schnaufer (centre) with radioman Fritz Rumpelhardt and mechanic Wilhelm Gansler. (Courtesy Ken Wright).
Image Source: historynet.com

The airfield at St Trudien/Brusthem, Belgium, blanketed by bomb craters after attack by the Allied Air Forces.
Image Source: awm.gov.au

Unit Relocated Due Heavy Allied Bombing

In early September 1944, NJG 1 was forced to abandon its airfields in the Netherlands and Belgium. Continuous heavy attacks by RAF and USAAF bombers and strafing by Allied fighter-bombers rendered the airfields unsuitable for operations. On 2 September, VI/NJG 1 relocated from Saint-Trudien (Trond) to Dortmund-Brackel.

Knight's Cross of the Iron Cross with Oak Leaves, Swords and Diamonds

Schnaufer achieved his 100th victory on 9 October 1944, when he claimed two bombers shot down from an attack force of 415 bombers targeting Bochum. He was mentioned in the Wehrmachtbericht (the daily Wehrmacht High Command mass-media communiqué and a key component of Nazi propaganda during World War II) on 10 October 1944 and awarded the Knight's Cross of the Iron Cross with Oak Leaves, Swords, and Diamonds on 16 October 1944. He was the 94th Luftwaffe pilot to achieve the century mark.

Wing Commander of Nachtjagdgeschwader 4: The Youngest at 22

Schnaufer was then appointed Wing Commander of NJG 4 (4th Night Fighter Wing), at Gütersloh, on 20 November 1944. He was the youngest Wing Commander in the Luftwaffe at the age of 22. Schnaufer and his crew flew to Berlin on 27 November 1944 for the official presentation of the Diamonds to the Knight's Cross of the Iron Cross with Oak Leaves and Swords by Hitler. Schnaufer and his crew were filmed for the German newsreels. Three days later they returned to Gütersloh.

Friedrich Lang, Erich Hartmann and Heinz-Wolfgang Schnaufer receive the Oak Leaves with Swords, Horst Kaubisch, Eduard Skrzipek and Adolf Glunz the Oak Leaves to the Knight's Cross from Adolf Hitler.
Image source: warhistoryonline.com

Becomes Leading Night Fighter

Schnaufer became the leading night fighter pilot on 9 November 1944. Schnaufer surpassed Colonel Helmut Lent's record of 102 night-time victories after he claimed three Lancaster aircraft shot down from a force of 235 Lancaster aircraft from No 5. Group which attacked the Dortmund-Ems Canal. Schnaufer, whose victory total stood at 106 at the end of 1944, failed to shoot down a single bomber in January 1945. It was his first month without filing a claim since April 1943.

Offered Post of Inspector of the Night Fighter Force: Declines

Carinhall, the country residence of Hermann Göring.
Image Source: iltalehti.fi

Schnaufer was ordered to Carinhall, the residence of the Reichsmarschall Hermann Göring, on 8 February 1945. Göring informed him about the intent to appoint him as Inspector of the night fighter force, a role held by Colonel Werner Streib, a friend and mentor of Schnaufer. He politely excused himself by saying that he would better serve the German cause fighting the enemy. Göring was convinced and Schnaufer remained in active flying.

British Honorary Title "The Spook of St. Trond"

The British propaganda radio station (Soldiers' Radio Calais) congratulated Schnaufer on his 23rd birthday on 16 February 1945. The radio station explicitly addressed the soldiers of NJG 4 stationed in Gütersloh followed by the song "The Bogeyman" praising him for the honorary title given to him by the British bomber crew "The spook of St. Trond".

Second Time Ace in a Day – 9 Victories in a Night

Schnaufer's greatest one-night success and the second time he became an Ace-in-a-day was on 21 February 1945, when he claimed nine Lancaster heavy bombers in the course of one day. Two were claimed in the early hours of the morning and a further seven, in just 19 minutes, in the evening between 20:44 and 21:03.

Avro Lancaster heavy bomber.
Image Source: forces.net

Operation Gisela

Joseph Schmid, the architect of *Gisela*.
Image Source: Wikipedia

Schnaufer was one of the influential figures that instigated a brief return to mass intruder operations over England named Operation *Gisela*. General of Night Fighters, Lt. General Schmid, and de facto command-in-chief of the German Night Fighter Force until November 1943, had long since desired to return to intruder operations over Bomber Command bases in England. The proposals met resistance from General Staff. Eventually, in October 1944, Schmid won support to begin planning an operation. Schnaufer voiced his support also. In his experience, he had regularly pursued RAF bombers to the English coast, or at least the other side of the frontline. In British airspace, and over territory the Germans did not control, he experienced a lack of radar interference. Schnaufer recalled that he could fly around as if it was peace time, since all British jamming and interference stopped immediately once he was in Allied airspace.

Last Victories: Further Combat Flying Stopped

On 7/8 March, he claimed three RAF four-engine bombers for victories 119 to 121. These were his last victories of the war. He was then banned from further combat flying and was given the task of evaluating the new Dornier Do 335, a twin-engine heavy fighter with a unique "push-pull" layout, for its suitability as a night fighter. Disobeying his ban from combat flying, he flew his last mission of the war on 9 April 1945. Attempting to chase a Lancaster, he took off from Faßberg Air Base at 22:00 and landed after 79 minutes at 23:19 without success.

New Dornier Do 335, a twin-engine heavy fighter with a unique "push-pull" layout.
Image Source: militaryfactory.com

Prisoner of War

Schnaufer was taken prisoner of war by the British Army in Schleswig-Holstein in May 1945. He was reportedly taken to England for interrogation. The British authorities were especially interested in knowing whether his achievements had been made under the influence of methamphetamine or other stimulating psychoactive drugs which induce temporary improvements in either mental or physical functions or both, as had been documented in widespread Wehrmacht use and made for the German military by the Temmler-Werke GmbH firm, under the name Pervitin. Schnaufer was released later that year in November following a bout of diphtheria. Some others have said that as per Rumpelhardt's testimony, Schnaufer was never taken to England. Rumpelhardt was released on 4 August 1945 and soon after Schnaufer was admitted to a hospital in Flensburg, ill with a combination of diphtheria and scarlet fever. The interrogation had begun in late May 1945 by a team of twelve officers from the Department of Air Technical Intelligence (DAT), led by Air Commodore Roderick Aeneas Chisholm. The German prisoners were brought to Eggebek. Here they conducted a number of interviews with various members of the night fighter force.

Schnaufer House (Altdeutsches Haus), Lederstraße 39.
Image Source: calw.de

The Winery Business

Following his release from the hospital and as a prisoner of war, Schnaufer took over the family wine business. He had never planned to run the family winery as his ambition had always been to pursue an officer's career in the Luftwaffe. However, in the immediate aftermath of World War II the business had virtually ceased to exist and Schnaufer was given the task of rebuilding it from scratch. He had to re-establish business links to suppliers and customers and to consolidate them. Then he had to make new contacts in order to facilitate the expansion and growth of the business. Lastly, he had to create an infrastructure that supported the growth of the business. His motto for business was clearly "Quality before Quantity".

Failed Attempt to Go to South America

Soon the wine business began to prosper, Schnaufer also gave thought to alternative employment possibilities in peacetime aviation. With his wartime friend Hermann Greiner, he travelled from Weil am Rhein to Bern in Switzerland to meet South American diplomats; the two hoped to find employment as pilots in South America. To get to Bern, they crossed the Swiss-German border illegally. The meeting was a failure. As they attempted to make a second illegal border crossing to return to Germany they were caught by Swiss border guards. The Swiss handed them over to the French occupation authorities and they were imprisoned in Lörrach, where they remained until Schnaufer managed to make contact with a French general, who was a customer of the Schnaufer winery and had them released. This misadventure kept him away from his business for about half a year.

Dies After a Car Accident

In July 1950, Schnaufer was on a wine buying visit to France. On the afternoon of 13 July, he was heading south on the Route Nationale No. 10 in his Mercedes-Benz 170 convertible with a registration number "AWW 44-3425". Just south of Bordeaux, at about 18:30, he was involved in a collision with a Renault 22 truck. The accident occurred at the intersection of road D1, present-day D211, and the N10, present-day D1010, in Cestas (44°422 043N 0°422203W). The truck, driven by Jean Antoine Gasc, was carrying 6 tonnes (6.6 short tons) of empty gas cylinders. The collision ruptured the fuel tank of the Mercedes and ignited the petrol. Witnesses to the accident quickly put out the flames. Alice Ducourneau gave first aid to Schnaufer, who was bleeding from a wound from the back of his head. The police appeared at the scene of the accident at about 19:30, followed by an ambulance shortly thereafter. Schnaufer had suffered a fractured skull, and was immediately taken to the Saint-André Hôpital in Bordeaux.

Heinz Wolfgang Schnaufer is buried in the local cemetery of Calw, along the wall.
Image Source: ww2gravestone.com

Schnaufer never regained consciousness and succumbed to his injuries at the hospital two days later on 15 July 1950. The investigation into the accident concluded that though the impact of the two vehicles was severe, it seemed unlikely that the collision itself was the cause of his injuries. It was speculated that at least one of the truck's cargo of 30 empty gas cylinders, which were thrown off by the collision, had struck Schnaufer on the head. Subsequently, the truck driver was charged with manslaughter and breach of traffic regulations before a court at Jauge, Cestas. The hearing began on 29 July 1950 and concluded with his conviction on 16 November 1950. Gasc was found guilty of not yielding the right of way, and his speed was considered too high. It was ruled that as a consequence of not observing the law, he involuntarily caused the death of Schnaufer. Heinz Wolfgang Schnaufer is buried in the local cemetery of Calw, along the wall.

Summary of Aerial Victories

Heinz-Wolfgang Schnaufer was the top-scoring night fighter pilot of World War II. He was credited with 121 aerial victories claimed in just 164 combat missions. His victory total includes 114 RAF four-engine bombers; arguably accounting for more RAF casualties than any other Luftwaffe fighter pilot and becoming the third highest Luftwaffe claimant against the Western Allied Air Forces. His flight book indicated 2,300 sorties and 1,133 flying hours. Matthews and Foreman, authors of "Luftwaffe Aces—Biographies and Victory Claims", researched the German Federal Archives and found documentation for 119 nocturnal aerial victory claims, plus three further unconfirmed claims.

Image Source: alchetron.com

Until late 1944, Schnaufer documented his aerial victories with detailed geographical locations. After this date, he claimed his victories over the territory occupied by the Allies, and his victories were logged in a Planquadrat (grid reference). The grid map was composed of rectangles measuring 15 minutes of latitude by 30 minutes of longitude, an area of about 360 square miles (930 km²).

Post-War Recognition

Schnaufer's Messerschmitt Bf 110 G-4/U 8 was brought to England after the war. The aircraft was displayed in London's Hyde Park. The port-side vertical stabiliser of this twin-tailed aircraft, tallying all his victories, is preserved at the Imperial War Museum in London. A fin from another Bf 110 flown by Schnaufer is at the Australian War Memorial in Canberra. The street "Heinz-Schnaufer-Straße" in Calw was named after him.

One of the tail fins of Heinz-Wolfgang Schnaufer's Bf 110. It displays all of his 121 victories, Imperial War Museum (2010).
Image Source: wikiwand.com

Summary of Awards

- Front Flying Clasp of the Luftwaffe for Night Fighters in Gold
- Combined Pilots-Observation Badge
- Wound Badge in Black
- Iron Cross (1939)
- 2nd Class (2 June 1942)
- 1st Class (19 October 1942)
- Honour Goblet of the Luftwaffe on 26 July 1943
- German Cross in Gold on 16 August 1943.
- Knight's Cross of the Iron Cross with Oak Leaves, Swords, and Diamonds
- Knight's Cross on 31 December 1943
- 507th Oak Leaves on 24 June 1944
- 84th Swords on 30 July 1944
- 21st Diamonds on 16 October 1944

REFERENCES

1. Heinz-Wolfgang Schnaufer, Luftwaffe, http://www.luftwaffe.cz/schnaufer.html
2. Ken Wright, The Luftwaffe's Devil in the Dark, November 2009 issue of Aviation History, History Net, https://www.historynet.com/devil-in-the-dark.htm
3. Donald Greyfield, Heinz Wolfgang Schnaufer, Find A Grave, https://www.findagrave.com/memorial/7958647/heinz-wolfgang-schnaufer
4. Major Heinz Wolfgang Schnaufer, Germany's Top-Scoring Night Fighter of World War II, The loss of LANCASTER ME 683 EM-W 21/22 June 1944, 207 SQUADRON ROYAL AIR FORCE HISTORY, http://www.207squadron.rafinfo.org.uk/wesseling/meeuwen_schnaufer.htm
5. Schnaufer, Ace of Diamonds: The Biography of Heinz Wolfgang Schnaufer – Germany's Top-Scoring Night Fighter of World War II by Peter Hinchliffe.
6. Heinz-Wolfgang Schnaufer, Wikipedia https://en.wikipedia.org/wiki/Heinz-Wolfgang_Schnaufer

17
Israel's Giora "Hawkeye" Epstein
Ace of Aces of Supersonic Fighter Jets

Giora "Hawkeye" Epstein.
Picture Source: iaforgil

Brigadier General Giora "Hawkeye" Epstein (also called Giora Even) of the Israeli Air Force (IAF), is a fighter ace credited with 17 victories, 16 against Egyptian jets, making Epstein the "Ace of Aces" of supersonic fighter jets. Epstein was an active IAF pilot from 1956 until May 26, 1997, when he retired at age 59. Like many retired IAF flyers, he later worked as a pilot for El Al Airlines.

Early Years

Born on 20 May 1938, in Negba, British-mandate Palestine, he grew up in the traditional frontier-farmer lifestyle of his home, Kibbutz Negba. Born Giora Epstein in the Holy Land in the decade before Israel won her independence. An avid aviation enthusiast as a young boy, he poured over flying books and studied the biographies of Britain's wartime aces.

Struggles to Become a Pilot Due Medical State

After the 1956 Suez War, the 18 year old joined the Israeli Defence Force (IDF) and applied for flight training but was rejected for medical reasons. Enraged, he went to his commander and told him that if he could not be a pilot, he wanted to be a paratrooper. He served in the Efah Para Battalion (890th). He eventually made over 700 jumps and was a para instructor and a member of the IDF's parachute demonstration team. Before travelling outside Israel to the world parachute championships, he changed his name in accordance with IDF regulations. (Israel's Names Law established that Israeli citizens must adopt Hebrew names). He chose the last name "Even" ("stone" in Hebrew). Epstein left active duty in 1959, but life on the kibbutz left him restless and unfulfilled. He re-joined active service in late 1961 and reapplied for flight training, trying again to attain his dream of becoming a fighter pilot. The doctors determined that he had a "sportsman's heart," a condition resulting in unusual indications during examinations, but not a medically disqualifying factor for flying.

Negba.
Image Source: Pinterest (thinglink.com)

Israeli Paratroopers Brigade Insignia.
Image Source: Wikipedia

Pushes Case for Fighter Flying

He cleared the first hurdle and entered flight school, but he was not yet cleared for fighters. Although he was at the top of his class in 1963, the doctor had apparently allowed him to enter flight training only on the condition that he fly helicopters. So he first served as an IAF helicopter pilot. After he filed another appeal to the IAF, he was assigned to a combat helicopter pilot position. Epstein was determined though, and while flying helicopters, he sent his medical records to an expert USAF cardiologist who confirmed that his medical condition should not keep him from flying fighter aircraft. Armed with this recommendation, he took his complaint all the way up the IAF chain of command until he ended up sitting in the outer office of Ezer Weizman, legendary IAF commander, later seventh President of Israel. Even put his career on the line, stating emphatically, "I'll sit in your reception hall

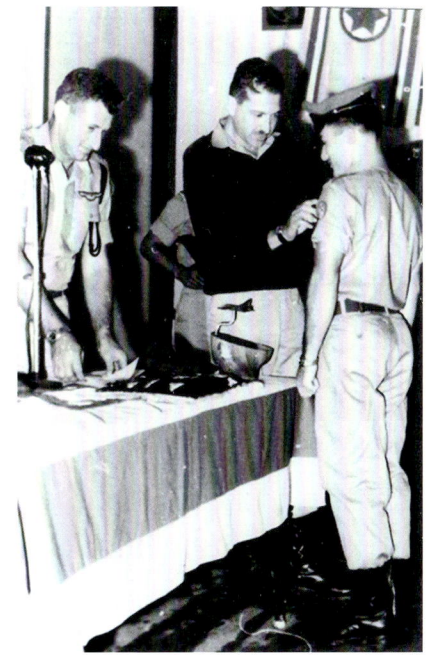

Receiving his wings from Ezer Weizman, legendary IAF commander, later seventh President of Israel.
Image source: Giora Epstein

until you send me to fighter training." It worked. After deliberating all night, Weizman called him and said, "You SOB, get your things and go to the fighter OTU. I don't want to hear another word." After seven years and countless obstacles, he was finally on his way to becoming a fighter pilot! Because of his stubbornness, Epstein received the necessary medical clearance and began training to become a fighter pilot. This stubbornness would not only prove to be good for the pilot, but also for the nation he served.

Begins Fighter Flying

Beginning 1963, he began to fly frontline fighters, including Dassault *Ouragons*, *Super Mysteres*, and *Mirage IIIs*, Israeli Aircraft Industries (IAI) *Neshers* and *Kfirs*; and General Dynamics F-16 *Fighting Falcons*. He soon gained the nickname "Hawkeye" due to his extraordinary eyesight. Epstein was allegedly able to spot aircraft at the distance of 24 miles (44 km) – nearly three times further than a normal pilot.

Israel Aircraft Industries Nesher "griffon vulture" the Israeli version of the French Dassault Mirage 5 multirole fighter.
Image: Wikipedia

Six-Day War – 1967

Epstein's first kill came on June 6, 1967, during the Six-Day War, when he downed an Egyptian Sukhoi-7 at El Arish. During the War of Attrition in 1969–70, Epstein served as deputy commander of the air force squadron, and downed a MiG-17, another Sukhoi-7, and two MiG 21s.

Recalls Beginning of Yom Kippur War – 1973

In an interview with Yaakov Lappin of Jewish News Syndicate on 08 October 2018, Epstein recalled that when the Yom Kippur War erupted in October of 1973, Epstein was an officer serving at the Israeli Air Force's headquarters in Tel Aviv, in the Operations Department. On Friday, the day before the war began on 06 October, he travelled south to Refidim air base in Israeli-controlled Sinai to be part of an emergency standby team for his 101 Squadron of Mirage jets. While at Refidim, Epstein began receiving phone calls from the IAF's headquarters in Tel Aviv, as it became clear that the war was drawing closer. Headquarters ordered Epstein to return, viewing his work in the IAF's Photography Branch, where he planned intelligence-gathering flights, as being of crucial importance.

Egyptian MiG-17s in Nile Valley camouflage.
Image Source: edokunscalemodelingpage.blogspot.com

Epstein returned to Tel Aviv, and the next day, began issuing orders for aerial photography flights over the Suez Canal. "Then, at around five minutes to 3 in the afternoon, an intelligence person tells me that the whole of the Egyptian and Syrian air forces are in the air, and that they are on the way to Israel," he recalled. "I immediately called the bureau of the IAF commander. He was in a commanders' meeting. They headed underground to the IAF's command room immediately and took control. I was appointed

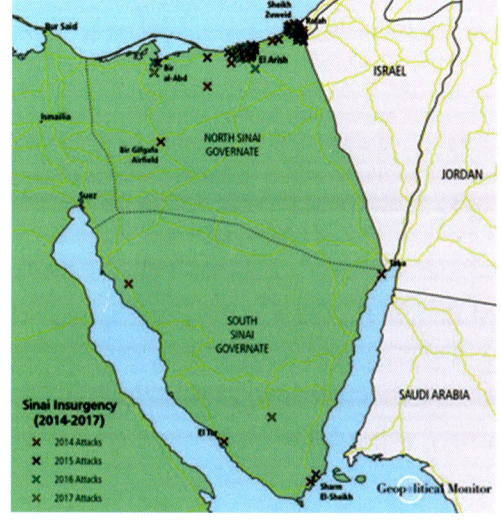
Bir Gifgafa Airfield (Refidim air base in Israeli-controlled Sinai).
Image Source: geopoliticalmonitor.com

to coordinate air-to-ground attacks, even though it wasn't my job. But at the time, nobody else was responsible for that."

Epstein was sent to the underground control room of the Artillery Corps, where he spent two days overseeing air strikes on advancing enemy forces. He then headed back to the Photography Branch, spending days in war planning and operations. On a few occasions, he also took to the air, joining his squadron for air patrols and a night flight. But after 10 days of raging conflict, Epstein had decided that the time had come to fully join the air battles that were raging between the Israeli and Egyptian air forces.

Seeks Permission to Leave HQs and Join War

Epstein in the Dassault Super Mystere Cockpit.
Image Source: Giora Epstein

"I went to my commanders and said I had to go to my squadron because I feel that I am not taking part in the war," Epstein recalled. By that time, the IAF's headquarters had filled up with enough people to operate at full capacity, and Epstein was given the green light to join his squadron.

In the war's first days, Egyptian forces advanced beyond the Bar-Lev Line and as far as nine miles into the Sinai. Then they were stopped.
Image Source: historynet.com

First Aerial Kill of Yom Kippur War: Mi-8 Helicopter

On 18 October, Epstein was in the air, patrolling the Suez Canal and about to get his first taste of the action. "We called this patrol 'turning off the sun' because it was the last patrol of the day. We would fly until the last light of day and then land, allowing Phantom planes to continue the patrols because they had radar. We in the Mirage jets did not have radars," he explained. "When it got really dark, the air controller told us to land at Refidim (front-line airbase in Sinai desert)." But suddenly, the controller came back on the radio, telling them to abort the landing and head straight back to the Canal, where the Egyptian air assault was underway. The IDF ground forces were preparing to cross the Suez Canal into Egypt, and the Egyptian planes had come to bomb them. Epstein flew into the area and spotted a napalm bomb explosion on the ground. "Against the background of the blast, I saw the silhouette of an Egyptian Mi-8 helicopter, with its back door open and the crew rolling off napalm barrels onto our forces," he said. "I did the tightest turn I could, placed the helicopter in front of my gun sight, and released a burst of cannon fire. The helicopter entered the burst and exploded." Epstein had just completed his first shoot-down of the war.

Mirage III CJ of 101 squadron.
Image Source: alchetron.com

First two Su-7 Kills and Wall of Fire

Egyptian Su 7.
Image Source: users.skynet.be

Epstein was mostly assigned the "filling holes" sorties. That meant that his mission was to fill in for planes that were on patrol, and then became engaged in air battles, before using up their fuel and having to land. On 19th October, Epstein and a second pilot were scrambled to the Suez Canal, after being told that an Egyptian attack was forming and a hole had formed in the Israeli patrol. "We sped to the Canal and when I got there, I saw to the west a convoy of four pairs of Sukhoi – 7 Egyptian planes, which were arriving to attack. We turned hard and went after them," he said. "I did not manage to catch the first pair, but I caught the second pair. As they were diving towards our ground forces on the canal, I fired a missile at the rear plane and struck it. It continued to dive straight into the canal, crashing into it. Then I turned and flew after the second plane that was fleeing west to Egypt. It released munitions west of Canal and escaped. I pursued it." Epstein was attempting a high-speed low-altitude pursuit, but he knew he was approaching a known trouble spot. "We called this area 'the wall of fire.' Because every Egyptian there with whatever weapon they had would raise their weapons and fire in the air, so it was very dangerous to cross that line. I said to myself, 'I'll pursue up to that line. If can reach him, OK, if not, then no.'" As they approached the "fire wall," Epstein fired a small burst from his cannon that exploded in front of the Egyptian plane. "This startled him, and he turned. As soon as he turned, I had a target, fired a burst, struck the plane, and he crashed into the ground. I returned to land at Refidim," he said.

Two More Kills

Hours later, Epstein scrambled to join a formation of Israeli Phantoms that were patrolling the Suez Canal. They spotted a convoy of Egyptian planes, attacking from the west, and Epstein engaged. He hit one enemy plane with a missile, causing it to blow up in mid-air. The Egyptian pilot ejected just before being hit. "He descended over the battle area, waving his arms and legs. Of course, none of us fired at him," said Epstein. He continued onwards, destroying another aircraft with a machine-gun burst.

When Epstein Fought off 11 Enemy Jets Alone

Epstein recalls that on 20th October, he was back in the air, once again on a "filling holes" mission. It was a four-aircraft formation. "We took off, flying towards the Canal. We had not yet been assigned a target," he said. "When we reached the Canal, the controller told us to fly at 20,000 feet and search for adversaries. I reported seeing nothing. The controller told me to look south, and sure enough, I saw a pair of MiG-21s coming from south to north, at our attitude. When they reached us, they turned west towards Egypt. We pursued and closed in on them, fast. I fired a missile at the rear plane, struck it, and it exploded. Then I went after the second plane, which was manoeuvring to avoid being shot down."

Painting of Air Action of the Great Israeli fighter pilot – Ace of Aces – Giora "Hawkeye" Epstein.
Image Source: onjewishmatters.com

Logan Nye of "We are the Mighty", wrote about Epstein's this encounter. The Israeli Air Force's Nesher was a highly-capable delta-wing fighter based on the French Mirage. Israeli Col. Giora Epstein, one of the world's greatest fighter aces of the jet era, was leading a flight of four planes during the Yom Kippur War when his team spotted two Egyptian MiG-21s. Epstein pursued the pair and quickly shot down the trail plane. But the Israelis got a surprise. The pair of MiG-21s were a bait. While the four Israeli planes were pursuing the surviving MiG they could see, approximately 20 more MiG-21s suddenly ambushed them. What followed was a very classic one sided victories of modern aerial combat. The four Israeli Neshers fought the approximately 21 MiGs, coordinating between them for tail clearance from Egyptian MiGs and identifying vulnerable aircraft to chase and shoot.

Epstein after 13 Kills.
Image Source: jewish.ru

Unfortunately for the Egyptian pilots, the MiG-21s flew at them from a low altitude at low speed, meaning that when they reached the Israelis' altitude, they "did not have a big advantage," stated Epstein. A major air battle had, however, begun. He recalled later.

Every plane in the Israeli formation began fighting, firing missiles and machine-gun bursts at the Egyptian aircraft. During the fight, Epstein's No.2 shot down a MiG but the missile firing resulted in his aircraft engine flame-out. Epstein talked him through a restart and sent him home. His No.3 shot a MiG but was running out of fuel and went back to base to land. The No. 4 pursued a plane and brought it down, but moved away from the main battle.

Epstein was left alone in the fight with 10 MiG-21s, and out for vengeance for their lost comrades. Epstein's immediate target pulled off a manoeuvre thought impossible in a MiG-21, a split S at approximately 3,000 feet. Normally the MiG should have crashed but it barely survived. It was so close to the ground that it raised the sand of the desert. "At this stage, I'm pursuing a plane, and he is manoeuvring like crazy to avoid getting shot down. He reached a low altitude and lost speed. When he finished his manoeuvre, I struck him with a (gun) burst, and the plane exploded," he said. The MiGs engaged in pairs against Epstein, firing bursts of machine-gun fire and missiles. Epstein outmanoeuvred them, killing two with 30mm cannon fire and forcing the rest to exit combat. The entire battle lasted 10 minutes. "I finally ran out of fuel so went to land at Refidim," said Epstein.

Egyptian MiG 21.
Image Source: idfmod.wixsite.com

Effectively, Epstein had shot down the lead MiG of the decoy pair, then managed to outduel the other five pairs of MiG-21s shooting down two of the Fishbeds. When he returned to base, having scored four kills that day, ground crew lifted him from the plane. Within 40 hours, he had destroyed a total of nine enemy aircraft.

What about Us? We want to join the Battle

Three days later on 23 October, Epstein was flying in a four-aircraft formation. They were replacing another patrol that was running low on fuel. As the planes that completed their patrol moved towards their base to land, the controller suddenly shouted out, "negative on going home," and directed them to fly towards an area where enemy activity had been detected. "They turned towards the battle. Over the radio, Epstein asked, 'What about us?' The controller said, 'You go on patrol.' Epstein said, 'You go on patrol. I'm going to the battle. Where is the battle?' He didn't want to tell us, but we knew where it was. As we approached, we saw blasts in the sky and aircraft crashing." Epstein had flown into the largest battle of the Yom Kippur War. Ten Israeli Mirage jets were up against 20 Egyptian MiG-21s. Once again, the Egyptians had planned an ambush.

Aviation Art. Mirage IIIC of Giora Epstein, the highest scoring ace in the Israeli Air Force. Commissioned by Rick Turner. *Image Source: aviation-art.net*

"They were very frustrated that we shot them down all of the time," said Epstein. During that battle, Epstein shot down three aircraft out of that sky. In total, the Israelis destroyed 11 Egyptian planes in the battle. On his way back to the landing, Epstein's base commander called in. "He received the details of the battle, congratulated us, and then informed us that the war had ended," recounted Epstein. "My feeling to this day hasn't changed. We did our job in the best way we could." Epstein bagged three more Fishbeds, giving him 11 fighter kills in less than a week. What a badass pilot said his colleagues.

Epstein sketch on the left, with the 101 pilots in the Squadron boys. *Image Source: Giora Epstein*

Summary of 1973 Yom Kippur War

Already an Ace with 5 kills from previous conflicts, on 18-20 October 1973, Epstein skilfully flew his 101 Squadron Nesher to eight victories over Egyptian aircraft in 26 hours. He downed a Mi-8 helicopter and eight jets: two Sukhoi-7s, two Sukhoi-20s and four MiG 21s. Then, on October 24, 1973, Epstein downed three more MiG-21s west of the Great Bitter Lake. Eight of these victories were with the French-built Mirage III, a delta wing fighter designed primarily as a high altitude interceptor. His other nine victories came in an IAI Nesher, an Israeli-built version of the Mirage 5. Five of his kills were downed using air-to-air missiles, the rest with cannon. Through this incredible accomplishment, he achieved legendary status in the IAF and became the top scoring Israeli ace of all time.

Medal of Distinguished Service – Israel. *Image Source: emedals.com*

Post War Career and Air Flying Summary

After the Yom Kippur War, Epstein received the Medal of Distinguished Service, one among Israel's high military honours. In 1974, he was appointed commander of the IAF 117th Squadron, also known as the First Jet Squadron. He went on to command Mirage and Kfir squadrons and was flying "ready" missions in the F-16 up until his 59th birthday. In 1977 Epstein retired from permanent service. As a reserve officer, he commanded the IAF's 254th Squadron and after he completed a retraining process he became an F-16 pilot. He accumulated over 5,000 hours of fighter flying and scored 17 victories making him Israel's "Ace of Aces". During the 1973 Yom Kippur War, he shot down 12 Egyptian aircraft, eight of them in 26 hours! Giora Epstein effectively shot down more enemy aircraft than any other pilot since World War II.

Ace of the Aces of supersonic jets, Colonel Giora Epstein in the cockpit of his F-16 decorated with his 17 kill marks, about to embark on his final flight as an active fighter pilot. 1998.
Image Source: 9gag.com/tag/giora-epstein/fresh

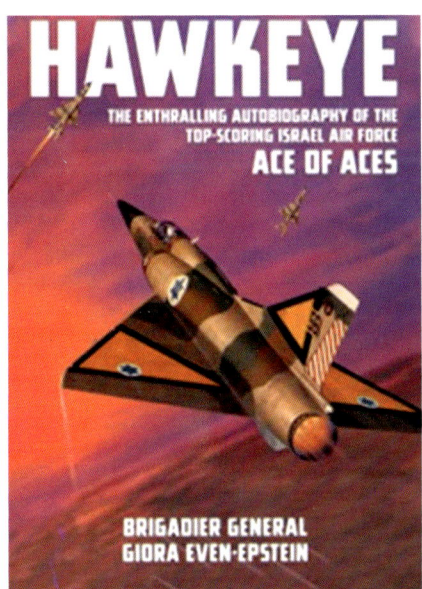

Hawkeye: The Enthralling Autobiography of the Top-Scoring Israel Air Force Ace of Aces by Giora Even-Epstein
Image Source: casematepublishers.com

Professional Modesty

Yaakov Lappin did an interview with Brig.-Gen (Retd) Giora Epstein on October 8, 2018, and talked of his battles with the Egyptian air force in the Yom Kippur War. Epstein, who took part in many life-or-death dog fights displayed remarkable skill and cool-headedness. But when asked, he very modestly said that it was simply the results of his training and of sticking to his mission, and that he was merely doing his job. "It does not give me any sense of elevation. I simply did what I had to do and took advantage of every situation that I had," Epstein said matter-of-factly.

Feelings as a Fighter Pilot

"We were not thinking about grand ideas, like defending the country during battles. The feeling during a battle of a fighter pilot is about how to conduct this battle in the best way. You're not thinking about spiritual stories or anything else. You're focused on what you're doing, on how to manage to take care of yourself in the best way and hit the enemy as much as possible. Those are the two things. Beyond that, there are no thoughts. It's about doing all of the things you trained for and knowing how to do them in the best way you know how."

Post Air Force Career

Epstein was a Captain for El Al Airlines from 1977 to 1997. He had sought, but was not allowed to participate in combat in the Bekaa Valley Turkey Shoot of June, 1982. He added another 4,000 flying hours post-retirement making an impressive total of 9,000. Epstein was the primary subject of the "Desert Aces," an episode of The History Channel series "Dogfights'. The episode was first aired on August 10, 2007.

Giora Epstein.
Image Source: iaf.org.il

Honour at 80: Brigadier General Rank

IDF Chief of General Staff Lt. Gen. Gadi Eisenkot and Israel Air Force (IAF) Commander Maj. Gen. Amikam Norkin awarded the rank of Brigadier-General to Giora "Hawkeye" Epstein, 80, former IAF pilot and the Ace of Aces of supersonic jets, in September 2018. Yoav Zitun wrote about the ceremony for Y Net News. The ceremony was held at the IAF headquarters in the presence of air force veterans, Epstein's family members, and other distinguished guests. Chief of Staff Eisenkot praised Epstein at the ceremony. "This modest ceremony is an expression of our great appreciation to you personally, and to the generation of founders who developed the extraordinary abilities of the IDF and the State of Israel," Eisenkot said. "The legacy that is passed over to pilots and air force commanders is a legacy of tremendous commitment, of winning every battle, of completing the mission, and of excellence as a way of life," he continued. "The spirit of man and the spirit of the fighter who carries the weight of the battle on his shoulders characterizes you, Giora, as a fighter pilot. I wish you good health, success, and wish us all a happy new year," Eisenkot concluded.

With Douglas Bader.
Image Source: Giora Epstein

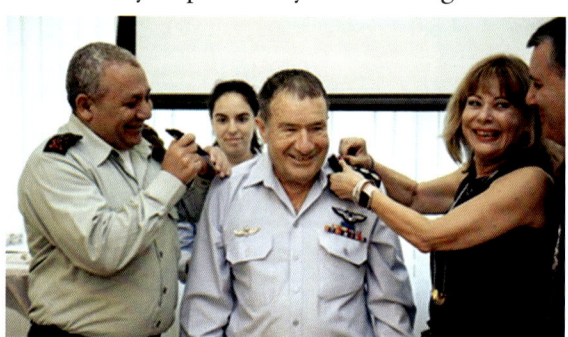

Epstein receiving his Brigadier-General Rank (Photo: IDF Spokesperson's Unit).
Image Source: ynetnews.com

"This rank is a clear statement. We appreciate those who fought for our homeland, we are aware of the size of your contribution throughout Israel's wars and we are proud to have a world champion in our midst," Israel Air Force commander Maj. Gen. Amikam Norkin said at the ceremony. " I am proud to be the commander of the air force, who awards the Brigadier General rank to a world champion fighter pilot," Norkin concluded.

Israeli Flying Aces

There are a total of 14 Israeli Air Aces. The next highest after Epstein were Abraham Salmon with 14.5, Amir Nachumi with 14, Asher Snir 13.5, Israel Baharav and Yiftah Spector with 12.

Epstein with Neil Armstrong.
Image Source: Giora Epstein

REFERENCES

1. Yaakov Lappin did an interview with Brig.-Gen (retd) Giora Epstein, Jewish News Syndicate, on October 8, 2018, https://www.jns.org/pilot-giora-epstein-1973-then-i-was-told-the-egyptian-and-syrian-air-forces-are-on-the-way/
2. Logan Nye of "We are the Mighty", wrote about Epstein's this encounter. https://www.businessinsider.com/israeli-ace-pilot-egypt-dogfight-2017-1?IR=T
3. Dog Fights (TV Series) https://en.wikipedia.org/wiki/Dogfights_(TV_series)
4. Yoav Zitun, IDF promotes 80-year-old world's fighter aces champion, September 13, 2018. ynetnews.com. https://www.ynetnews.com/articles/0,7340,L-5349290,00.html
5. Giora Epstein, Wikipedia, https://en.wikipedia.org/wiki/Giora_Epstein

Living Legend Giora "Hawkeye" Epstein's Message

I am very happy to hear that an Air Marshal of the Indian Air Force is writing a book on air aces. There are many books written about individual air aces, but it will be really unique to read a book covering 25 air aces, spanning many wars and many regions. I am looking forward to reading it. I recall with pleasure having met most of the then living Aces in 1976 at the Le Bourget airshow in Paris.

The times have changed since I engaged in air combat. The airplanes, the systems, the radar, and the missiles today will prevent any close dogfight of the kind we engaged in. Modern day engagements in the air will be finished even before the pilot sees the adversary. Therefore, there is a little air combat advice I can give to the current and future fighter pilots. The Air Marshal has very nicely recounted some of my engagements. For more details, the readers can read my book "HAWKEYE" and also see the "Dogfights of Desert Aces" on History Channel.

Wishing all the aviators, safe flying and happy landings, and combat aviators success in their missions.

—**Brigadier General Giora "Hawkeye" Epstein**

18

Iranian Air Ace Jalil Zandi

Highest-Scoring F-14 Tomcat Pilot

Image Source: Wikimedia Commons

Brig. General Jalil Zandi was a fighter pilot in the Islamic Republic of Iran Air Force who served during all of the Iran–Iraq War. His combat record qualifies him as one of the most successful pilot of that conflict in air-to-air combat, as well as one of the best Iranian Aces ever. It also made him the highest-scoring pilot in the history of the F-14 Tomcat.

Service Career

Born on 2 May 1951 at Garmsar in Iran, (Garmsar is located about 95 kilometres southeast of Tehran. It lies on the edge of Dasht-e Kavir, Iran's largest desert) Jalil Zandi began in the Imperial Iranian Air Force (IIAF) and stayed on to serve in the Islamic Republic of Iran Air Force (IRIAF) when it was somewhat dangerous for pilots to continue their military service. While a Major, he often clashed with his superior Lt. Col. Abbas Babaei. Abbas Babaei was "notorious for his merciless treatment of the pilots and officers considered disloyal to the new regime.

Image Source: military.wikia.org

Iranian F-14s.
Image Source: erospaceweb.org

Iranian Purchase of F-14s

Back in the early 70s, the pro-western Shah of Iran had a big problem with the Soviets flying reconnaissance flights over his airspace, so he sought from the USA some aircraft capable of engaging the MiG-25 Foxbat. The U.S. offered F-14 Tomcats and F-15 Eagles. The Shah bought two billion dollars' worth of F-14s and missiles and the other stuff that goes with it. Iran trained its best pilots, engineers and technicians with the US Navy personnel in the operation and maintenance of the aircraft. By 1979, Iran had almost 80 F-14s and over 120 qualified pilots to fly them.

Ouster of the Shah: Jalil Put Behind Bars

Iranian Air Force Major Jalil Zandi was a daredevil fighter pilot and in awesome love of his aviator sunglasses. Trained with the US Navy, he was pro-American on many counts, so naturally, when the Ayatollah came to power and declared a new ultra-conservative Islamic republic, Jalil Zandi was one of the first guys arrested at gunpoint in the middle of the night, thrown in jail, and sentenced to ten years in prison for disloyalty to the new regime. Loyalty outweighed the benefits of having a highly-trained F-14 Tomcat pilot. Many of the F-14 pilots and technicians who had trained in the US were arrested or killed, but not before some of the techs sabotaged the aircraft and missiles on their way out. When he was in prison, he was threatened to be sentenced to death, but by demand of the then air force commander and many other air force pilots, he was released after six months. It was also because in 1980, when Saddam Hussein's Iraqi army launched a full-scale invasion of Iran, the Ayatollah had to change his mind and released them from prison to save their country.

Mohammad Reza Pahlavi, Shah of Iran, and Ayatollah Khomeini.
Image Source: middleeasteye.net

Eight Years of Action: Many Times Single Handed

For the next eight years, Jalil Zandi was in constant combat against the MiGs and Mirages of the Iraqi Air Force. He was flying the ever-shrinking operational force of F-14 Tomcat fighters, as the United States refused to ship replacement parts, missiles, or weapons for the Tomcat. By 1986 the operational strength of the Iranian Tomcat fighter wing had dropped from 80 to 25. Jalil often had to engage much larger swarms of enemy fighters that vastly outnumbered him. He was tasked primarily with defending Iranian oil fields. He sometimes had to engage nearly the entire Iraqi squadron, but he brought down many enemy MiGs without mercy.

Iraqi MiG-23.
Image Source: theaviationgeekclub.com

The Early Aerial Victories

In one of his first engagements, Jalil attacked a pair of MiG-23s on his own, nailing one with a long-range Phoenix missile and the other with a Sidewinder. In another engagement he escorted a massive Iranian oil tanker plane on a dangerous run deep through Iraqi airspace and found himself engaged with a squadron of MiGs. The plane he was escorting survived.

The Final Engagement

Jalil's final fight took place in October of 1988 when he found himself going up against eight French-built Mirage F1 fighters. Zandi fought hard, scoring two unconfirmed kills in the engagement, but eventually was shot. He somehow managed to limp the aircraft across the Iranian border, then ejected when his second engine also shut down. For the rest of the war he was running things from the ground.

Iraqi Air Force (IrAF) Sukhoi Su-22M-4 observed at Saddam International (BGW / ORBI) in April 1989.
Photo by T. Laurent. Source: airliners.net

Iran–Iraq War: Aerial Victories

The Iran–Iraq War was an armed conflict between the Islamic Republic of Iran and the Republic of Iraq lasting from September 1980 to August 1988. He earned his fame as an F-14 Tomcat pilot during the Iran–Iraq War. He has been reliably credited with shooting down 11 Iraqi aircraft. Eight victories have been confirmed through examination with US intelligence documents released according to Freedom of Information Act (FOIA) inquiry and three probable victories. The victories include four MiG-23s, two Su-22s, two MiG-21s, and three Mirage F1s. This makes him one of the most successful Iranian fighter ace ever, & the most successful F-14 Tomcat pilot worldwide.

Counting air-to-air kills has always been a complex exercise, and it was much worse in the case of a war-like the Iran-Iraq War where each side was running an intense propaganda machine. Iran officially credits Major Jalil Zandi with destroying eight enemy aircraft, with three more listed as "probable" to bring his total to 11. Iraq claims he only shot down three, which, incidentally, would still be enough to tie the number of kills earned by the United States' top-scoring fighter pilot of the supersonic age. What was great was that Jalil survived eight years of non-stop war, and flew the F-14 despite limited repair facilities. Undoubtedly, Jalil was the most successful pilot of the war in air combats, with 11 aerial victories. The most successful F-14 Tomcat pilot worldwide.

Iraqi Mirage F1.
Image Source: flying-tigers.co.uk

Jalil Zandi.
Image Source: badassoftheweek.com

Jalil's Personal Impressions

In a story about Jalil in the Badass of the week Jalil has reportedly said, "The MiG-23 was not the fighter the Iraqis had hoped for. It could not outmanoeuvre any of our fighters and we have had very little respect for them on a one-to-one basis. We were concerned only when facing large numbers of Iraqi MiG-23s, later during the war. The most impressive thing about the MiG-23 was its ability to rapidly accelerate when we chased them—but it could not outrun an F-14."

The Other Aces of the War

There were two more Iranian Air Aces of the war. Shahram Rostami scored six victories flying F 14 Tomcat, and Yadollah Sharifirad who scored five victories flying the Northrop F-5. There was only one Iraqi Air Ace Colonel Mohammed Rayyan. Nicknamed "Sky Falcon," who scored five air combat kills, making him the most successful Iraqi fighter pilot of that war. Rayyan, while only a Flight Lieutenant, and flying a MiG-21MF, claimed two (later confirmed) kills against Iranian F-5 Freedom Fighters in 1980. Later he qualified on the MiG-25P in 1981 and claimed 5 more victories (3 verified by western sources.) Most of his victories were F-4 Phantoms. In 1986, having attained the rank of Colonel, Rayyan was shot down and killed by IRIAF Grumman F-14 Tomcats. His 5 to 8 air combat victories make Rayyan the most successful MiG-25 fighter pilot ever. Another notable Iranian pilot, though not an Ace, was Major Rahnavard, who on 16 February 1982 is reputed to have shot down four Iraqi fighter jets in two separate encounters over Kharg Island. Records indicate that two of his confirmed kills were Mirage F1s.

Iranian F-14 Air Ace Shahram Rostami.
Image Source: Wikipedia

Iraqi Colonel Mohommed Rayyan (died 1986), nicknamed "Sky Falcon", the most successful MiG 25 Pilot ever.
Image Source Wikipedia

Book on Iranian Tomcat Units in Combat

In 2004, Tom Cooper published "Iranian F-14 Tomcat Units in Combat", based mainly on primary interviews with Iranian pilots. The book makes many claims that contradict previous reports. In particular, Cooper claims that Iran's F-14s had up to 159 kills, and that in one incident, four Iraqi aircraft were shot down with one AIM-54 (The missile's warhead exploded between them and severely damaged them). By 1980, with the prospect of war with Iraq becoming ever more likely, most of the 77 surviving F-14 airframes were found to be in non-operational condition, or at least had non-functioning radars. As a result, F-14 pilots were forced to rely on ground control for their first wartime patrol and intercept missions. Within a few days of the start of the war, a dozen or so F-14s were made operational. Tom Cooper, wrote in the "Persian Cats", Smithsonian Air & Space.

The first confirmed kill by an F-14A during the Iran–Iraq War occurred before the formal start of hostilities on 7 September 1980, an IRIAF F-14A destroyed an Iraqi Mil Mi-25 (export variant of the Mil Mi-24) Hind helicopter using

Iranian Tomcat Pilots. Jalil's face is circled.
Image Source: badassoftheweek.com

its 20mm Vulcan cannon. Six days later, Major Mohammad-Reza Attaie shot down an Iraqi MiG-21 with an AIM-54 Phoenix while flying a border patrol. A single AIM-54 fired in July 1982 by Captain Hashemi may have destroyed two Iraqi MiG-23s flying in close formation.

F-14 Spares and Modifications

Cooper claims the AIM-54s were used only sporadically during the start of the war, most likely because of a shortage of qualified radar intercept officers, and then more frequently in 1981 and 1982—until the lack of thermal batteries suspended the missiles' use in 1986. There were also rumours that suggested that Iran's Tomcat fleet would be upgraded with avionics derived from the MiG-31 "Foxhound". However, IRIAF officials and pilots insist that the Soviets were never allowed near the F-14s, and never received any F-14 or AIM-54 technology. Also, the AIM-54 missile was never out of service in the IRIAF, though the stocks of operational missiles were low at times. Clandestine deliveries from US sources and black-market purchases supplied spares to top up the Phoenix reserves during the war, and spares deliveries from the US in the 1990s have also helped. Furthermore, an attempt was made to adapt the MIM-23 Hawk surface-to-air missiles that were also a carry-over from the pre-revolution period, to be used as air-to-air missiles for the F-14. At least two F-14s had been successfully modified to carry the hybrid weaponry.

Major Mohammad-Reza Attaie.
Image Source: airspacemag.com

Tomcats with AIM-54A missiles.
Image Source: badassoftheweek.com

AIM-54A Targets

All in all, the IRIAF was said to have launched possibly 70 to 90 AIM-54A missiles, and 60–70 of those scored. Of those, almost 90 percent of the AIM-54A missiles fired were used against Iraqi fighters and fighter-bombers. Only about a dozen victories by AIM-54s were claimed to be against fast, high-flying targets such as the MiG-25 or Tu-22 'Blinder'.

Spares Shortage and Depleting F-14s

By the close of the war, both sides were unable to obtain new aircraft or parts, and aerial combat had become rare, since neither side could afford to lose aircraft they could not replace. In particular, the IRIAF F-14 fleet suffered from a lack of trained technicians,

Tomcat F-14 with its weapons.
Image Source: nationalinterest.org

and by 1984 only 40 F-14s were still in service. By 1986, that number had dropped to just 25. The F-14 was relegated to protecting Iran's vital oil refining and export infrastructure; in this role, they often encountered French-built, Iraqi Dassault Mirage F1EQ fighters attempting to attack Iranian oil pipelines.

A Tomcat Shoots Three MiG 23s with One Missile

Blake Stilwell writes in the "Mighty History" a story about "That time an F-14 killed three MiGs with a single missile." A lot of crazy things happened in the Iran-Iraq War. The backbone of the Iranian Air Force at the time was the beloved F-14 Tomcat, a plane the Iranians still fly. It gave them an edge in the air war against Iraq. But technology can only take you so far. And it was the skills of Iranian pilots that allowed the IRIAF to claim three kills with one missile. Iranians are really good behind the stick of the Tomcat. In fact, the highest-scoring Ace in a Tomcat is the Iranian Jalil Zandi. The U.S. Air Force has also confirmed Jalil Zandi's 11 kills flying F-14s, an amazing achievement for any fighter pilot. Before the Iraqi ground troops crossed the border, Saddam's air forces attempted to destroy the Iranian Air Force while it was still on the ground. They missed and it cost them big time. In the opening days of the war, Tomcats took their toll on the Iraqi Air Force, downing fighters and bombers alike. Their most deadly weapons, Phoenix missiles, carried an explosive payload that was much larger than other anti-aircraft missiles. They were designed to take down Soviet-built Tupolev bomber aircraft, the same kind the Iraqis were trying to fly over Tehran.

Iranian pilot Asadullah Adel. 3 Kills with one Missile.
Image Source: blog.museumofflight.org

On 07 January 1981, Iranian pilot Asadullah Adeli and his Radar Intercept Officer Mohammed Masbough responded to reports of unidentified aircraft headed toward Kharg Island in the Persian Gulf. The Tomcat realised that the intruders were actually three Iraqi MiG-23s. Iranian ground radar couldn't see all three, but authorised Adeli and Masbough to engage the MiGs anyway. "They were flying really low," Adeli recalled. "Even though it was night, they were flying at around 2,000 feet." Masbough told him to target the one in the middle, just hoping to damage the other two enough that they might break off. That's almost what happened. The American-built Phoenix missile's explosive delivery was so powerful, it downed all three enemy aircraft. The wreckage of all three MiGs was found on Kharg Island the next day.

Aviator sunglasses and the Top Gun which was released in 1986. Jalil standing on the left.
Image Source: YouTube

Iran and the Tomcat F-14

The F-14 "Tomcat" made its combat debut during Operation Frequent Wind, the evacuation of American citizens from Saigon, in April 1975. F-14As from Fighter Squadron 1 (VF-1) and VF-2, operating from the USS Enterprise (CVN-65), flew combat air patrols over South Vietnam to provide fighter cover for the evacuation route. Tomcats of the U.S. Navy shot down two Libyan Su-22's on 19 August 1981. But Iranian F-14's had been blasting Iraqi MiGs out of the sky since September 1980. The F-14 is an example of Iran's so-called "resistance economy" wherein Iran stretched its resources to the limit, getting the most mileage out of its planes as possible by domestically-built or internationally-acquired parts. The world is likely to see Iran's F-14s until they're shot out of the sky.

Brigadier General Jalil Zandi.
Image Source: Wikipedia

Jalil's Post-War Days

His last official post was deputy for planning and organization of the Iranian Air Force, in the rank of Brigadier General. He was awarded the Fath Medal, Grade II On February 4, 1990. He died with his wife Zahra Moheb Shahedin in 2001 in a car accident near Tehran. He was 49 years old. He is buried in Behesht-e Zahra cemetery in the south of Tehran. He had three sons: Vahid, Amir, and Nader.

REFERENCES

1. Jalil Zandi, Bad Ass of the Week, https://www.badassoftheweek.com/zandi
2. Blake Stilwell, That time an F-14 killed three MiGs with a single missile, Mighty History, December 10, 2018 https://www.wearethemighty.com/history/iranian-tomcat-kills-3-jets
3. DARIO LEONE, The legendary F-14 Tomcat's first victory in air-to-air combat, The Aviationist, August 19, 2013. https://theaviationist.com/2013/08/19/tomcats-2-libya-0/#.Uj3kAxCE6B4
4. Tom Cooper, Persian Cats, AIR & SPACE MAGAZINE, September 2006, https://www.airspacemag.com/military-aviation/persian-cats-9242012/
5. Iranian Air Force F-14, Aerospace Web, http://www.aerospaceweb.org/question/planes/q0077.shtml
6. Iran-Iraq War 1980-1990 http://aces.safarikovi.org/victories/iran-iraq.html
7. Samurai in the skies, http://www.iiaf.net/iiafmisc/announcements/announcements.ht
8. THE F-14 TOMCAT IN COMBAT, November 8, https://luckypuppy.net/the-f-14-tomcat-in-combat/
9. Jalil Zandi, Wikipedia, https://en.wikipedia.org/wiki/Jalil_Zandi
10. F-14 Tomcat operational history, Wikipedia, https://en.wikipedia.org/wiki/F-14_Tomcat_operational_history

19

Finnish Ilmari Juutilainen

Highest Scoring Non-German Ace

Ilmari Juutilainen.
Picture Source: sotapolku.fi

Eino Ilmari "Illu" Juutilainen was a fighter pilot of the Ilmavoimat (Finnish Air Force), and the top scoring non-German fighter pilot of all time. This makes him the top flying ace of the Finnish Air Force, leading all Finnish pilots in score against Soviet aircraft in World War II (1939–40 and 1941–44), with 94 confirmed aerial combat victories in 437 sorties, though he himself claimed 126 victories. Juutilainen flew Fokker D.XXI, Brewster Buffalo, and Messerschmitt Bf 109 fighters. He achieved 34 of his victories while flying the Brewster Buffalo fighter. He finished the war without a single hit to his plane from enemy fighter airplanes. Only once he was forced to land after a friendly anti-aircraft gun fired at his Bf 109. Like Japanese fighter ace Saburō Sakai, Juutilainen never lost a wingman in combat. He also scored the first radar-assisted victory in the Finnish Air Force on 24 March 1943, when he was guided to a Soviet Pe-2 by a German radar operator, who was testing out the freshly-delivered radar sets that officially became operational three days later. He was one of the four double recipients of the Mannerheim Cross 2nd Class.

Ilmari Juutilainen.
Image Source: Wikipedia

Early Years: Interest in Flying

Juutilainen was born 21 February 1914 in Lieksa, a town located in the North Karelia region of Finland. He spent his childhood in Sortavala. As a teenager, he was a member of the Volunteer Maritime Defence Association and loved sailing at the Laatokka Sea. There was a Finnish Air Force base in the middle of their town, and most youngsters were enamoured and developed an interest in flying. Many became pilots. Also, another inspiration was a book about the Red Baron, Manfred von Richthofen, which his elder brother gave him. He remembered reading it and dreaming about aerial manoeuvres.

Juutilainen.
Image Source: aminoapps.com

Initial Training and Service

He began his national service as an assistant mechanic in the 1st Separate Maritime Squadron from 1932 to 1933, and then got a pilot's license in a civilian course. He joined the Finnish Air Force as a non-commissioned officer and got military pilot training in the Air Force Academy at Kauhava from 1935 to 1936. His first assignment was on 04 February, 1937, in *LeLv 12* (12 air squadron) at Suur-Merijoki Air Base near Viipuri. In 1938 he went to Utti Air Base and got one year of fighter flying and weaponry exposure. Then, on March 3, 1939, he was assigned to *LeLv 24*, a fighter unit equipped with Dutch-built Fokker D.XXIs, at Utti Air Base, north east of Helsinki.

Fokker DXXI.
Image Source: mediastorehouse.com

His Brother: The Terror of Morocco

His brother was the Finnish Army Captain Aarne Juutilainen, nicknamed "The Terror of Morocco", who served in the French Foreign Legion in Morocco between 1930 and 1935. After returning to Finland, he served in the Finnish army and became a national hero in the Battle of Kollaa during the Winter War with the Soviet Union. He was wounded three times during World War II. He fought in several battles against the Berber rebels in the Atlas Mountains. In November 1939, the Soviet Union attacked Finland, starting the Winter War. Aarne served in the Finnish army during this war, notably during the Battle of Kollaa. When Lt. General Woldemar Hägglund's questioned, "Will Kollaa hold?" Lieutenant Aarne Juutilainen famously answered: "Kollaa will hold, unless the orders are to run." During his command at Kollaa in December 1939, Aarne negotiated with Hägglund about the strategy for the Kollaa Front. The Battle of Kollaa was strategically important. A week earlier, he had received a regimental order to withdraw, which he disregarded. Aarne was honoured with the term "Creator of the Kollaa Spirit". Aarne's men called him "papa". He used the guerrilla warfare skills he learned with the French Foreign Legion to train his men. By this time, Aarne had lost one finger of his right hand as a result of Russian shrapnel.

Finnish Army Captain Aarne Juutilainen, the "The Terror of Morocco".
Image Source: Wikipedia

Soviet–Finnish War

When Hitler was embroiled in a war against Britain and France, Stalin grabbed what he considered strategic territories adjacent to Russia. One concession Stalin sought was part of Finland's Karelian Isthmus on which he wanted to build air and naval bases. Stalin's real plan was to occupy the entire Finland just like the Baltic countries. When Finland refused to give up her lands the Soviets bombed Helsinki and launched an invasion on November 30, 1939. The ensuing conflict, known as the Winter War, ended on March 13, 1940, with the Soviet occupation of 10 percent of Finnish land, but not before the Red Army had suffered several humiliating defeats. The *Voyenno Vozdushny Sily* (Red Army air force, or VVS) had suffered even more, disproportionate to the outnumbered but highly skilled pilots of the Finnish air force.

Finland's territorial concessions to the Soviet Union displayed in red.
Image Source: Wikipedia

Initial Tactics

Ilmari "Illu" Juutilainen, was interviewed many years later by Jon Guttman, for *Military History*. He explained about prevailing tactics that the international trend in the early 1930s was to use a tight, three-plane formation, or "Vic", as a basic fighter element. However, Finland had very few fighters, and they considered the tight formations ineffective. Finns preferred a loose two-aircraft section as the basic fighter element. Divisions (four fighters) and flights (eight aircraft) were made of loose sections, but always maintaining the independence of the section. The distance between the fighters in the section was 150-200 meters, and the distance between sections in a division was 300-400 meters. The principle was always to attack, regardless of numbers. This way the larger enemy formation was broken up and combat became a sequence of section duels, in which the better pilots always won. Finnish fighter training heavily emphasized the complete handling of the fighter and shooting accuracy. Even basic training at the Air Force Academy included a lot of aerobatics with all the basic combat manoeuvres and aerial gunnery.

Juutilainen.
Image Source: century-of-flight.freeola.com

Initial Reaction to Outbreak of War

Juutilainen was mentally ready when the war broke. But it seemed real only when he took off for his first intercept mission. Finns were angry about Stalin's demands that Finland give the Soviet Union certain areas to improve Leningrad's security. The nation's reaction to the war was not analytical – it was emotional. The feeling was, "When I die, there will be many enemies dying, too." All fighters and weapons were prepared. In October 1939, with the situation worsening, the Squadron moved further north-

Soviet Polikarpov I-16.
Image Source: Pinterest. Credit: Don and Christina

east to Immola, closer to the Finnish-Soviet frontier. Shelters were built for the fighters. They began flying combat air patrols – careful to stay on their side, so that they didn't provoke the Soviets. The younger pilots got additional training in aerial combat and gunnery. During bad weather, they indulged in sports, pistol shooting, and discussions about fighter tactics.

The Fokker D.XXI

That was the best available fighter with Finns in 1939, but the opposing Soviet Polikarpov I-16 was faster, had better agility, and also had protective armour for the pilot. Illu got a chance to fly the I-16 many years later as part of war booty, and he found it did 215 knots at low level and turned around a dime. In comparison, the Fokker could make about 175. The D.XXI also lacked armour, but it had good diving characteristics and it was a steady shooting platform. The gunnery training made the Fokker a winner in the Winter War.

Winter War: Juutilainen Initial Air Success

Soviet Ilyushin DB-3 2M-87.
Image Source: armedconflicts.com

The Winter War began three months after the outbreak of World War II, and ended three and a half months later with the Moscow Peace Treaty on 13 March 1940. Despite superior military strength, especially in tanks and aircraft, the Soviet Union suffered severe losses and initially made little headway. During the Winter War, Juutilainen flew the Fokker D.XXI. December 19, 1939, was the first real combat day after a long period of bad weather. When he was close to Antrea, he got a radio message of three enemy bombers approaching. Soon, he made contact with three Ilyushin DB-3s. He was about 1,500 feet higher than them and started the attack. The DB-3s immediately dropped their bomb loads in the forest and turned back. He shot the three rear gunners, one by one. Then he started to shoot the engines. He followed them a long way and kept on shooting. One of them nosed over and crashed. The two others continued in a shallow descent, smoking down. When he finished all of the ammunition, he turned back. There was no real air combat. It was more like firing on a training target.

Meaning of 1/6 Shared Victory

The Soviet bombers flew without fighter escort, and that was a typical situation when the Finnish flight attacked a formation of Tupolev SB-2s. Several fighters shot at several targets, and the kills were then shared, because it was impossible to distinguish a decisive attack. Later, the system was changed and kills were assigned to some particular pilot.

First Encounter with I-16

As per Illu, 31st December was a classic air combat. He was at a very good initial position behind that Soviet aircraft, but he saw the attacker and began a hard left turn. Illu followed, shooting occasionally, to test his nerves. Both lost speed as they circled tightly under the 600 feet low cloud base. The I-16 was

Juutilainen.
Image Source: Pinterest. Saved By Grzegorz Gembala

much more agile, and was gradually gaining advantage. As he was getting into Illu's rear sector, Illu pulled into the cloud, and continuing his hard left turn. Once inside it, he reversed right and came out of the cloud. He was soon behind the opponent. Before he could see him, he was already close at about 100 yards range. Illu fired with tracers a few yards in front of him, and eased the stick pressure to lower the aiming point. Burst hit the engine, and it began to belch smoke. The target pitched over and went into the forest. At the end of the Winter War, Juutilainen had achieved one shared and two individual victories.

Other Types of Missions by Finns

Finnish reconnaissance aircraft were obsolete, so they flew many daytime reconnaissance missions. They occasionally carried out some ground-attack missions until the last days of the war. For the fighter pilots, they were boring missions. The Soviets mostly massed their fighters to cover the ground troops. Finns tried to achieve surprise by using the weather and attacks from different directions every time.

Between the Wars

At the end of the Winter War, the Soviets gained part of Finland's land area, albeit at significant cost and losses. The intervening period was used by Finland to overhaul the fighter fleet. Also new aircraft like Gloster Gladiators, Fiat G.50s and Morane-Saulnier MS 406s, got inducted. Finland did not have enough resources to continue a prolonged campaign alone, but they never surrender. A new fighter, the American Brewster 239 Buffalo got inducted. These were acquired during the Winter War despite the U.S. law which prohibited the sale of war material to the combatant countries. The loophole which permitted the acquisition of the Brewster

Brewster Buffalo.
Image Source: asisbiz.com

239s was a clause in the law that permitted the sale of "rejected" equipment. It was "arranged" that the U.S. Navy rejected 44 Brewster Buffaloes which were then sold to Finland at a "nominal price". 43 were finally released. Juutilainen a joined Brewster flight in the beginning of April 1940. Brewster was a good aircraft, agile, had over 4 hours endurance, one 7.62 mm and three 12.7 machine guns, and armoured plating around pilot's seat.

Ilmari's Tactics

Ilmari Juutilainen's autobiography, "Double Fighter Knight", describes the general tactics followed by him and the FAF. A FAF fighter formation consisted of eight planes in two divisions, with two sections in each division. The forward division attacked. The rear division flew at a higher altitude and "a little behind and off to one side," going into combat only when the situation demanded. What's most astonishing about Sgt. Juutilainen's fighting style was the extreme close range at which he preferred to fight. He regularly recalled shooting at 50 yards, and speaks of following a MiG-3 plane close that his Brewster was "drafting" on the slipstream. Even a Hawker Hurricane left him undaunted. He would come in at high speed from above and behind and pull the throttle back to idle. As the target grew in the gun-sight, and looked really big, he

Ilmari Juutilainen (right), Joppe Karhunen, Pekka Kokko briefing for combat 1941.
Image Source: facebook.com/Ilmari-Juutilainen

would think he is in perfect firing range. Checked his tail once again. Now he could count rivets on the target and the range was about thirty yards.

Finnish fighter Pilots Jorma Karhunen and Ilmari Juutilainen (pictured right), and Brewster Buffalo in the background.
Image Source: historiavareissa.blogspot.com

Continuation War

The Continuation War was a conflict fought by Finland and Nazi Germany, as co-belligerents, against the Soviet Union (USSR) from 1941 to 1944, during World War II. Russians call it the Finnish Front of the Great Patriotic War. Germany regarded its operations in the region as part of its overall war efforts on the Eastern Front and provided Finland with critical material support and military assistance, including economic aid. The Continuation War began 15 months after the end of the Winter War, also fought between Finland and the USSR. By the time of the Continuation War, the Finns had acquired open-cockpit Fiat G.50 Freccias from Italy, Morane-Saulnier 406s from Vichy France, and war-booty export models of the Curtiss P-36 from Germany (captured in France), in addition to the Brewsters. During the Continuation War, Juutilainen served in 3rd Flight of 24 Squadron, flying a Brewster B-239 "Buffalo".

Becomes Ace on Brewster Buffalo

On 21 July 1941, Juutilainen and five other Buffaloes scrambled to intercept Soviet fighters from 65th ShAP that were strafing Finnish troops near Käkisalmi. During that sortie, he destroyed a Polikarpov I-153 'Chaika', making him an "ace" in the Brewster Buffalo. A few days later, on 1 August, seven fighters under the command of First Lieutenant Karhunen destroyed six I-16s near Rautjarvi, and Juutilainen (having been promoted to Warrant Officer in the meantime) claimed two of them.

Polikarpov I-153 and I-153 DM 'Chaika'.
Image Source: forum.worldofwarplanes.eu

Fighter ace Emil Vesa (29 victories) receiving advice from Ilmari Juutilainen (94 victories), June 1942.
Image Source: reddit.com

Major Encounter: Three Victories

On the morning of 6 February 1942, while reconnoitring the Petrovkiy-Jam region of Russia with other LeLv 24 pilots, Juutilainen intercepted seven Tupolev SB bombers escorted by 12 MiG-3s. Juutilainen claimed two SBs. Juutilainen later recalled: I noticed the bombers at 3,000 metres, and radioed the boys about them. As we intercepted the Soviet aircraft, I spotted a formation of three SBs heading for a nearby railway line and dived after them. Targeting the aircraft to the left of the formation, my fire set its port wing aflame. The SB crashed next to the railway line. Just as I started after the lead bomber, I observed a MiG fighter closing in on me. In spite of the threat posed by the latter, I managed to hit the bomber in the starboard engine, which poured out smoke and oil. Moments later the aeroplane rolled over to the right and plunged into the forest close to the railway line. Turning my attention to the MiG, which was above me, I managed to shoot at it as we raced towards each other. My aim was good and the fighter started to trail black smoke from the engine. He banked away to the east, losing altitude as it went.

At Immola: Completes 22 Victories

On 27–28 March 1942, 3/LeLv 24 moved to Immola in preparation for a Finnish Army offensive on Suursaari, in the Gulf of Finland. Although grossly outnumbered

Ilmari Juutilainen in Cockpit Buffalo, BW-364 in 1942.
Image Source: facebook.com/Ilmari-Juutilainen

over the Gulf of Finland, LeLv 24 pilots were more experienced than their Soviet opponents from Red Banner Baltic Fleet. Even when they had the advantage of surprise and height, Soviet pilots did not succeed in shooting down Finnish pilots. On 28 March, W.O. Juutilainen, in patrol with Sgt Huotari, attacked some "Chaikas" of 11 IAP over the Suurkyla shoreline, at Gogland, and shot down two of them. These air victories took Juutilainen's tally to 22. A month later, on 26 April, he became his unit's first recipient of the Mannerheim Cross.

In June 1942 Adolf Hitler visited Immola Air Base in a four-engined Focke Wulf Condor escorted by two Brewsters. His moustache, "Illu" recalls, was dark brown rather than the expected black.

Finnish flying ace Ilmari Juutilainen demonstrating his dog fights to Lieutenant Jori Alanko. Römpötti, Finland. September 1942.
Image Source: mgur.com

34 Victories in Brewster B-239

On 20 September, he took off with Capt Jorma Karhunen and 3/LeLv 24 pilots for a patrol of the Kronstadt-Tolbukhin - Seiskari region, an island in the Gulf of Finland, part of the Leningrad Oblast of Russia. Near the Estonian coast, they were bounced by ten Soviet fighters. But the Finnish quickly reacted and managed to down three of their opponents. Juutilainen was credited with two kills. All in all, Juutilainen scored 34 victories in Brewster B-239, 28 of them (including three triple kills) between 9 July 1941 and 22 November 1942, in his BW-364 "Orange 4".

Eino Ilmari Juutilainen.
Image Source: militaryimages.net

Ace in a Day Flying Bf 109G-2

In February 1943, Juutilainen was transferred to *LeLv 34*, which used new Messerschmitt Bf 109G-2s. This was fortunate, because the Russians were now flying better aircraft. With the Bf 109, he shot down a further 58 enemy planes. He shot down six Soviet airplanes on 30 June 1944 (all confirmed on Soviet loss

Jorma Kalevi Sarvanto was a Finnish Air Force pilot and the foremost Finnish fighter ace of the Winter War, and Ace in a Day.
Image Source: Wikipedia

records), becoming an ace in a day and paralleling Jorma Sarvanto's score on 6 January 1940 in the Winter War. The war ended for Finland in 1944, by which time Illu had earned two Mannerheim crosses (making him the "double knight" of the title) and was the FAF's top ace, credited with 94 air-to-air kills, including 34 on his pet Brewster, BW-364.

Love of Flying and Aerial Victories

Juutilainen refused an officer commission, fearing it would keep him from flying. In the two wars, Ilmari Juutilainen, and his fellow pilots helped preserve their country's independence and taught the Soviet Union a lesson: "If you threaten Finns, they do not become frightened, they become angry. And they never surrender." Ilmari Juutilainen scored more than 94 victories in two wars, flying Fokker D.XXIs, Brewster B-239s, and Messerschmitt Me-109Gs, making him the Finnish Ace of Aces. His 94th and last victory was a Lisunov Li-2, the Russian version of the Douglas C-47, shot down on 3 September 1944 over the Karelian Isthmus. Summary of his victories was:

Aircraft	Victories
Fokker D.XXI	2 1/6
Brewster B-239	34
Messerschmitt Bf 109G	58
Total	94 1/6

Summary of His Flying Qualities

Ilmari Juutilainen's autobiography, "Double Fighter Knight", was translated by General Heikki Nikunen of the Finnish Air Force and Rear Admiral Paul Gillcrist (retd) of the U.S. Navy. In the foreword, Adm. Gillcrist writes the qualities that in his opinion made Illu such an exceptional fighter pilot. These included, aggressive – always the one to attack; superb situational awareness; good eyesight, always looking around; good at estimating deflection, a natural shooter; understood his aircraft, both strengths and weaknesses, as well as

Staff sergeant Jouko Huotari, Sergeant Major Ilmari Juutilainen and Sergeant Emil Vesa from Flying Squadron 24 at Hirvas airfield, June 27, 1942. Airplane is Brewster Buffalo BW-364.
Photo: Sa-kuva.fi, colour by jhlcolorizing

those of the plane he was attacking; a natural pilot; pushed his aircraft to the edge, pressed close (taking the fight to 20 yards from the enemy); physical endurance; self-confidence; and coolness under fire.

Eino Ilmari Juutilainen's Awards

1. April 26th, 1942 Vapaudenristin 2.luokan Mannerheim-risti (MR 2) (Knight of Mannerheim Cross)
2. June 28th, 1944 Vapaudenristin 2.luokka (VR 2)

Ace of Aces. A Book on Eino Ilmari Juutilainen.
Image Source: lukuhetki.fi

3. Cross of Liberty, 3rd Class with Oak Leaves and Swords
4. Cross of Liberty, 4th Class with Oak Leaves and Swords (decorated twice)
5. Medal of Liberty, 2nd Class
6. Iron Cross, 1st Class
7. Iron Cross, 2nd Class

Continues to Fly Till Old

After the wars, Juutilainen served in the air force until 1949. He worked as a professional pilot until 1956, flying people in his De Havilland Moth. His last flight was in 1997 at age 83, in a two-seat F-18 Hornet of the Finnish Air Force.

Quiet Death: Great Legacy

Juutilainen died at home in Tuusula (Tusby), a small town near Helsinki on his 85th birthday on 21 February 1999. Eino Ilmari "Illu" Juutilainen was the top-scoring non-German fighter pilot of all time. This makes him the top flying Ace of the Finnish Air Force, leading all Finnish pilots in score against Soviet aircraft in World War II (1939–40 and 1941–44), with 94 confirmed aerial combat victories in 437 sorties.

July 1996: Ilmari Juutilainen.
Image Source: www.is.fi

REFERENCES

1. Ilmari Juutilainen Interview by Jon Guttman, Century of Flight, Military History. http://www.century-of-flight.freeola.com/Aviation%20history/WW2/aces/Ilmari%20Juutilainen.htm
2. A Finnish pilot's view of the Brewster excerpts from Ilmari Juutilainen's autobiography, "Double Fighter Knight", was translated by General Heikki Nikunen of the Finnish Air Force and Rear Admiral Paul Gillcrist (ret) of the U.S. Navy. https://www.warbirdforum.com/illu.htm
3. Ilmari Juutilainen, Wikipedia, https://en.wikipedia.org/wiki/Ilmari_Juutilainen
4. Knights of the Mannerheim Cross, THE JEWISH Eino Ilmari, http://www.mannerheim-ristinritarit.fi/ritarit?xmid=35
5. Ilmari Juutilainen, Finish Fighter Ace, Facebook, https://www.facebook.com/Ilmari-Juutilainen-65891828784/photos/

20

GERMAN GERHARD BARKHORN

Second Highest Scoring Ace of All Time

Image: airpowerasia.com

Gerhard "Gerd" Barkhorn (Gerd being short form for Gerhard) was the second most successful fighter ace of all time after fellow *Luftwaffe* pilot Erich Hartmann. Other than Hartmann, Barkhorn is the only fighter Ace to ever exceed 300 claimed victories. Barkhorn flew his first combat missions in May 1940, during the Battle of France and then the Battle of Britain without shooting down any aircraft. His first victory came in July 1941 and his total rose steadily against Soviet opposition. In March 1944 he was awarded the second highest decoration in the 'Wehrmacht' when he received the Knight's Cross of the Iron Cross with Oak Leaves and Swords for 250 aerial victories.

Despite being the second-highest scoring pilot in aviation history, Barkhorn was not awarded the Diamonds to his Knight's Cross with Oak Leaves and Swords after achieving his 300th victory on 5 January 1945. Barkhorn flew 1,104 combat sorties and was credited with 301 victories on the Eastern Front against the Soviet Red Air Force piloting the Messerschmitt Bf 109 and Focke-Wulf Fw 190D-9. He flew with the famed Jagdgeschwader 52 (JG 52–52nd Fighter Wing), alongside fellow Aces Erich Hartmann and Günther Rall, and in JG 2 (2nd Fighter Wing). Later he left JG 52 on the Eastern Front and joined JG 3 (3rd Fighter Wing), defending Germany from Western Allied air attack.

From the Left. German Aces Gerhard "Gerd" Barkhorn (301), Erich Hartmann (352), Johannes "Macky" Steinhoff (176), Günther Rall (275).
Image Source: historynet.com

Barkhorn surrendered to the Western Allies in May 1945 and was released later that year. After the war, Barkhorn joined the German Air Force of the Bundeswehr, serving until 1975. On 6 January 1983, Barkhorn was involved in a car crash with his wife Christl. She died instantly and Gerhard died two days later on 8 January 1983.

Gerhard Barkhorn's Messerschmitt Bf 109 G-6 of Stab II/JG52, November 1943, with his wife Christl's name on it.
Image Source: military.wikia.org

Early Life and Career

Barkhorn was born on 20 March 1919 in Konigsberg in the Free State of Prussia of the Weimar Republic. Today it is Kaliningrad in Kaliningrad Oblast, the Russian exclave between Poland and Lithuania on the Baltic Sea. He was one of four children of an urban design inspector Wilhelm and his wife Therese. Barkhorn had two brothers, Helmut and Dieter, and a sister Meta. Following four years of primary school, Barkhorn attended the Wilhelms-Gymnasium, a secondary school, where he graduated with a diploma. After his compulsory Reich Labour Service Barkhorn joined the military service in the Nazi German Luftwaffe in November 1937 as a Cadet. He started his flight training in March 1938.

Image Source: donhollway.com

Selected For Fighter Stream

World War II in Europe began on Friday 01 September 1939 when German forces invaded Poland and Barkhorn was selected for specialized fighter pilot training. Upon completion of his training, he was commissioned as a Second Lieutenant and posted to 3rd squadron in JG 2 "Richthofen". This unit had an old tradition and was named after the World War I fighter pilot Manfred von Richthofen. He was flight trained by Franz Stigler, who would later become a *Luftwaffe* Ace himself. He was then transferred to the 6th squadron of JG 52 on 1 August 1940.

First Combat: World War II

Barkhorn flew his first combat sorties over Belgium and France during the Battle of France and later over southern England during the Battle of Britain, flying the Messerschmitt Bf 109E. On 1 August 1940 Barkhorn was transferred to 6/JG 52. He did not have any success, although he flew some 21 combat sorties and was shot down in the English Channel on 29 October by RAF fighters, but rescued. Shortly thereafter he was rewarded with the Iron Cross 1st Class. In the Squadron, he flew alongside another promising pilot, later an Air Ace, Hans-Joachim Marseille, "Star of Africa".

Winged sword unit emblem of JG 52
Image Source: Wikipedia

Air Ace, Hans-Joachim Marseille, "Star of Africa"
Image Source: alchetron.com

Moved to Eastern Front: Opens Air Victory Account

In 1941, JG 52 was transferred to the east and participated in Operation *Barbarossa*, the invasion of the

Ilyushin DB-3 Bomber.
Image Source: airpages.ru

Knight's Cross of the Iron Cross.
Image Source: Wikipedia

Soviet Union, on 22 June 1941. Subsequently, Gerhard Barkhorn scored his first victory by shooting down a Red Air Force DB-3 bomber on 2 July, flying his 120th combat sortie. By November his tally had reached 10 victories and he was promoted to Lieutenant on 11 November 1941. On 21 May 1942, Barkhorn was appointed Squadron Commander of 4/JG 52.

Ace in a Day

He continued to add to his score over the next year, until on 19 July when he became "ace-in-a-day" by shooting down six aircraft in his Bf 109F. He was wounded on 25 July and put out of action for two months, returning to combat in October. During July 1942 alone, Barkhorn had destroyed 30 Soviet aircraft. On 23 August he received the Knight's Cross of the Iron Cross for having shot down a total of 59 aircraft. After a two-month break from the front he returned to action in early October.

Crosses Century Mark

On 19 December 1942, Barkhorn had raised his score to 101 victories. That day, he became the 32nd *Luftwaffe* pilot to achieve the century mark. Barkhorn came to respect the Soviet pilots. On one occasion he was involved in a forty-minute dogfight with a Lavochkin-Gorbunov-Gudkov LaGG-3. "Sweat was pouring off me just as though I had stepped out of the shower," he recalled: despite having a faster aircraft he was simply unable to get a bead on the Russian pilot.

Lavochkin-Gorbunov-Gudkov LaGG-3.
Image Source: aviastar.org

Shoots Hero of Soviet Union

On 9 January 1943, Barkhorn claimed his 105th. His victims included Lieutenant Vasiliyev, and Hero of the Soviet Union Podpolkovnik (Lt Col) Lev Shestakov of the 236 IAP Fighter Regiment. Barkhorn strafed their Yakovlev Yak-1 fighters until they caught fire. Both pilots survived. Barkhorn was awarded the Knight's Cross of the Iron Cross with Oak Leaves on 11 January 1943.

Podpolkovnik (Lt Col) Lev Shestakov.
Image Source: Wikipedia

First German to Cross 1,000 Missions

Barkhorn, now a Captain, was appointed Group Commander of II./JG 52 on 1 September 1943. On 5 September he shot down another Hero of the Soviet Union and Soviet fighter ace Nikolay Klepikov, Ace with 10 personal and 32 shared victories. This was offset by the loss of II./JG 52's 173-victory German Ace Heinz Schmidt. The two Lavochkin La-5s shot down by Barkhorn were his 165th and 166th aerial victories. Barkhorn reached the 200 mark on 30 November 1943. On 23 January 1944 Barkhorn became the first German pilot to fly 1,000 combat missions.

Gerhard Barkhorn's 1,000 Combat Sortie.
Image Source: Twitter @DownedWarbirds

Second German to Cross 250 Victories

The main German fighter unit covering the Crimea and Kuban was his II/JG 52 and in the three months between December 1943 and 13 February 1944 the unit claimed 350 victories, of which 50 were claimed by Barkhorn personally. Earlier he had claimed 15 victories in September, 23 in November and 28 in December, including seven on 28 December alone. On 13 February 1944 he reached 250 victories. Barkhorn was the second to reach this total.

Gerd Barkhorn receiving congratulations from his Gruppe after his 250th victory.
Image Source: luftwaffe.cz

Best Man for Fellow Ace Erich Hartmann's Wedding

On 2 March 1944, he was awarded the Knight's Cross of the Iron Cross with Oak Leaves and Swords He attended the wedding of fellow pilot Erich Hartmann as best man. Barkhorn was promoted to Major on 1 May 1944.

Barkhorn as the Best man at Erich Hartmann's wedding.
Image Source: donhollway.com

Hospitalised: Erich Hartmann Overtakes Total

On 31 May 1944, on 274 victories, Barkhorn was flying his sixth mission of the day and, being fatigued, was not concentrating on keeping a good look-out when he was bounced by a Russian Airacobra fighter and shot down in Bf 109 G-6 (W.Nr. 163 195) "Black 5". He received severe wounds to his right arm and leg which put him out of action for four months as he was hospitalised. With Barkhorn side-lined, Hartmann was to overtake his total. Eventually returning to his unit the psychological damage and combat stress on Barkhorn became apparent; sitting in his cockpit he became overcome with anxiety, and even when flying with friendly aircraft behind him he felt intense fear. It took several weeks for him to overcome this condition. Returning to combat in October he claimed his 275th victory on 14 November. Over the next few weeks Barkhorn added another 26 victories, scoring his 301st (and final) victory on 5 January 1945.

German Ace fighter pilots Erich Hartmann and Gerhard Barkhorn.
Image Source: warfarehistorynetwork.com

Defence of the Reich: Eased From Command Due Health

On 16 January 1945 Barkhorn was made Wing Commander of JG 6 Fighter Wing, a unit assigned to defend the Reich and equipped with the Focke-Wulf Fw 190D. Barkhorn led this unit until the 10th of April 1945. During this tenure, he did not claim any aerial victories. JG 6 was a unit consisting mostly of new recruits and former Bf-110 pilots; it suffered heavy losses against the American air fleets. Barkhorn did not last long in this position and was forced to take a medical absence because of severe physical and mental strain and eventually relinquished command for another spell in hospital.

Moved To New Me-262 Jet Fighter: Last Missions

After his hospitalisation, Barkhorn was invited by Adolf Galland to join the elite JV 44 – 44th Fighter Detachment flying the Messerschmitt Me 262 jet-fighter. He found flying the Me 262 over the western front difficult and he did not score any victories in it. On 21 April 1945, he flew his 1,104th and last mission. One of the engines of his aircraft flamed out as he was approaching an enemy bomber formation and he was forced to make an emergency landing. As he approached the airfield, his jet was attacked by several prowling North American P-51 Mustang fighters.

Barkhorn in his Fw 190D-9 on 17 February, 1945 in Welzow.
Image Source: luftwaffe.cz

Barkhorn managed to land his burning plane though he received a slight wound as a result of this action when the cockpit canopy, which on the Me 262A flipped open to starboard like a Bf 109's did prior to a crash landing, slammed shut on his neck. He was hospitalised. Barkhorn was one of the few noted *Luftwaffe* Experts who escaped being imprisoned by the Russians. However, he did become a prisoner of war of the Allies, finally being released by them in September 1945.

German Aces meet Hitler 04 April 1944. From the left Major Werner Streib, Major Gerhard "Gerd" Barkhorn, Major Erich Walther, Major Kurt Bühligen, Hauptmann Hans-Joachim Jabs, Major Bernhard Jope, Major Reinhard Seiler, Major Erich Hartmann (hidden by Hitler), Major Horst Ademeit, Major Johannes Wiese, Wachtmeister Fritz Petersen, Major Dr. Maximilian Otte, Hauptmann Walter "Graf Punski" Krupinski.
Image Source: thevintagenews.com (Photo credit: Walter Frentz).

Initial Civil Life and Re-joins Service

After Barkhorn was released as a prisoner of war, he then found work in the automobile industry. Following the decision of the Cabinet of Germany to rearm the Federal Republic of Germany, Barkhorn joined the military service of the West German Air Force in 1956. From 1 April 1957 to 31 December 1962, he commanded the 31 Fighter-Bomber Wing.

Flight Test Assignment

VSTOL Kestrel, the Hawker Siddeley P1127.
Image Source: acesflyinghigh.wordpress.com

In 1964 he was posted to the staff of Air Force Test Command holding the rank of Colonel. From October 1964 until November 1965, Colonel Barkhorn headed the six-man West German Air Force contingent of the Tripartite Kestrel Evaluation Squadron at Royal Air Force Station West Raynham, Norfolk, England. The squadron's mission was to evaluate the military capabilities of the very short take-off and landing, VSTOL Kestrel, the Hawker Siddeley P1127 and forerunner of the Harrier VSTOL aircraft. The squadron consisted of military pilots and ground staff from three nations: Great Britain, USA and West Germany. In addition to being one of the squadron pilots, Barkhorn also served as one of the squadron's two Deputy Commanders. At the conclusion of the evaluation, Barkhorn then accompanied the American contingent to the US, where he assisted in that nation's continuing trials of six of the Kestrels that had been shipped to the US and renamed the XV-6A.

Lieutenant General Gerhard Barkhorn.
Image Source: Wikipedia.

With his Wife Cristl, a few days before the accident.
Image Source: donhollway.com/gerhardbarkhorn

Higher Appointments

Barkhorn was promoted to Brigadier general in 1969 and to Lieutenant General in 1973. His last position was Chief of Staff of the Second Allied Tactical Air Force, a NATO military formation under Allied Air Forces Central Europe. He retired from active service on 30 September 1975.

Death in Car Accident

On 6 January 1983, during a winter storm on an autobahn near Cologne, Barkhorn and his wife, Christl, were involved in a serious automobile accident. His wife died instantly and Barkhorn, without regaining consciousness, died in hospital in Frechen/Cologne on 8 January 1983, at age 63. They were buried in Tegernsee, Bavaria.

Summary of Operational Career

Barkhorn claimed 301 victories in 1,100 combat missions, mostly piloting Messerschmitt Bf 109 and Focke-Wulf Fw 190D-9 fighters. He flew in the famous Jagdgeschwader 52 (JG 52) fighter unit, along with companions aces, such as Hartmann and Günther Rall.

On His Grave.
Image Source: alifrafikkhan.blogspot.com

He also served in the Jagdgeschwader 2 hunting wing.

Barkhorn's success had not come without some cost. He was shot down nine times, bailed out once and was wounded twice. Matthews and Foreman, authors of *Luftwaffe Aces – Biographies and Victory Claims*, researched the German Federal Archives and found records for 300 aerial victory claims, plus one further unconfirmed claim. Each claim is well documented with time and place. He was "Ace in a Day" on five occasions, on 22 June 1942 (5 Victories), 19 July 1942 (6 Victories), 20 July 1942

Shaking Hands with Adolf Hitler.
Image Source: alifrafikkhan.blogspot.com

Image Source: Pinterest (Credit: Catran)

(5 Victories), 02 December 1942 (6 Victories), 28 December 1943 (7 Victories). He shot down a very wide variety of aircraft including, Ilyushin DB-3, Polikarpov I-16, I-18 (MiG-1), Ilyushin Il-2, I-61 (MiG-3), Hurricane, I-153, P-40, R-5, Yak-1, LaGG-3, Pe-2, Spitfire, P-39, La-5, Yak-7, and Yak-9 aircraft.

Critical Milestones

- 5 January 1942 – his 30th victory
- 1 March 1942 – Squadron Commander of 4./JG 52
- 25 July 1942 – wounded while flying Bf 109 F-4 'White 5'
- 23 August 1942 - awarded the Iron Cross, with 64 victories
- 19 December 1942 – his 100th victory
- 11 January 1943 – awarded the Oak Leaves, with 105 victories
- 8 August 1943 – his 150th victory
- 1 September 1943 – Group Commander of II./JG 52
- 30 November 1943 – his 200th victory
- 23 January 1944 – 1,000 combat missions (the first fighter pilot in history to do so)
- 13 February 1944 – his 250th victory
- 2 March 1944 – awarded the Swords
- 5 January 1945 – his 301st, and last, victory

Image Source: YouTube (Das Vergessene As – Gerhard Barkhorn)

Honours and Awards

- Honorary Cup of the *Luftwaffe* on 20 July 1942
- German Cross in Gold on 21 August 1942
- Iron Cross (1939)
- 2nd class (23 October 1940)
- 1st class (3 December 1940)
- Knight's Cross of the Iron Cross with Oak Leaves and Swords
- Knight's Cross on 23 August 1942
- 175th Oak Leaves on 11 January 1943
- 52nd Swords on 2 March 1944

The Bilingual Book – The Forgotten Ace: Fighter Pilot Gerhard Barkhorn

The Book: The Forgotten Ace: Fighter Pilot Gerhard Barkhorn

Despite Gerhard Barkhorn scoring the second-highest 301 kills while flying with Jagdgeschwader 52 on the Eastern Front (making him the second most successful fighter pilot of the Second World War after Erich Hartmann) but today he is almost unknown. He was also not awarded the Diamonds to the swords and Oak leaves of his Iron Cross. Maybe he was not ardent enough a Nazi. This book describes Barkhorn's wartime experiences with JG 2 on the Channel, JG 52, JG 6 and finally JV 44. Excerpts from original combat reports enable the reader to feel the drama of the events that happened back then. Published in this book for the first time are numerous colour slides taken by Barkhorn during his service with JG 52, as well as many black and white photos. It also contains colour profiles of every aircraft flown by Barkhorn. "The Forgotten Ace" is a book that examines all aspects of this fighter ace.

REFERENCES

1. German Knight on the Russian Front By Don Hollway, Try Try Try Again, Aviation History September, 2012 Issue. http://donhollway.com/gerhardbarkhorn/
2. Gerhard "Gerd" Barkhorn, Wikipedia, https://en.wikipedia.org/wiki/Gerhard_Barkhorn
3. Gerhard Barkhorn, *Luftwaffe*, http://www.luftwaffe.cz/barkhorn.html
4. Stephen Sherman, Gerhard Barkhorn - Second-highest scoring ace of all time, December 2008, http://acepilots.com/german/barkhorn.html
5. Gerhard Barkhorn, Find the Grave https://www.findagrave.com/memorial/13812330/gerhard-barkhorn
6. Barkhorn, Gerhard "Gerd", World War II Graves, https://ww2gravestone.com/people/barkhorn-gerhard-gerd/
7. Gerhard Barkhorn - German Messerschmitt Bf 109 Ace, YouTube, https://www.youtube.com/watch?v=RDyUzT8lhlU

21

GÜNTHER RALL

Third Highest Scoring Air Ace of All Time

Günther Rall.
Image Source: griffonmerlin.com

General Günther Rall was a highly decorated German military aviator, whose military career spanned nearly forty years. Rall was the third most successful fighter pilot in aviation history, behind Gerhard Barkhorn, who is second, and Erich Hartmann, who is top-scoring air ace of all times. Rall flew combat missions in the Battle of France, and Battle of Britain. He was later moved to the Eastern front. Rall claimed his first successes in the air defence of Romania. In November 1941, he was shot down, wounded and invalidated from flying for a year. At this time Rall had claimed 36 aerial victories.

Rall returned in August 1942 and by September 1942 had 65 aerial victories. By 22 October Rall had claimed 100. He reached 200 in late August 1943. By the end of 1943, Rall had achieved over 250, the second flier to do so after Nowotny (final total 258) did in October 1943. He left the eastern front and was sent for the Defence of the Reich where he was wounded thrice. In November 1944 Rall was appointed as an

Günther Rall. Knight's Cross portrait, 1942.
Image Source: Wikipedia

Günther Rall.
Image Source: commons.wikimedia.org

instructor and flew captured Allied fighter aircraft in order to prepare instruction notes on their performance for German fighter pilots. Rall ended the war with an unsuccessful stint commanding JG 300 near Salzburg, Austria where he surrendered in May 1945. He was credited with 275 air victories in 621 missions. He was shot down five times and wounded on three occasions. Rall claimed all of his victories in a Messerschmitt Bf 109, though he also flew the Focke-Wulf Fw 190 operationally. All but three of his claims were against Soviet opposition.

Günther Rall, During a Lecture Tour to Finland in June 2003.
Image Source: virtualpilots.fi

After the war, Rall again joined the West German Air Force in 1956, served as Inspector of the Air Force from 1971 to 1974, and as the German representative to the NATO Military Committee until 1975. After his retirement, Rall became a consultant.

Early Life

Rall was born on 10 March 1918 in Gaggenau, at that time in the Grand Duchy of Baden of the German Empire (1871–1918) during World War I. He was the second child of merchant Rudolf Rall and his wife Minna, née Heinzelmann. His sister Lotte, was four years older than Rall. Rall stated that his father was a member of *Der Stahlhelm, Bund der Frontsoldaten* (The Steel helmet, League of front-line Soldiers) and had an affiliation with the German National People's Party. He grew up in the Weimar Republic.

Nazi Schooling

Young Rall.
Image Source: luftwaffe39-45.historia.nom.br/ases/rall.htm

In 1922, the Rall family moved to Stuttgart. There, in 1928, Rall joined the Christian Boy Scouts. In 1934, the *Gleichschaltung* (Nazi terminology the process of Nazification) converted the Christian Boy Scouts into the *Deutsches Jungvolk* (a term for German Youngsters in the Hitler Youth). He attended the *Volksschule* (compulsory school) in Stuttgart. For his secondary education, he first attended the humanities-oriented *Karls-Gymnasium* in Stuttgart and then in 1935 transferred to the National Political Institutes of Education (Napola) in Backnang, a secondary boarding school founded under the recently established Nazi state. The goal of the Napola schools was to raise a new generation for the political, military and administrative leadership of Nazi Germany. There he received his university entry qualification.

Rall attended the Napola in Backnang.
Image Source: wikiwand.com/en/Günther_Rall

Decides Military Career: Later Luftwaffe

In 1933 the Nazi Party seized power. Following graduation, Rall volunteered for military service and joined the *Wehrmacht* (Nazi German Armed Forces) in December 1936 to train as an infantry soldier. On 4 December 1936, Rall joined the 13 Infantry Regiment in Ludwigsburg as a non-commissioned officer. From 1 January to 31 June 1938, he attended the military school in Dresden. In the summer of 1938, Rall requested

to be transferred to the *Luftwaffe*. He was initially trained as a pilot at Unterbiberg airfield, and then transferred to the *Luftwaffe* and then trained to be a fighter pilot. On 1 September 1938, he was promoted to Second Lieutenant. Rall then attended the *Jagdfliegerschule Werneuchen* (fighter pilot school) from 15 July to 15 September 1939 and was then posted to No. 4 Squadron of *Jagdgeschwader* 52 (JG 52—52nd Fighter Wing) on 16 September 1939 where he served as a flight leader of a *Rotte* (two-aircraft section).

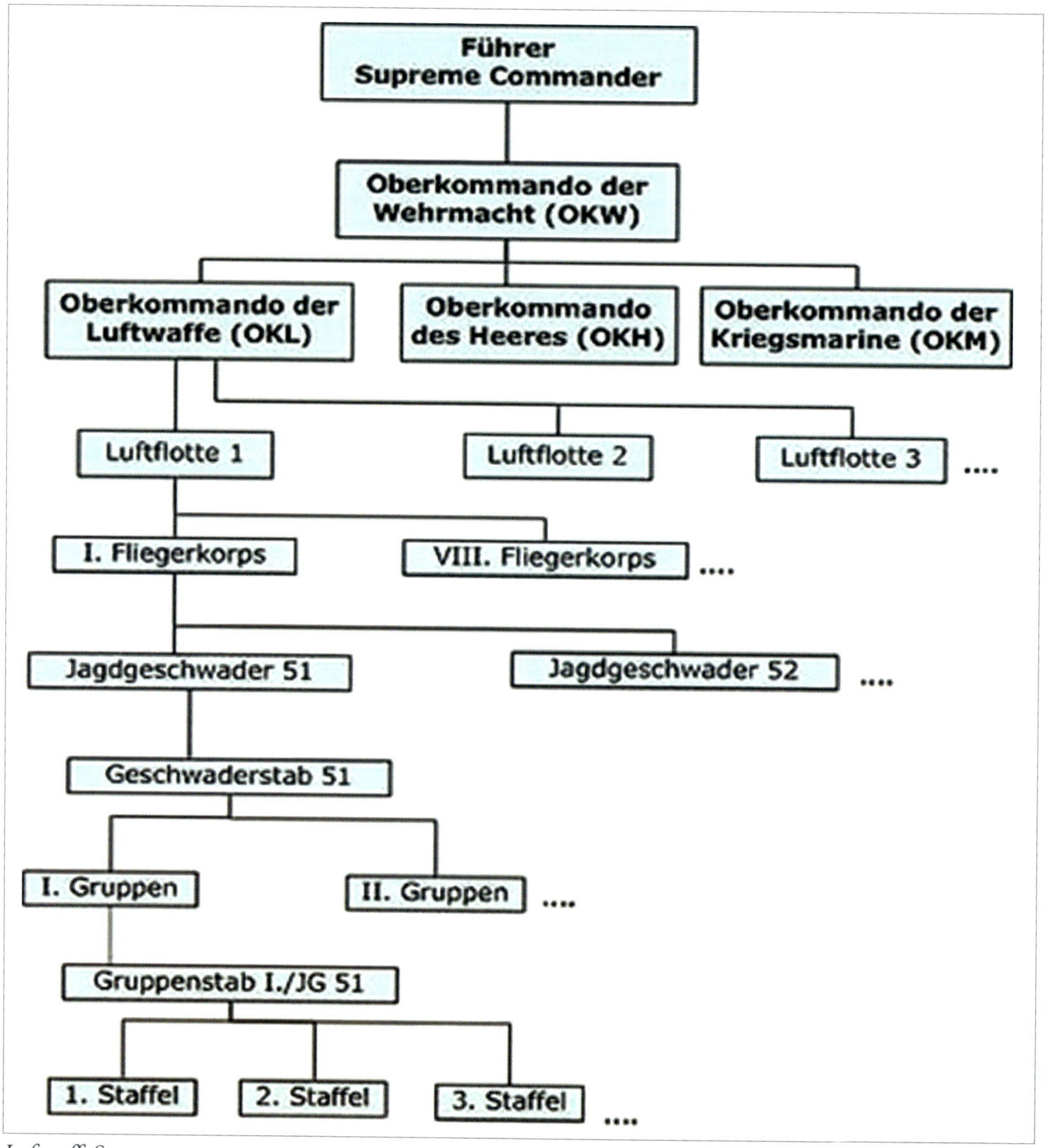

Luftwaffe Structure.
Image Source: Wikipedia

World War II: Initial Action

World War II in Europe began on Friday, 01 September 1939 when German forces invaded Poland. JG 52 was not directly involved in the invasion. It was positioned in western Germany, for Defence of the Reich duties and Rall did not see combat. Rall flew combat patrols in the Phoney War which was an eight-month period at the start of World War II, during which there was only one limited military land operation on the Western Front, when French troops invaded Germany's Saar district. The Phoney period began with the declaration of war by the United Kingdom and France against Nazi Germany on 3 September 1939, and ended with the German invasion of

Image Source: luftwaffe39-45.historia.nom.br/ases/rall.htm

France and the Low Countries on 10 May 1940. On 7 March 1940, he was transferred to No. 8 Squadron when JG 52 was augmented by the newly created 3rd Group. On 10 May 1940 *Fall Gelb* (The Invasion of France and the Low Countries) began, and JG 52 supported German forces in the invasion of Belgium and Battle of France. Rall flew combat missions in the Battle of France and Battle of Britain.

First Air Victory: Lots of Bullets but Renewed Self Confidence

On the third day of the campaign, 12 May 1940, 22 years old Rall achieved his first victory. Three French Curtiss H75-C1 fighters were attacking a German reconnaissance aircraft at a height of 26,000 feet (7,900 m). Rall attacked them and shot down one, stating: "I was lucky in my first dogfight, but it did give me a hell of a lot of self-confidence...and a scaring, because I was also hit by many bullets." The victory was his only success on the Western Front. Rall's Wing sustained heavy casualties.

French Curtiss H75-C1.
Image Source: asisbiz.com

Battle of Britain: Faulty Tactics

Alfred Grislawski. 109 Victories.
Image Source: Wikipedia

JG 52 was later moved to Peuplingues and Coquelles, on the French channel coast where it fought in the Battle of Britain. Due to heavy losses, Rall was given command of 8. *Staffel* (unit) of JG 52 on 25 July 1940 and was promoted to Lieutenant a week later, on 1 August 1940. Rall replaced Lieutenant Lothar Ehrlich, who was killed in action against No. 610 Squadron of RAF the previous day during the convoy battles. He was one among the three pilots killed that day. Rall said of the battle, "probably no one even had time to shout a warning. Suddenly a flock of Spitfires were on us like hawks on a bunch of chickens." Rall placed the blame for losses on faulty tactics; such as tying the Bf 109s to close escort of the slow Junkers Ju 87 *Stuka* dive-bombers. On the day he was appointed, JG 52 lost another four pilots, including two men of unit commander status. Rall's Unit lost one pilot missing in action with No. 65 Squadron RAF over Dover in the early afternoon. Rall and his unit achieved little. Several of the highest claiming pilots of *JG* 52, Gerhard Barkhorn, Alfred Grislawski, and Adolf Dickfeld (1,072 combat missions and 136 claims) were not successful over England.

Transferred to Greece and Romania

Rall was then transferred to Greece, and participated in the final phase of the Balkans Campaign in April to May 1941 without success. Based at Athens, he flew combat missions in support of the airborne invasion and subsequent Battle of Crete in June 1941. JG 52 was transferred back to Romania to help defend their recently acquired allies' Ploieiti oil fields.

Luftwaffe pilots in Balkans 1941-02.
Image Source: asisbiz.com/Battles/Balkans-Campaign

Eastern Front: Air Victories Begin

On 22 June 1941 Operation Barbarossa, the Axis invasion of the Soviet Union, began the war on the Eastern Front. The majority of JG 52 were supporting Army Group South, and the invasion of the Ukrainian SSR. Rall's contingent remained in eastern Romania. The Red Air Force (VVS) immediately began a campaign to destroy the Romanian oil fields. Major General Pavel Zhigarev, commanding the VVS ChF (Air Command Crimea), committed the 63 BAP (63rd Bomber Aviation Regiment) and 40 SBAP (40th High Speed Bomber Aviation Regiment). The attacks met with some success, although heavy losses forced the switch to night bombing from mid-July. Rall scored his second, third and fourth victories in interceptions of Soviet bombers. During a five-day period, III/JG 52 claimed between 45 and 50 Soviet aircraft. Rall remarked the reasons for the success was the Soviets did not provide fighter escort for their bombers.

June 1941 Operation Barbarossa.
Image Source: oedeboyz.com

Becomes an Ace

Rall claimed his fifth victory on 4 August thus becoming an "Ace". While providing escort for 77th Dive Bomber Wing on 13 August 1941, flying with 3rd Fighter Wing, Rall claimed a Polikarpov I-16 as did JG 3's Günther Lützow (later Ace with 110 Victories). The Soviet pilots were from the 88 IAP and identified as Lieutenants Yakov Kozlov and Ivan Novikov. III/JG 52 supported the encirclement Battle at Kiev in August.

First Battle of Kharkov

Rall claimed 12 victories in October 1941 as III/JG 52 fought for air superiority during the First Battle of Kharkov; an autumn offensive to seize the industrialized regions of Eastern Ukraine. On 14 October there was heavy air fighting. Rall claimed an Ilyushin IL-2 over his group's Poltava airfield after being scrambled in the midst of a Soviet air attack. The Germans had failed in the race for the Ukrainian industrial heartland. After the capture of Kharkov and Stalino, the Germans found 54 medium and 223 large factories, all empty. Some 1.5 million wagonloads had been evacuated.

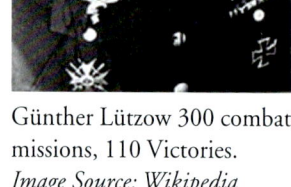

Günther Lützow 300 combat missions, 110 Victories.
Image Source: Wikipedia

Ilyushin IL-2.
Image Source: Wikipedia

Crimean Campaign: 28th Victory

On 23 October III/JG 52 moved to Chaplinka in the Crimea. With II/JG 3 and JG 77 it was ordered to clear the skies. The Crimean Campaign lasted into the following year. The German fighter units claimed 140 aircraft from 18 to 24 October over Perekop. Rall had reached 28 victories by this date.

Shot Down and Crashes: Survives With Broken Back

In November the Red Army regrouped and conducted a well-orchestrated recapture of Rostov on Don. The victory denied the Germans access to the Caucasus. On 28 November 1941, Rall claimed his 36th victory near the contested city, but as he watched the burning I-16 fall in the fading light Rall relaxed his vigilance and was shot down. He tried to fly back to German lines with a damaged engine, but crash-landed and was knocked unconscious. A German tank crew rescued him from the wreck. His Bf 109 F-4 (Tail No. 7308) came down in the vicinity of Rostov. X-rays revealed he had broken his back in three places. Doctors told him he was finished as a pilot, and transferred him to a hospital in Vienna in December 1941.

Günther Rall Bf 109 F-4 that Crashed.
Image Source: flickr.com/photos/farinihouseoflove

One Year Hospitalisation: Meets His Future Wife

Despite the prognosis, Rall defied odds and returned to combat a year later. During his treatment, he met Hertha Schön, whom he married in 1943. In Rall's absence, third group claimed 90 of the 135 aircraft claimed shot down by Luftflotte 4 in December. This was achieved without loss; making it the most successful of the German fighter groups. The VVS Southern Front admitted the loss of 44 aircraft from 1 to 22 December. The losses for the remaining nine days are not stated.

Günther Rall and his Jewish wife Hertha Schön.
Image Source: reddit.com

Back to Action after One Year

Rall came back to 8/JG 52 in August 1942. From 2 to 30 August Rall claimed victory 37 through to 62; a run of 26 aerial victories in a four-week period. On 6 August Rall claimed four in one day. At this time Rall's unit was operating in support of the Battle of the Caucasus, deep in southern Russia. His 61st claim was for a victory achieved in the vicinity of Grozny. German forces reached the Terek River in late August 1942 and erected pontoon bridges. The Soviets began air attacks on the crossings, and Rall's III/JG 52 claimed 32 aerial victories in their defence.

Knight's Cross holders of JG 52 during a mission break in late 1942 in front of Rall's Bf 109 G-2 "Black 13". From left: Uffz. Karl Gratz (138 victories), Oblt. Günther Rall and Uffz. Friedrich Wachowiak (about 120 victories).
Image Source: luftwaffe.cz/rall.html

Rall is Second from Left.
Image Source: defensemedianetwork.com

Awarded Knight's Cross of the Iron Cross

On 3rd September, Rall was awarded the Knight's Cross of the Iron Cross. The pilots of JG 52 opposed the Soviet 4th Air Army (4 VA) effectively; and with pilots such as Rall, Dickfeld and Grislawski, they dominated the airspace whenever they appeared in strength. The 4 VA reported the loss of 149 aircraft in September 1942. On 30 September 1942 Rall claimed his 90th aerial victim, bringing his total for the month to 28.

100th Victory: Oak Leaves Presented by Hitler

On 22 October, Rall was credited with his 100th aerial victory. He was the 28th Luftwaffe pilot to achieve the century mark. On 2 November 1942, Rall was required to meet Adolf Hitler and was personally awarded the Knight's Cross of the Iron Cross with Oak Leaves. Rall took the opportunity to ask Hitler when the war would be over. To Rall's surprise, Hitler replied that he did not know. After the ceremony, Rall was granted leave. Rall travelled by train to Vienna on 11 November and married Hertha. Upon completion of his leave, Rall returned to the front as III/JG 52 was ordered to cover the retreat after the Battle of Stalingrad in which several Axis field armies were destroyed.

Adolf Hitler personally awarded the Knight's Cross of the Iron Cross with Oak Leaves.
Image Source: luftwaffe39-45.historia.nom.br/ases/rall.htm

Kuban Bridgehead: First Real Test for German Fighters in the East

The Kuban bridgehead was the main area of operations for Rall in early 1943. Hitler wished to maintain a foothold in the Caucasus to defend the Crimea and retain the captured, and now operational, facilities at Maykop town near Krasnodar. Hitler harboured a forlorn hope he could use the region as a staging area for a renewed offensive against the Soviet oilfields. The Luftwaffe was rushed to the Kuban to support the German 17th Army's defences. StG 2, StG 77, SG 1, and the fighter wings JG 3, JG 52 were sent to the region as powerful close support just as the Soviet Front began its offensive. The fighter units were able to inflict heavy losses to Soviet aviation. Rall, who was not impressed by the latest Bell P-39 Airacobra, now in use by Soviet pilots, observed that Soviet fighter aviation displayed a newly aggressive posture in late 1942 and early 1943.

Messerschmitt Bf 109G-2 "Black 13" Gunther Rall 1942.
Image Source: Pinterest (Small Scale Art)

Rall achieved his first successes over the Kuban on 21 March 1943 and by 30 April had claimed 126. In April 1943, he was promoted to Captain and on the 20th of that month scored the Geschwader's (British Group equivalent) 5,000th victory. In the first week of May, Rall claimed a Soviet-flown Supermarine Spitfire. After filling out and submitting the combat report Rall was told by his superiors to keep the encounter to himself lest it lower morale. Three weeks later he was credited with a 145th victory. Rall noted the improvement of Soviet pilot training and regarded the Kuban as the first serious test of the German fighter force on the Eastern Front.

Group Commander

The German defences held in the Kuban in 1943, until the autumn. JG 52 moved north in preparation for Operation Citadel and the Battle of Kursk. Rall, who had already served as acting group commander of III. *Gruppe* in February and March 1943, officially moved to this position on 5 July 1943. Rall had 42 aircraft under his command, two over the full complement of machines and pilots, and 35 were operational. Rall continued to claim enemy aircraft. On 8 July a two-man patrol with Erich Hartmann resulted in two claims, and a third for Rall. A Soviet after-battle analysis mentioned this specific engagement: "Eight Yak-1s in the Provorot region observed two Me 109s off their flight path. Paying no attention to the enemy aircraft our fighters continued. Seizing a convenient moment, the German fighters attacked our aircraft and shot down three Yak-1s."

Gunther Rall with his Unit Crew Luftwaffe Aces dog unit mascot JG 52 Russian front.
Image Source: ww2images.blogspot.com

Yakovlev Yak-1.
Image Source: Wikipedia

Force Landing and a Mid Air Collision

On 9 July, following combat with Soviet fighters, Rall made a forced landing in his Bf 109 G-6 (Tail No. 20019) near Petrovka, north of Belgorod (North of Kharkiv). Four days later, a mid-air collision with a Lavochkin-Gorbunov-Gudkov LaGG-3 fighter resulted in another forced landing at Ugrim airfield. Rall claimed 21 air victories in July but the German offensive rapidly bogged down. The Red Army began a counteroffensive in the region to contain the German operation and destroy its forces (Operation Kutuzov and Operation Polkovodets Rumyantsev). Rall stated that after Kursk, his pilots no longer believed the "Endsieg" (German for ultimate victory), though the German army managed to stabilise the front somewhat over the following weeks. On 3 August Rall's group had only 22 operational Bf 109s from a total of 29; from its designated strength of 40 aircraft.

Lavochkin-Gorbunov-Gudkov LaGG-3.
Image Source: Wikipedia

Gunther Rall being received after a flight Bf109 G4.
Image Source: Archive: Messerschmitt Bf 109 flickr.com/photos/farinihouseoflove

Claim Disputes

The claims of fighter pilots on each side have often been disputed. The 2nd Air Army, responsible for defending the airspace opposite *Stab.* I and III/JG 52 at the start of the battle, lost 153 fighters from 5 to 10 July 1943, representing 40 per cent of initial strength. The Soviets admitted the loss of 1,000 aircraft in their "defensive" phase of the battle. In the first three days, till 8 July, Soviet records admit the loss of 566 aircraft while the Germans claimed 923; not all of the German claims were confirmed by their own side. The 17th Air Army, opposite II./JG 52, were reduced to 706 aircraft from 1,052. At the beginning of the offensive, the only fighter support for JG 52 came from II and III/JG 3.

Crosses 200 Victories: Third to Reach

Walter Nowotny. (258 Air Victories). Image Source: Wikipedia

In August 1943 Rall claimed 33 aircraft shot down as JG 52 fought over Central Ukraine through the late summer. Rall claimed two Lavochkin-Gorbunov-Gudkov LaGG-3 fighters shot down on 29 August 1943 in the vicinity of Kuybyshev, present-day Samara, taking his total to 200 aerial victories. Following Hermann Graf and Hans Philipp, Rall was the third fighter pilot to reach the double century mark. This achievement earned him a named reference in the *Wehrmachtbericht* (Nazi propaganda communiqué) that day and on 12 September was also honoured with the Knight's Cross of the Iron Cross with Oak Leaves and Swords. He was the 34th member of the German armed forces to be so honoured. The presentation was made by Hitler at the Wolf's Lair, Hitler's headquarters in Rastenburg on 22 September 1943. Three other Luftwaffe officers were presented with awards that day by Hitler, *Major* Hartmann Grasser and *Hauptmann* Heinrich Prinz zu Sayn-Wittgenstein were awarded the Oak Leaves, and *Hauptmann* Walter Nowotny also received the Swords to his Knight's Cross with Oak Leaves. Following the award ceremony, Rall went on vacation until the end of September.

Back to Air Victories after Short Leave – 250th Mark

On his return Rall immediately began where he left off, claiming Soviet aircraft consistently. Over the course of October 1943, he claimed exactly 40 aircraft; his first coming on the 1 October. The majority were claimed in Southern Ukraine. With few exceptions, the enemy aircraft claimed were fighters. On 1 November 1943, Rall was promoted to the rank of Major, a rank he retained until the end of the war. In November claimed 12 aircraft and on the twenty-eighth day became the second fighter pilot after Nowotny to reach 250 aerial victories mark.

Victories Start Slowing

Rall filed his last claim of the year on 30 November. It was credited as his 252nd aerial victory. On 11 January 1944 Rall received the certification for the Oak Leaves and Swords, along with the medals from Hitler. In 1944 Rall continued to claim but at a slower rate. The Soviet Crimean Offensive opened on 8 April

28 November 1943, Congratulations Rall after his 250th aerial victory. Image Source: ww2gravestone.com

and five weeks later ended the German occupation in the Crimea. Rall claimed his 273rd and last aerial victory on the Eastern Front on 16 April 1944 over a Lavochkin La-5 fighter aircraft in the vicinity of Sevastopol.

Defence of the Reich: Heavy German Losses

On 19 April 1944, Rall was transferred to JG 11(11th Fighter Wing), as a Group Commander II, flying on operations in Defence of the Reich. Rall led his unit against the bomber fleets of United States Army Air Force's (USAAF) Eighth Air Force. The purpose of his Bf 109-equipped group was to engage the American escorting fighters, to allow the slower, heavier, and well-armed Focke-Wulf Fw 190 "Sturmbock" (Battering Ram) aircraft to intercept the bombers. Within two weeks since Rall was in combat, on 29 April, he claimed a Lockheed P-38 Lightning shot down north of Hanover.

Rall.
Image Source: neunundzwanzigsechs.de/jungmaenner

That day the USAAF targeted Berlin with 679 heavy-bombers escorted by 814 fighter aircraft. The German day fighter force was beginning to falter under the pressure. General of Fighters, Adolf Galland reported that for January-April 1944, 1,000 German pilots had been killed or wounded; Rall would soon become one of them (wounded).

American "Zemke fan" Tactic

On 12 May 1944, Rall was leading a formation of Bf 109s and bounced a flight of three P-47 Thunderbolts led by USAAF Colonel Hubert Zemke. Zemke was experimenting with a new tactic, the "Zemke fan", in which independent flights scattered in front of the bombers in order to cover as much sky as possible, thereby maximising the chance of intercepting German fighters. Zemke's flight had strayed too far in front of the bomber stream, and the fighters of JG 11 spotted an opening. The Zemke tactic left the four flights of P-47s isolated when a large numbers of Bf 109s were encountered.

Hartmann Grasser (103 victories), Walter Nowotny (258 victories), Günther Rall (275 victories), Heinrich Prinz zu Sayn, Wittgenstein (83 victories).
Image Source: http://www.flying-tigers.co.uk

Rall Gets Hit: Bails Out – Injured – Hospitalised

Rall was flying at 36,000 ft (11,000 meters) without cabin heating or pressurization, and 10,000 ft above the Fw 190s. Rall attacked claiming a Thunderbolt. His formation was then ambushed by other P-47s. Rall dived to escape, but his Bf 109 could not out-dive the Thunderbolts, which were attacking in line-abreast, preventing him from turning left or right. Rall was near to 620 kmph, but took hits in the engine and radiator by pilots of the 56th Fighter Group. Rall's left thumb was hit, and after he cleared the ice from his windshield with his remaining good hand, he decided there was no escape, and bailed out. He landed in a tree on a steep slope, then rolled down it into a gully after releasing his parachute harness. By luck, he avoided aggravating his earlier back injury and was tended to by

Many Years Later. General Rall with USAAF Colonel Hubert Zemke.
Image Source: planejunkie.com

Rall's damaged flying glove, which he wore when shot down in 1944 by American fighters, is now on display at the National Air and Space Museum in Washington, D.C.

farmers. Rall was hospitalised for many months in Nassau. Doctors found his thumb was attached only by skin and could not be saved. Rall credited the wound with saving his life as the Eighth Air Force established air superiority over Germany through the remainder of the war.

High Losses for the Unit

Rall's unit succeeded in this battle, but at a high cost. Besides Rall's claim of one P-47, two P-51 Mustangs were also claimed by other pilots. The group lost 11 Bf 109s, with two pilots killed and five wounded, and many aircraft were shot down.

Tall guy on the right is Gunther Rall. This was in the winter of 1944/1945.
Image Source: Pinterest upload by Tim Effler

Moves as Instructor

In the autumn, 1944, Rall moved to Bad Wörishofen and became an instructor at the Training School for Unit Leaders. Part of this training involved flying captured Allied aircraft and preparing notes for student pilots on their capabilities and deficiencies. Rall flew in mock-combat with Bf 109s. He specifically, flew the Supermarine Spitfire, Lockheed P-38 Lightning, P-47 Thunderbolt, and the P-51 Mustang.

Appointed Head of Fighter Wing

On 20 February 1945, he was appointed *Geschwaderkommodore* of 300th Fighter Wing, operating from airfields in southern Germany during the last months of the war. On his arrival, Rall found 15 burning German fighters on the airfield, courtesy of a low-level P-51 attack. Rall reported that the Wing was in chaos, with no radar while fuel and food had to be sought from day to day. JG 300 withdrew to Salzburg in Austria as American and French forces advanced deep into southern Germany. Rall did not claim an enemy aircraft during his time with his Wing.

On 2 March 1945, JG 300 flew with all four groups for the last time supported by JG 301. The two units sent 198 fighters to contest an American air raid. Only small groups reached the bombers but successes had no effect. JG 300 continued to fly and fight into 1945. On one mission the pilots claimed an optimistic total of 50 to 60 aircraft at the cost of 24 killed to the ever-present USAAF fighter escorts while Rall was hospitalised again due to his wound. On May 7, 1945, Germany unconditionally surrendered to the Allies in Reims, France, ending World War II and the Third Reich. Rall was taken as a prisoner of war.

Gunther Rall.
Image Source: Pinterest, uploaded by Luis Gimenez

Prisoner of War

Rall remained in a prisoner of war camp for a few weeks. He was approached by the Americans who were recruiting Luftwaffe pilots who had experience with the Messerschmitt Me 262 fighter. He was transferred to Bovingdon near Hemel Hempstead, and then based at RAF Tangmere, where he met the RAF fighter pilot Robert Stanford Tuck, with whom he became close friends.

Wife' Jewish Connection

After his release as POW, Rall settled back into civilian life. In 1948 he visited England again. He accompanied Hertha Rall and stayed in Grosvenor Square with Dr Paul Kaspar and Jewish acquaintants, whom she had helped to escape from the Nazis. Rall knew of Hertha's wartime Jewish connections and was concerned it would attract the attention of Nazi authorities. In 1943, Hertha was suspected of Jewish sympathies by the Gestapo, but no action was taken. By 1954 Hertha was a physician at the *Schule Schloss Salem*, near Lake Constance. Rall became PA to the dean of Prince Georg Wilhelm of Hannover School.

About Nazi Crimes

Of Nazi crimes, Rall acknowledged the pilots at the front knew of Nazi concentration camps but didn't know exactly what they were used for. When he first heard of Auschwitz and the Holocaust, initially he believed it to be propaganda. Rall could not believe that Germans would do such things. The criminal nature of the Nazi Party did not occur to Rall when Hitler came to power; "The fact that we did not explore the essence of the Nazi regime when it came to power is, of course, one of our great failings."

Re-joins German Air Force

Rall re-joined the newly established West German military in 1956 and became one of the first cadre of officers in the German Air Force. Around 6,000 air veterans survived the war but only 160 were fit to fly through years of idleness. The Bundesluftwaffe was ten years behind the times in modern aviation experience. The German military cadre knew they would have to spend years as pupils before they could stand on their own.

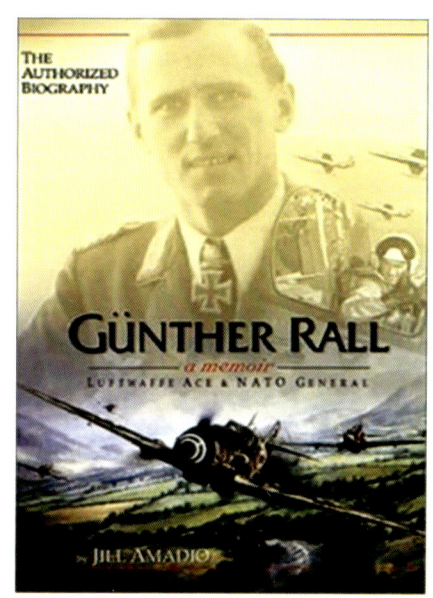

Gunther Rall: Luftwaffe Ace and NATO General. The Authorised Biography
Image Source: http://www.bookdepository.com

Trained with USAF. This one after German F-104 Starfighter flight.
Image Source: luftwaffe39-45.historia.nom.br/ases/rall.htm

Trains with USAF

Rall was sent to the United States to train on modern jets. Rall, and the former Luftwaffe officers he trained with, aspired to make the Bundesluftwaffe a carbon copy of the United States Air Force (USAF). The future chief of staff commented on modern USAF training methods compared to the old, highly individualistic training program of the Nazi Luftwaffe: "The systematic and consistent American training methods were impressive. All in all, these methods were better, more efficient in view of the aircraft we were being trained to fly. Indeed, we were going to fly jets. For most of us, this was a new era. The memories of flying the Me 262 were nostalgic for some of us but not a secure foundation you could build on."

Rebuilding German Air Force. Gunther Rall in F-104 cockpit.
Image Source: http://www.wsj.com/articles/SB125548213064683963

World War II German Aces who got a second chance in Bundesluftwaffe service included (from left) Gerhard Barkhorn (301), Erich Hartmann (352), Johannes Steinhoff (176) and Günther Rall (275). (Courtesy of Wolfgang Meuhlbauer)
Image Source: historynet.com

Rebuilding German Air Force

One of his tasks was to oversee modifications to the Lockheed F-104 Starfighter to comply with the requirement of the Bundeswehr, leading to the F-104G version. The accident rate of the new version was alarming when introduced in 1960. The machine was nicknamed the Witwenmacher (widow maker) after 292 crashes and 116 deaths. Officers like Erich Hartmann and Johannes Steinhoff believed the type too advanced for German pilots. Rall and Steinhoff thought it was a matter of training. They visited the United States to receive further training which reduced accidents when introduced to the German program. In particular, the training sought to address the fundamental change in role from high-altitude interceptor in the United States to fighter-bomber in Germany; and the radically different climate and weather conditions experienced at low altitudes by German pilots over Germany.

Senior Command

Rall received recommendations for senior commands by his then superior General Kurt Kuhlmey. Following his promotion to Brigadier General, he was appointed commander of the 3rd Air Force Division in Münster. Rall was then promoted to Major General on 15 November 1967 and on 1 April 1968 was given command of the 1st Air Force Division in Meßstetten. From 1 January 1971 to 31 March 1974, he held the position of Inspector of the Air Force and from 1 April 1974 to 13 October 1975, he was a military attaché with NATO.

A pair of Aces: Chuck Yeager (left) and famed World War II German ace Gunther Rall (2nd from right; 275 combat victories) during a meeting at Ramstein Air Base, West Germany, 1970.
Image Source: chuckyeager.com

India Visit

In May 1973, as part of a West German delegation, he was sent by the designer Kurt Tank to witness a demonstration of the Marut. He was reportedly quite impressed by the machine even as he was saddened by the fact that it never achieved its original supersonic dream.

Forced Retirement

Rall's forced retirement in 1975 was as a result of a controversial visit to apartheid-governed South Africa. Rall received a request from a German journalist, and former Bundesluftwaffe pilot, to attend a veterans meeting there. When news of the general's ill-

Image and Information Source: marutfans.wordpress.com/2010/04/23/german-ace-gunther-rall/ By Polly Singh

advised visit to Cape Town broke, German weekly magazine *Stern* claimed Rall held high-level meetings with South African officials and emphasised the personal nature of the trip. Despite its policy of apartheid, South Africa was seen as strategically important to NATO and the South Africans exploited Rall's visit. The political embarrassment, following a concerted press campaign, encouraged Defence Minister Georg Leber to retire Rall in October 1975. Rall subsequently resigned as military attaché to NATO. By the end of his career, he had attained the rank of Lieutenant General.

Summary of Service and Death

Lieutenant General Günther Rall with a Messerschmitt Bf 109 G-6.
Image Source: Pinterest Image Upload by Hernan Claudio

Rall served from 1936 to 1945 (Last Rank Major) and later from 1956 to 1975 (Last rank Lt. General). He Commanded 8/JG 52, III.JG 52, II.JG 11, and II.JG 300. Of his wartime service, he said simply: "We fought for our country, and to stay alive". Rall died at his home in Bad Reichenhall on 4 October 2009, aged 91, after suffering a heart attack two days earlier. He is buried at Friedhof Sankt Zeno, Bad Reichenhall, Berchtesgadener Land, Bavaria (Bayern), Germany

Burial Stone and Cemetery.
Image Source: ww2gravestone.com

Memoir "Flight Log Book"

Book Cover Image Source: goodreads.com

His memoir *Mein Flugbuch* ("My Flight Logbook") was released in 2004. Rall was interviewed in documentaries such as Thames Television's *The World at War*, and was a contributor to the *Wings* documentary television series produced by the Discovery Channel.

Aerial Victories

Matthews and Foreman, authors of *Luftwaffe Aces — Biographies and Victory Claims*, researched the German Federal Archives and found records for 274 aerial victory claims, plus one further unconfirmed claim. This number includes one victory over a French P-36, one victory over a U.S. P-38, and 272 Soviet-piloted aircraft on the Eastern Front. Victory claims were logged to a map-reference. During World War II Rall was credited with the destruction of 275 enemy aircraft in 621 combat missions. He was shot down five times and wounded on three occasions. Rall claimed all of his victories in a Messerschmitt Bf

109, though he also flew the Focke-Wulf Fw 190 operationally. All but three of his claims were against Soviet opposition. He was Ace-in-day on 03 May 1943, 20 August 1943 and 10 October 1943 (5 Victories each). There was only one unconfirmed claim of a 12 May 1944 against a P-47 for which he did not get credit. Over half his victims were in LaGG-3. His last credited Victory was on 29 April 1944, a P-38.

Image Source: Art Work by Pinterest (Upload by small scale art)

Awards

- Iron Cross (1939)
- 2nd Class (23 May 1940)
- 1st Class (July 1940)
- Wound Badge (1939) in Gold
- "Crete" Cuff band
- Front Flying Clasp of the Luftwaffe for fighter pilots in Gold with pennant "600"
- Honour Goblet of the Luftwaffe on 17 November 1941 as Lieutenant
- German Cross in Gold on 15 December 1941 as Lieutenant in the 8. JG 52
- Knight's Cross of the Iron Cross with Oak Leaves and Swords
- Knight's Cross on 3 September 1942 in 8. JG 52
- 134th Oak Leaves on 26 October 1942 as Lieutenant in the 8. JG 52
- 34th Swords on 12 September 1943 as Captain in the III/JG 52
- Commander's Cross of the Order of Merit of the Federal Republic of Germany (1973)

REFERENCES

1. Luftwaffe History, Gunther Rall http://www.luftwaffe39-45.historia.nom.br/ases/rall.htm
2. Rall Günther, Griffon Merlin, https://www.griffonmerlin.com/wwii-interview/gunther-rall/
3. Rall, Günther. World War II Graves, https://ww2gravestone.com/people/rall-gunther/
4. Günther Rall, Wikiwand, http://www.wikiwand.com/en/G%C3%BCnther_Rall
5. Günther Rall, Luftwaffe History, http://www.luftwaffe.cz/rall.html
6. World War II Pictures in Detail, 24 July 2013, http://ww2images.blogspot.com/2013/07/gunther-rall-and-his-men-with-unit.html
7. Glory. The largest archive of German WW II Images. https://www.flickr.com/photos/farinihouseoflove
8. Flying Tigers – The Aeroplane People https://www.flying-tigers.co.uk/
9. Stephen Miller, Nazi-Era Pilot Helped Lead Germany's Postwar Military, The Wall Street Journal October 15, 2009 https://www.wsj.com/articles/SB125548213064683963
10. Polly Singh, German Ace Gunther Rall, Marut Fans, https://marutfans.wordpress.com/2010/04/23/german-ace-gunther-rall/

22

FRENCH COLONEL RENÉ PAUL FONCK

The Highest Scoring All-time Allied Ace of Aces

"I put my bullets into the target as if I placed them there by hand." – **Rene Fonck**

René Paul Fonck.
Image Source: historynet.com

Colonel René Paul Fonck (27 March 1894 – 18 June 1953) was a French aviator who ended the First World War as the top fighter Ace and, when all succeeding aerial conflicts of the 20th and 21st centuries are also considered, Fonck still holds the title of "all-time Allied Ace of Aces". He received confirmation for 75 victories (72 solo and three shared) out of 142 claims. In the confusion of war, however, his tally could actually be much higher with some commentators estimating that at least 100 aircraft were victims of his gunnery and piloting skills. He was made an Officer of the Legion of Honour in 1918 and later a Commander of the Legion of Honour after the war, and raised again to the dignity of Grand Officer.

Early Life

Fonck was born on 27 March 1894 in the small village of Saulcy-sur-Meurthe in the Vosges region of north-eastern France. During his formative years, he reportedly received an engineering education at École Nationale Superieure des arts et metiers (The Grand School of Technology), but still there is a doubt about

René Paul Fonck.
Image Source: Pinterest (Danny's History & Ancient Cash Coins)

Image Source: commons.wikimedia.org

this aspect. Fonck actually left school when he was 13. Jon Guttman of HistoryNet says René Paul Fonck grew to be a rather short, unremarkable-looking young man whose own self-serving writings suggest ambitions at least partially driven by an inferiority complex. He claimed that his upbringing in the Alsace-Lorraine region, seized by the Germans after the humiliating Franco-Prussian War of 1870-71, had imbued him with a desire for revenge. When World War I broke out, he was mobilized on August 22, 1914, and assigned to the 2nd *Groupe d'Aviation* (Aviation Group) at Dijon. Although he had been interested in aviation from his youth, he was rejected for the air service when conscripted on 22 August 1914. Instead, he underwent five months of basic training for the role of the French Army's combat engineer because of his engineering skills. His training duties included first digging trenches near Épinal, and later bridge repairs on the Moselle River. Throughout this time, however, he harboured dreams of becoming a pilot.

Finally Joins Military Aviation

After numerous applications, he was finally selected for pilot training on February 15, 1915. The training was undertaken at centres at St. Cyr and later at Le Crotoy where he would practice flying on a Bleriot Penguin. It was a flightless version (simulator) of the Bleriot XI aircraft that gave the sensation of flight whilst firmly attached to the ground. In June 1915, having successfully earned his pilot's

Caudron G III observation planes. Image Source: YouTube

brevet and passed the final examinations, he was posted as a pilot to Escadrille (squadron) C47 flying Caudron G III observation planes, based at Corcieux, not far from his hometown. Fonck considered the unit's Caudron G.3s "slow and cumbersome," and after encountering a German plane while returning from reconnaissance over Colmar, he wrote that he "no longer took off without carrying a good carbine."

Transferred to an Ex Bomber Unit

Fonk with his senior and Ace French pilot Georges Guynemer. Image Source: aviadrix.blogspot.com

Early after training, Fonck was inducted into WW I. Jon Guttman writes that on March 17, 1917, Fonck and his observer helped bring down an Albatros north of Cernay-en-Laonnais. Fonck was clearly more fighter than recon pilot material. At age of 23, on 15 April 1917 ("Bloody April"), Fonck received a coveted invitation to join the famous *Escadrille les Cigognes*. Group de Combat 12, with its four escadrilles (or squadrons), was the world's first fighter wing. The then leading French Ace, Georges Guynemer, was serving at the time in one of its escadrilles, N3, and had just scored his 36th victory. But actually, on April 25, Fonck was transferred to N.103 of *Groupe de Combat* 12. Also known as "*Les Cigognes*" for the stork emblems that graced the sides of its Nieuport 17s and Spad VIIs. GC.12 was the elite group in the French air service, boasting such renowned fighters as Alfred Heurteaux, Albert Deullin, René Dorme, and Georges Guynemer. When Fonck arrived, however, N.103, a bomber unit recently turned into a fighter squadron, had yet to boast an Ace of its own. Fonck aimed to be the first.

World War I: First Victory

"I had obtained a new plane, a brand-new Spad with which I promised myself to do a great job," wrote Fonck. It took him and his mechanics two days to get the aircraft performing to his satisfaction, but his careful preparations paid off on May 5, when he and three comrades encountered five Albatros D.IIIs over Laon. Sergeant Pierre Schmitter's plane was hit, and Sergeant Claude Haegelen and Lieutenant Pierre Henri Hervet were hard-pressed when Fonck intervened and fired point-blank at a German who suddenly emerged from a cloud in front of him. "His plane immediately nose-dived to a crash at the corner of a wooded area," Fonck wrote. His victim, Warrant Officer Anton Dierle of *Jagdstaffel* (fighter squadron, or *Jasta*) 24, was killed.

More Actions – 1917

His second victory came on 17 May 1917, when Fonck downing an Albatros in conjunction with his observer, Sergeant Huffer. By this time, Fonck had amassed over 500 hours of flight time, an incredible amount in those early days of aviation. On 25 May 1917, Fonck's observer was killed by an anti-aircraft shell burst, a fate that almost befell Fonck a few weeks later. Fonck claimed his first enemy aircraft in July 1916, but his victory was

German Albatros D.III. The first type of aircraft shot down by Fonck.
Image Source: commons.wikimedia.org

unconfirmed. On 6 August, he and fellow pilot Lieutenant Thiberge engaged a German Rumpler CIII, and by manoeuvring over and around the reconnaissance plane, staying out of its fields of fire, forced it lower and lower until the German crew landed behind French lines. "For twenty minutes at least, from bank to bank and spiral to spiral," he wrote, "we descended from an altitude of 4,000 meters until we landed on a grassy field where, with their Will broke, and two Boche officers surrendered. They were the only prisoners I ever took." German records noted that 2nd Lt. Hermann von Raumer and Reserve 1st Lt. Adam Brey were taken prisoner that day. It was his first verified victory. It brought him the Médaille militaire (Military Medal) in late August 1917.

German Rumpler CIII Reconnaissance Aircraft.
Image Source: Wikipedia

Unit Switches to G.4s

In October C.47 switched from G.3s to twin-engine Caudron G.4s, and Fonck flew 13 long-range recon missions and 24 artillery-spotting flights during the month. Some of the G.4s carried cameras, which as Fonck noted in his autobiography, *Mes Combats*, "gives a clearer and more exact map, once corrected and adjusted for scale, than the work of the best professional geographer." He also observed that German anti-

Caudron G.4 Aircraft.
Image Source: tumblrgallery.xyz

aircraft fire was intensifying. During a photo-reconnaissance mission in June 1917, a shell tore through Fonck's right wing, missing his nacelle by less than a yard. "If the projectile had exploded on contact with my wing, my fate would have been sealed," he wrote. "I am not ashamed of the slight case of shivers that I still experience at this memory."

Becomes a Flying Ace

Fonck was instead assigned to another escadrille in the group, Spa 103. Flying the SPAD VII, he quickly made a name for himself, achieving fifth aerial victory and attaining flying Ace status by 13 May. He picked off another target on 12 June, then went on hiatus until 9 August. In late July, when GC.12 moved to Dunkirk in the Flanders sector, to face some of the best fighter squadrons in the German air service, the aerial action heated up considerably.

Interesting Air Challenges

Shortly after GC.12's arrival in their sector, some British pilots arrived to familiarise their personnel with their aircraft. There was a difference in technique between the renowned Guynemer and the rising star Fonck. There was a Canadian Ace, who offered to have a mock dogfight with Fonck and Guynemer. It was with the senior, Guynemer first. The two aircraft would cross in the air and the 'combat' would begin at once. Immediately, Guynemer was on his tail and he could not get him off. Fonck reportedly said, "Send me three pilots, and I will attack them. They will

Rene Paul Fonck.
Image Source: /malcolmmarshauthor.com

never see me." Three English pilots started, and were over the field, where they had lost sight of Fonck. Suddenly, there was a Spad flying through the three Englishmen. It was Fonck. That was the difference between the two schools. Fonck was a very good pilot, of course, but he never made a dog-fighting manoeuvre in the air. He always flew flat. Not to be seen by anybody...that was his style.

French Air Ace Georges Guynemer.
Image Source: musee-aeroscopia.fr/

Winning Streak

On August 19, Fonck embarked on a winning streak, downing an enemy plane daily until the 22nd. On September 14 he attacked a German observation aircraft and quickly killed the pilot. The plane suddenly and violently inverted and threw the observer through the wing of Fonck's aircraft. Such was the Frenchman's determination to accurately report his victories that he went to the crash site and ripped out the German aircraft's barograph to confirm, his twelfth, so its readout would confirm his combat report. Fonck ended the year with nineteen confirmed kills as well as a commission and had been awarded the highest French military honour the Légion d'honneur.

Avenges Guynemer Death

On September 11, 1917, Captain Georges Guynemer, victor over 53 German aircraft since 1915, did not return from a patrol. Everyone in GC.12 swore revenge, including Fonck. On September 14, he destroyed a two-seater in flames over Langemarck. "Such was the funeral of Guynemer to me," he later wrote. On 30 September, he and Adjutant Dupre jointly shot down a German two-seater Rumpler CIV 6787/16 of FA 18. The news reported

Reputation as an Excellent Shooter.
Image Source: albindenis.free.fr

30 Sept 1917; French Ace René Fonck and Adj. Dupre jointly shot down a German two-seater. On the return of his mission, Fonck was wounded in the head. Here with a piece of the plane he shot down. Image by Eva Conyne Sterner
Sourced From: Wikiwand

the killed pilot to be Lieutenant Kurt Wissemann, who had allegedly shot down Guynemer, and that Fonck had boasted of avenging the death of his "good friend" Guynemer. This story is put into question by German records, indicating that Kurt Wissemann of *Jasta 3* had been killed two days before in a different fight, in which he was flying a single-seater, probably against No. 56 Squadron.

Legion D'Honneur

September and October 2017 added four victories to Fonck's score. Thus, by year's end, he had raised his tally to nineteen. He was commissioned as an officer, and had received the *Légion d'honneur*.

Clinical Professionalism

Fonck got only better. Fonck's confidence in his ability was matched by his excellence in the air. He was a studious man and known for his clinical professionalism. He applied mathematical principles to combat flying, and his engineering knowledge regarding the capabilities of the aircraft he flew was unsurpassed among his fellow pilots. Fonck took few chances, patiently stalking his intended victims from higher altitudes.

The French Ace stands next to S452, one of two Spad XIIs armed with a 37mm cannon that were assigned to Spa.103 in May 1918. (Louis Risacher Album via Jon Guttman).
Image Source: historynet.com

Legion D'Honneur.
Image Source: Wikipedia

He then used deflection shooting with deadly accuracy at close range, resulting in an astonishing economy of ammunition expended per kill. More often than not, a single burst of fewer than five rounds from his Vickers machine gun was sufficient. Many commentators have compared his tactics as less a dogfight than surgical merciless executions. He was also reputed to be able to spot enemy observation aircraft from very far away, where most other pilots would have perceived nothing.

SPAD XII

Fonck, like France's leading Ace, Captain Guynemer, flew a limited-production SPAD XII fighter, distinguished by the presence of a hand-loaded 37mm Puteaux cannon firing through the propeller boss. He is apparently credited with downing 11 German airplanes with this type of weapon, called a "moteur-canon".

René Fonck Spad XIII Spa 103 in 1918.
Image Source: commons.wikimedia.org

This was made possible by the gear-reduction version of the Hispano-Suiza V8 SOHC engine first used in that model of SPAD fighter. It offset the now-hollow propeller shaft above the crankshaft axis, and the 37mm cannon was mounted in the V space between the two rows of cylinders.

Fonck would later fly the highly successful SPAD XIII, the first SPAD fighter model to use twin Vickers machine guns. Elegant looking on the outside, the Spad XIII was decidedly different inside the cockpit, where the cannon breech protruded between the pilot's legs, necessitating Deperdussin-type elevator and aileron controls on either side of his seat instead of a central control column. A highly skilled pilot like Guynemer could master such a system, but he was also forced to deal with the heavy recoil of a single-shot weapon that filled the cockpit with smoke upon firing and had to be reloaded by hand. In spite of the Spad XIII's shortcomings, Fonck found its speed and sturdiness in a dive ideal for his stalking tactics.

Spectacular 2018: "Ace-in-a-Day"

For Fonck 1918 started with something of a fallow period of nineteen days without a single kill to his name but a double victory on 19 January ended this drought. Adapting to it readily, he downed two opponents on January 19, and by March 17 had raised his score to 30. February added another five, March seven more, and another three in April. Then came a spectacular performance on 9 May. Rene Fonck was a serious character and a heated disagreement between the Frenchman and two American squadron-mate pilots, Edwin C Parsons and Frank Baylies led to perhaps the single most spectacular day in his career. Perturbed by Fonck's lectures on aerial success, the two Americans bet Fonck a bottle of champagne that one of them would shoot down an enemy plane before Fonck. Baylies took off despite hazy weather and shot down a Halberstadt CL.II. Back at the airfield, he expected Fonck to honour the bet. He did not. Rather than pay off the bet, a sulky Fonck badgered the Americans to change the terms of the bet so that whoever shot down the most Germans that day would win. The lingering

French soldiers examine the recovered remains of a Rumpler C.IV brought down by Sub-Lieutenant Fonck in the spring of 1918. (SHAA B76.32).
Image Source: historynet.com

fog kept Fonck grounded most of the day. It was well into the afternoon before it cleared enough for him to take off at 1500 hours. Between 1600 and 1605 hours, he shot down three enemy two-seater reconnaissance planes. A couple of hours later, he repeated the feat. Thus becoming "Ace-in-day" for the first time. Understanding the importance of reconnaissance planes, with their potential to direct intensive artillery fire onto French troops, Fonck concentrated his attentions upon them; six shot down within a three-hour span proved it. He added a double victory on 19 May and five more in June. By now, he was shooting doubles frequently, and with 49 on his score sheet, he was rapidly closing in on Guynemer's record.

Surpasses Legendary Guynemer

On 18 July 1918, he achieved another double, to bring his total to 53 and into a tie with Guynemer. The following day, he shot down three more enemy aircraft and surpassed the score of the legendary Guynemer, who had remained the leading French Ace after his death on 11 September 1917.

Becomes Leading Allied Ace

He added four more victories in August, raising his total to 60. Fonck was becoming a legend amongst the Allied forces and even in Germany, his reputation was well known and respected. Then, on 26 September, he

A reunion of Aeroclub of Paris in 1918 to honour legends of WW I. Roland Garros (Died 5 October 1918, aged 29) is standing first from the left. Fonck is third in Uniform.
Image Source: albindenis.free.fr

repeated his feat of knocking down six enemy airplanes in a day, although this time three of his six victories were over Fokker D.VII fighters. "I now had sixty-six official victories to my credit." He had also become the only World War I Ace with two six-victory days in his combat log. Another success two days later and two on 5 October put his score at 69, very close to the 72 of Major William Avery Bishop, then the leading Allied Ace. On 30 October, he matched Bishop with three more victories. He shot down two more the following day, and another the day after that, finishing with 75 confirmed victories. By end of the war, Rene Fonck had become France's highest-scoring combat Ace. His victory over a Halberstadt on 01, November 2018, was also the last for GC.12 before the armistice was signed on the 11th, bringing Group's total to 286 aircraft and five balloons, although if victories prior to the Group's formation are counted, the collective wartime total of its component squadrons came to 411 planes and 11 balloons. The group's top-scoring squadron had been Spa.3 with 175 victories, but Spa.103 ranked second with a wartime total of 111, with 73 of them scored by one individual: René Fonck.

Fonck of Spa.103 (left) shares the limelight with Lieutenant Gustave Lagache, Commander of Spa.3, and Lieutenant Bernard Barny de Romanet, 18-victory Ace and Commander of Spa.167. (SHAA, B88.3570).
Image Source: historynet.com

Summary of WW I

To summarize, he was officially awarded 75 confirmed victories. With 75 confirmed, and 52 unconfirmed, Fonck was the undisputed Allied Ace of Aces, yet he never received the adulation bestowed upon Guynemer and Nungesser. He had 56 victories during the whole of 1918. His 1918 list by itself would have made him France's leading Ace. Unlike many leading French Aces, Fonck's score contained only three shared victories. Also unlike most Aces, he remained unwounded; indeed, only a single enemy bullet had ever hit his aircraft. He had also forgone the most hazardous air-to-air combat: he had shot down no balloons.

Image Source: totalcontentdigital.com

Fonck: Ascetic and Withdrawn

Yet for all his skill and success, Fonck never captured the heart of the French public as Guynemer had. Fonck was ascetic and withdrawn. Instead of drinking or socialising with the other pilots, he planned his flying missions and tactics, ironed his uniforms, and stayed physically fit through calisthenics. He seemed to overcompensate for his shyness by constantly mentioning his exploits. As a result, he seemed distant, arrogant, and even abrasive. His comrades respected his skills, but even one of his few friends, Marcel Haegelen, considered him a braggart and shameless self-promoter. Fonck may have resented the fact that Guynemer remained more popular in the French press even after he surpassed him in victories. Fonck also seemed to lack insight into the effect his personality had upon his image or career. However, he and he alone carried the flag of the French Air Force at the victory parade on the Champs-Elysées.

Rene Fonck carried the flag of the French Air Force at the victory parade on the Champs-Elysées on 14 July 1919.
Image Source: rene.fonck.free.fr/

Civil Life after WW I: Member of Parliament

Fonck returned to civilian life after World War I, and published his war memoirs *Mes Combats*, prefaced by Marechal Foch, the French General, Marshal of France and Allied Supreme Commander in World War I, in 1920. The fame he got from the war allowed him to be elected Member of Parliament representing the Vosges from 1919 to 1924.

Transatlantic Air Race

During the 1920s, Fonck persuaded Igor Sikorsky to redesign the Sikorsky S-35 for the transatlantic race for US$ 25,000 Orteig Prize. On 21 September 1926, Fonck crashed on take-off when the landing gear collapsed, killing two of his three crew members. Charles Lindbergh won the prize seven months later in 1927.

With Igor Sikorsky in front S-35 aircraft for Trans-Atlantic attempt.
Image Source: Sikorsky Archives

Returns to Military Aviation & Nazi Links Controversy

Fonck eventually returned to military aviation and rose to Inspector of French fighter forces from 1937 to 1939. His inter-war contact with the likes of former World War I foe Hermann Göring and Ernst Udet cast a shadow upon Fonck's reputation and independence during the German occupation of France in 1940, as did allegations of collaboration with the Nazis and the Vichy regime. On 10 August 1940, Vichy Foreign Minister Pierre Laval announced that Fonck had recruited 200 French pilots to fight on the Nazi side. However, the truth was more complicated. Marshal Philippe Pétain wished to exploit Fonck's relationship to Göring in order to meet Adolf Hitler. He ordered Colonel Fonck to talk to Göring. A meeting was planned at Montoire, but after discovering evidence about the pro-Nazi politics of Pierre Laval, Fonck tried to convince Pétain not to attend. Initially, Pétain appeared to heed Fonck's advice, but for some reason, he eventually decided to disregard Fonck's warnings and met Hitler at Montoire on 24 October 1940. Fonck's loyalties were thus questioned by the Vichy regime, and he returned home to Paris, where he was eventually arrested by the Gestapo and imprisoned in the Drancy internment camp. Effectively he could not be part of WW II.

Lindbergh with French WWI Ace Capt. Rene Fonck in 1927.
Image Source: flickr.com/photos/sdasmarchives

René Fonck with German pilot Ernst Udet ten years after World War 1.
Image Source: Saved by Paul Maerten on Pinterest

Post-World War II: Cleared For Loyalty

After the war, a French police inquiry about his supposed collaboration with the Vichy regime completely cleared Fonck. The conclusion was that his loyalty was proved by his close contacts with recognised resistance leaders such as Alfred Heurtaux during the war. Additionally, he was awarded the Certificate of Resistance in 1948. The citation reads "Mr. Fonck, René, a member of the fighting French forces without uniform, took part, in the territory occupied by the enemy, to glorious fights for the liberation of the nation".

Tombe de René Fonck, Saulcy-sur-Meurthe.
Image Source: aerosteles.net

Last Days: Dies at 59

Fonck remained in Paris but also visited frequently his native Lorraine, where he had business interests. He died of a stroke in his Paris apartment, Rue du Cirque, on June 18, 1953, at the age of 59, and is buried in the cemetery of his native village of Saulcy-sur-Meurthe.

Great Way to Sum Up

Jon Guttman of History Net wrote based on an interview with Historian Ian Toll "Twilight of the Gods", covering the story "Allied Ace of Aces: René Fonck." He wrote, "When Germans, Americans, Italians or Belgians think of World War I aviation, the first names that come to mind are usually their highest-scoring fighter pilots, Manfred Freiherr von Richthofen, Edward Rickenbacker, Francesco Baracca, and Willy Coppens. An exception is France, which most reveres its second-ranking ace, Georges Guynemer, among its martyred heroes, while the higher-scoring René Fonck settles for posterity's grudging respect for his wartime achievements. A less romantic, more practical mind might note that Guynemer literally burned himself out in his single-minded patriotism, making his death, in September 1917, almost inevitable. Fonck, in contrast, flew, fought, and lived by a philosophy that dying for one's country was less desirable than making one's opponent die for his. Cynical though that outlook seemed

Pilots of the Group, Squadrons 3, 26 and 103 in August 1918. Fonck in the centre in Uniform
Image Source: airpowerasia

at the time, it was arguably more mature and better suited for a fighter pilot's success—and survival. But perhaps Fonck's biggest problem compared to Guynemer was that he survived.

"He is not a truthful man," said Haegelen, who was nevertheless one of Fonck's best friends. "He is a tiresome braggart, and even a bore, but in the air, a slashing rapier, a steel blade tempered with unblemished courage and priceless skill....But afterwards, he can't forget how he rescued you, nor let you forget it. He can almost make you wish he hadn't helped you in the first place." Swiss volunteer Jacques Roques summed up Fonck by saying, "As a fighter pilot, in one word, the best...but he was not a very sympathetic character."

In seeming contradiction to his grating personality, Fonck's lifestyle was arguably among the most sensible for a fighter pilot of his time. While Guynemer flew relentlessly, and third-ranking French Ace Charles Nungesser alternated between fighting, womanising, and drinking, getting barely two hours of sleep at night, Fonck rested between missions, drank moderately, and spent much of his leisure time practicing his marksmanship.

Book on Fonck

While uncounted volumes have been written about Guynemer, the only author who wrote a book devoted to Fonck was Fonck himself. The Storks, *by Norman Franks and Frank Bailey, and* Ace of Aces, *by René Fonck.* An unrequited seeker of glory whose deeds could easily have spoken for him eloquently enough by themselves—if only he had let them. Fonck is also remembered for his two famous quotes.

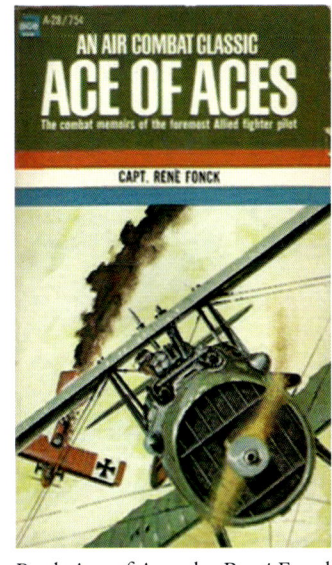

Book Ace of Aces, by René Fonck

- On his closing into the target aircraft during combat, he said *"I put my bullets into the target as if I placed them there by hand."* – Rene Fonck
- On his preference of flying alone, he said, *"I prefer to fly alone... when alone, I perform those little coups of audacity which amuse me..."* – Rene Fonck

Citations

Médaille Militaire citation, 1916 – "A pilot of remarkable bravery and skill, having already engaged in a great number of aerial combats. On 6 August 1916, he resolutely attacked two strongly armed enemy planes, took on one in pursuit, and by a series of bold and skillful manoeuvres, forced it to land uninjured within our lines. He has been cited in orders twice."

Légion d'honneur – "A fighting pilot of great value, combining outstanding bravery and exceptional qualities of skill and sang-froid. He came to pursuit aviation after 500 hours of flight on army corps aircraft and became, in a short time, one of the best French combat pilots. On 19, 20 and 21 August 1917, he shot down his 8th, 9th, and 10th enemy aircraft. He has already been cited seven times in orders, and has received the Médaille Militaire for feats of war."

Commandeur de la Légion d'Honneur, awarded on 16 June 1920. Decorated by General Fayolle Dans at Invalides, Paris, on 27 July 1922.
Image Source: albindenis.free.fr

Légion d'honneur chevaliership in 1917. He was raised to the Grade of Commander in 1920, and to the Dignity of Grand Officer in 1936.

René Fonck Grand Officer de la Légion d'Honneur presented by President de la République Albert Lebrun, on 14 July 1936. Photo collection du CRI de Nancy.
Image Source: albindenis.free.fr

Image Source: Pinterest (samll scale art)

One of the Most Decorated French War Heroes

A remarkable officer from every point of view; of admirable fighting ardour. Pilot of the highest order, for reconnaissance missions and artillery range intelligence, as well as for surveillance service that he completed many times despite very unfavourable atmospheric conditions. He demonstrated, during the course of an uninterrupted series of aerial combats, an exceptional strength and will to win, which sets an example for the French chasse pilots of today.

Rene Fonck was also awarded the British Military Cross and the British Distinguished Conduct Medal. He had twenty eight army citations ("palmes"), and one bronze regimental citation ("étoile de bronze") attached to his War Cross.

REFERENCES

1. Find A Grave https://www.findagrave.com/memorial/51798788/rene-paul-fonck
2. Vocal Media. https://vocal.media/serve/aerial-combat-ace-rene-paul-fonck
3. Aviation History *research director Jon Guttman,* History Net https://www.historynet.com/allied-ace-of-aces-rene-fonck.htm
4. Kennedy Hickman, Thought Co. https://www.thoughtco.com/world-war-i-colonel-rene-fonck-2360477
5. René Paul Fonck http://albindenis.free.fr/Site_escadrille/Rene_Fonck.htm
6. Aeroscopia Musee, France. http://www.musee-aeroscopia.fr/
7. Memoire de Rene Fonck http://rene.fonck.free.fr/spip/index.php
8. SDASM Archives https://www.flickr.com/photos/sdasmarchives
9. Steles, monuments, plaques commemoratives, https://www.aerosteles.net/
10. The Construction of an Image in Aviation: the Case of René Fonck and the French Press (1917-1926), Dossier thématique / Thematic Section, http://revues.univ-tlse2.fr/pum/nacelles/index.php?id=654
11. Colonel René Paul Fonck, Wikipedia https://en.wikipedia.org/wiki/Ren%C3%A9_Fonck

23

BILLY BISHOP

The Top Canadian and British Empire Ace of all Time

Billy Bishop.
Image Source: theworldwar.org

The most important thing in fighting was shooting, next the various tactics in coming into a fight, and last of all flying ability itself.

– Billy Bishop

William Avery Bishop, VC, CB, DSO & Bar, MC, DFC, ED (8 February 1894 – 11 September 1956) was a Canadian flying Ace of the First World War. He was officially credited with 72 victories, making him the top Canadian and British Empire ace of the war. He was a Victoria Cross recipient who rose to become an Air Marshal. During the Second World War, Bishop was instrumental in setting up and promoting the British Commonwealth Air Training Plan.

Early Life

Bishop was born on 8 February 1894 in Owen Sound, Ontario. He was the third of four children born to William Avery Bishop Sr. and Margaret Louisa (Green) Bishop. His father, a lawyer and graduate of Osgoode Hall Law School in Toronto, Ontario, was the Registrar of Grey County. Attending Owen Sound Collegiate and Vocational Institute, Bishop earned the reputation of a fighter,

William Avery 'Billy' Bishop (1894–1956), VC painting by S. J. Payne Royal Air Force Museum. *Image Source: artuk.org*

Bishop as an officer cadet of the Royal Military College of Canada, c. 1914.
Image Source: Wikipedia

defending himself and others easily against bullies. He avoided team sports, preferring solitary pursuits such as swimming, horse riding, and shooting. Bishop was less successful at his studies; he would abandon any subject he could not easily master, and was often absent from class.

Initial Love for Aviation

At the age of 15, Bishop built an aircraft out of cardboard, wooden crates and string, and made an attempt to fly off the roof of his three-story house. He was dug, unharmed, out of the wreckage by his sister. In 1911, Billy Bishop entered the Royal Military College of Canada (RMC) in Kingston, Ontario, where his brother Worth had graduated in 1903. At RMC, Bishop was known as "Bish" and "Bill". Bishop failed his first year at RMC, worked hard his second year but in his third year was caught cheating.

The Mackenzie Building of Royal Military College of Canada (RMC) in Kingston, Ontario.
Image Source: Wikipedia

Billy Bishop as RFC Observer.
*Image Source: Billy Bishop Home: Museum, Archives and National Historic Site.
canadiankidsactivities.com*

Joins Horse Cavalry: Excelled on the Firing Range

When the First World War broke out in 1914, Bishop left RMC and joined The Mississauga Horse cavalry regiment. He was commissioned as an officer but was ill with pneumonia when the regiment was sent overseas. After recovering, he was transferred to the 7th Canadian Mounted Rifles, a mounted infantry unit, then stationed in London, Ontario. Bishop showed a natural ability with a gun, and excelled on the firing range: he put bullets in a target placed so far away, others saw only a dot, due to his seemingly "super-human" eyesight. His service record at the time of attestation shows that he was 5 feet, 6¾ inches tall, weighed 138 lbs., and had a fair complexion, blue eyes, light brown hair, and a scar on the inner part of the right leg, and was a Presbyterian. The unit left Canada for England on 6 June 1915 onboard the requisitioned cattle ship *Caledonia*. On 21 June, off the coast of Ireland, the ship's convoy came under attack by U-boats. Two ships were sunk and 300 Canadians died, but Bishop's ship was unharmed, arriving in Plymouth harbour on 23 June.

Switches to RFC: As an Air Observer

Bishop quickly became frustrated with the mud of the trenches and the lack of action. In July 1915, after watching a Royal Flying Corps (RFC) aircraft return from a mission, Bishop said "it's clean up there! I'll bet you don't get any mud or horse shit on you up there. If you die, at least it would be a clean death." While in France in 1915 he transferred to the RFC. As there were no places available for pilots in the flight school, he chose to be an observer. On 1 September, he reported to 21 (Training) Squadron at Netheravon for elementary air instruction. The first aircraft he trained in was the Avro 504, flown by Roger Neville (WWI Flying Ace). Bishop was adept at taking aerial photographs, and was soon in charge of training other observers with the camera. The squadron was ordered to France in January 1916 and arrived at Boisdinghem airfield, near Saint-Omer, equipped with R.E.7 reconnaissance aircraft.

R.E.7 reconnaissance aircraft.
Image source: 3squadron.org.au

First Combat Mission

Bishop's first combat mission was as an aerial spotter for British artillery. At first, the aircraft could not get airborne until they had offloaded their bomb load and machine guns. Bishop and pilot Neville flew over German lines near Boisdinghem and when the German howitzer was found, they relayed coordinates to the British, who then bombarded and destroyed the target. In the following months, Bishop flew on reconnaissance and bombing flights, but never fired his machine guns on an enemy aircraft.

Injured: Misses Battle of the Somme

During one take-off in April 1916, his aircraft engine failed, and he badly injured his knee. The injury was aggravated while on leave in London in May 1916, and Bishop was admitted to the hospital in Bryanston Square. While there he met and befriended socialite Baroness Lady St Helier, who was a friend to both Winston Churchill and Secretary for Air Lord Hugh Cecil. After Bishop's father suffered a small stroke, St Helier arranged for Bishop to recuperate in Canada, and he thereby missed the Battle of the Somme.

William Avery Bishop in a Nieuport aircraft, circa 1917.
Photo Credit: Library and Archives Canada.

Bishop returned to England in September 1916, and, with the influence of St Helier, was accepted for training as a pilot at the Central Flying School at Upavon on Salisbury Plain. His first solo flight was in a Maurice Farman "Shorthorn". In November 1916 after receiving his Wings, Bishop was attached to No. 37 Squadron RFC at Sutton's Farm, Essex flying the BE.2c. He was officially appointed to flying officer duties on 8 December 1916. Bishop disliked flying at night over London, searching for German airships, and he soon requested a transfer to France.

Bishop and a Nieuport 17 fighter in Filescamp, 1917.
Image Source: Wikipedia

Initial Combat Mess-ups

On 17 March 1917, Bishop arrived at 60 Squadron at Filescamp Farm near Arras, where he flew the Nieuport 17 fighter. At that time, the average life expectancy of a new pilot in that sector was 11 days, and German Aces were shooting down British aircraft 5 to 1. Bishop's first patrol on 22 March was less than successful. He had trouble controlling his run-down aircraft, was nearly shot down by the anti-aircraft fire, and became separated from his group. On 24 March, after crash-landing his aircraft during a practice flight in front of General John Higgins, Bishop was ordered to return to flight school at Upavon. Major Alan Scott, the new commander of 60 Squadron, convinced Higgins to let him stay until a replacement arrived.

First Aerial Victory

The next day, 25 March, Bishop claimed his first victory when his was one of the four Nieuports that engaged three Albatros D.III Scouts near St Leger. Bishop shot down and mortally wounded a Lieutenant Theiller, but his engine failed in the process. Bishop landed in no man's land, 300 yards (270 m) from the German front line. After running to the Allied trenches, Bishop spent the night on the ground in a rainstorm. There Bishop wrote a letter home, stating, "I am writing this from a dugout 300

German Albatros D.III Scout.
Image Source: Wikipedia

yards from our front line, after the most exciting adventure of my life." General Higgins personally congratulated Bishop and rescinded his order to return to flight school. On 30 March 1917, Bishop was named a Flight Commander with a temporary promotion to Captain a few days later.

Painting of Captain Albert Ball's Nieuport 17 fighter with red Spinner.
Image Source: arizonaskiesmeteorites.com

Becomes an Air Ace

On 31 March, he scored his second victory. Bishop, in addition to the usual patrols with his squadron comrades, soon flew many unofficial "lone-wolf" missions deep into enemy territory, with the blessing of Major Scott. As a result, his total of enemy aircraft shot down increased rapidly. On 8 April, he scored his fifth victory and became an Ace. To celebrate, Bishop's mechanic painted the aircraft's nose blue, the mark of an Ace. Former 60 Squadron member Captain Albert Ball, at that time the Empire's highest-scoring Ace, had had a red spinner fitted.

Leading From the Front: Bounty on His Head

Bishop's no-holds-barred style of flying always had him "at the front of the pack," leading his pilots into the battle over hostile territory. Bishop soon realized that this could eventually see him shot down. After one patrol, a mechanic counted 210 bullet holes in his aircraft. He then switched to a new method of using the surprise attack. This proved successful. He claimed 12 aircraft in April alone, winning the Military Cross for his participation in the Battle of Vimy Ridge. The successes of Bishop and his blue-nosed aircraft were noticed by the Germans, and they began referring to him as "Hell's Handmaiden". Ernst Udet called him "the greatest English scouting Ace" and one *Jasta* had a bounty on his head.

A replica of Canadian fighter ace Billy Bishop's Nieuport 17 biplane. The blue painted nose of this plane signifies "Ace" status.
Image Source: diecastaircraftforum.com

"Billy Bishop and the Red Baron" Air Encounter by Composer Jeremy David Hiebert.
Image Source: YouTube

Survives Encounter with Red Baron

On 30 April, Bishop survived an encounter with *Jasta 11* and Manfred von Richthofen, the Red Baron. In May, Bishop received the Distinguished Service Order for shooting down two aircraft while being attacked by four others.

Victoria Cross and Controversy

On 2 June 1917, Bishop flew a solo mission behind enemy lines to attack a German-held aerodrome, where he claimed that he shot down three aircraft that were taking off to attack him and destroyed several more on the ground. For this feat, he was awarded the Victoria Cross (VC), although it has been suggested that he may have embellished his success. His VC, awarded on 30 August 1917, was one of two awarded in violation of the warrant requiring witnesses (the other being the Unknown Soldier), and since the German records have been lost and the archived papers relating to the VC were lost as well, there is no way of confirming whether there were any witnesses. It seems to have been common practice at this time to allow Bishop to claim victories without requiring confirmation or verification from other witnesses.

Bishop in the cockpit of his Nieuport 17, c. August 1917. During this period, Bishop became the highest scoring flying Ace in the Royal Flying Corp.
Image Source: britannica.com/biography/William-Avery-Bishop

Highest Scoring RFC Ace: Third Highest of War

In July, 60 Squadron received new Royal Aircraft Factory S.E.5s, a faster and more powerful aircraft with better pilot visibility. In August 1917, Bishop passed the late Albert Ball in victories to become (temporarily) the highest-scoring Ace in the RFC and the third top Ace of the war, behind only the Red Baron and René Fonck. At the end of August 1917, Bishop was appointed as the Chief Instructor at the School of Aerial Gunnery and given the temporary rank of Major.

Wedding: Assignment in Washington

Bishop returned home on leave to Canada in fall 1917, where he was acclaimed a hero and helped boost the morale of the Canadian public, who were growing tired of the war. On 17 October 1917, Bishop married his long-time fiancée, Margaret Eaton Burden. After the wedding, he was assigned to the British War Mission in Washington, D.C. to help the Americans build an air force. While stationed there, he wrote his autobiography entitled *Winged Warfare*.

No. 60 Squadron RFC Lts Billy Bishop (left) & Graham Conacher Young at Filescamp Farm, summer 1917. Photo featured in "Osprey Aviation Elite Units 41: No 60 Sqn RFC/RAF" by Alex Revell.
Image Source: Pinterest uploaded by Mirela Nica

Return to Europe: Squadron Command – More Victories

Upon his return to England in April 1918, Bishop was promoted to Major and given command of No. 85 Squadron, the "Flying Foxes". This was a newly formed squadron and Bishop was given the freedom to choose any of the pilots. The squadron was equipped with SE5a scout planes and left for Petit Synthe, France on 22 May 1918. On 27 May, after familiarizing himself with the area and the opposition, Bishop took a solo flight to the Front. He downed a German observation plane in his first combat since August 1917, and followed with two more the next day. From 30 May to 1 June Bishop downed six more aircraft, including German ace Paul Billik, bringing his score to 59 and reclaiming his top-scoring Ace title from James McCudden, who had claimed it while Bishop was in Canada, and he was now the leading Allied Ace.

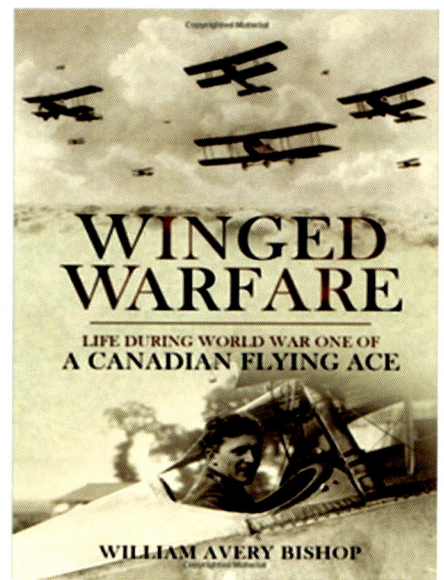

His memoire '*Winged Warfare: Life during the War of a Canadian Flying Ace*' was penned shortly after he'd been awarded the Victoria Cross in 1917.
Image Source: puttingdownuprising.com

Becomes an Icon: Asked to Return to Canada

The Government of Canada was becoming increasingly worried about the effect on morale if Bishop were to be killed, so on 18 June he was ordered to return to England to help organize the new Canadian Flying

Billy Bishop, 1918.
Image Source: Wikipedia

Corps. Bishop was not pleased with the order coming so soon after his return to France. He wrote to his wife: "This is ever so annoying." The order specified that he was to leave France by noon on 19 June. On that morning, Bishop decided to fly one last solo patrol. In just 15 minutes of combat, he added another five victories to his total. He claimed to have downed two Pfalz D.IIIa scout planes, caused another two to collide with each other, and shot down a German reconnaissance aircraft.

Promoted Lieutenant-Colonel

On 5 August, Bishop was promoted to Lieutenant-Colonel and was given the post of "Officer Commanding-designate of the Canadian Air Force Section of the General Staff, Headquarters Overseas Military Forces of Canada." He was on board a ship returning from a reporting visit to Canada when news of the armistice arrived. Bishop was discharged from the Canadian Expeditionary Force on 31 December and returned to Canada.

With His Wife Margaret Eaton Burden.
Image Source: dundurn.com

William Avery "Billy" Bishop in 1917 – by Jim Bruce
An illustration commissioned by the Reader's Digest for their 1977 book "Heritage of Canada".
Image Source: aviationartists.ca

Final Score

By the end of the war, he had claimed some 72 air victories, including two balloons, 52 aircraft shot and two shared "destroyed" and 16 aircraft "out of control". The 72 enemy aircraft, included 25 in one 10-day period. During WWI in many cases it was difficult to authenticate all victories because when Germans withdrew from France, they destroyed many records. Historians including Hugh Halliday and Brereton Greenhous (both of whom were official historians for the Royal Canadian Air Force) suggested that the actual total may have been far lower. Brereton Greenhous felt the actual total of enemy aircraft destroyed was only 27. But yet the Canadian Government and others kept honouring Bishop with the highest awards. It was thus presumed that the Victories were officially granted. He continues to be seen as the highest-scoring Allied Ace.

Post-WW I Career

After the war, Bishop toured the principal cities in the United States and lectured on aerial warfare. He established an importing firm, Interallied Aircraft Corporation, and a short-lived passenger air service with fellow Ace William Barker, but after legal and financial problems, and a serious crash, the partnership and company were dissolved. In 1921, Bishop and his family moved to Britain, where he had various business interests connected with flying. In 1928, he was the guest of honour at a gathering of German Air Aces in Berlin and was made an Honorary Member of the Association. In 1929 he became chairman of British Air Lines. However, the family's wealth was wiped out in the crash of 1929 and they had to move back to Canada, where he became Vice President of the McColl-Frontenac Oil Company.

Second World War: Promoted Air Marshal

In January 1936, Bishop was appointed the first Canadian Air Vice Marshal. Shortly after the outbreak of war in 1939, he was promoted to the rank of Air Marshal in the Royal Canadian Air Force (RCAF). He served during the war as Director of the Royal Canadian Air Force and was placed in charge of recruitment. He was so successful in this role that many applicants had to be turned away. Bishop created a system for training pilots across Canada and became instrumental in setting up and promoting the British Commonwealth Air Training Plan, which trained over 167,000 airmen in Canada during the Second World War.

William Avery ('Billy') Bishop by Bassano Ltd.
Image Source: National Portrait Gallery, London

Books, Movies and Civil Aviation

In 1942, he appeared as himself in the film *Captains of the Clouds*, a Hollywood tribute to the RCAF. By 1944 the stress of the war had taken a serious toll on Bishop's health, and he resigned from his post in the RCAF to return to private enterprise in Montreal, Quebec, before retiring in 1952. His son later commented that he looked 70 years old on his 50th birthday in 1944. However, Bishop remained active in the aviation world, predicting the phenomenal growth of commercial aviation post-war. His efforts to bring some organization to the nascent field led to the formation of the International Civil Aviation Organization (ICAO) in Montreal. He wrote a second book at this time, *Winged Peace*, advocating international control of global airpower.

Air Marshal Billy Bishop stands on parade during the filming of "Captains of the Clouds" at Uplands in Ottawa, Ontario, on April 5, 1943. During the film, Air Marshal Bishop, was playing himself.

Image Source: puttingdownuprising.com

Last Days

With the outbreak of the Korean War, Bishop again offered to return to his recruitment role, but he was in poor health and was politely refused by the RCAF. He died in his sleep on 11 September 1956, at the age of 62, while wintering in Palm Beach, Florida. His funeral service was held with full Air Force Honours in Toronto, Ontario. The body was cremated and the ashes interred in the family plot in Greenwood Cemetery, Owen Sound, Ontario. A memorial service for Air Marshal Bishop was held in St Paul's Church, Bristol, England, on 19 September 1956.

Family of Aviators

On 17 October 1917, at Timothy Eaton Memorial Church in Toronto he married Margaret Eaton Burden, his long-time fiancée, and daughter of Mr. C.E. Burden (a granddaughter of Timothy Eaton and sister of Ace Henry John Burden). They had a son, William, and a daughter, Margaret. Both of the Bishop children became aviators. William Arthur Christian Avery Bishop (1923 London, England – 2013 Toronto) was presented with his Wings by his father during the Second World War. Arthur went on to become a

Spitfire pilot and served with No. 401 Squadron RCAF in 1944. After the war, he became a journalist, advertising executive, entrepreneur, and author. Margaret Marise (Jackie) Willis-O'Connor (1926 London – 2013 Ottawa) was a wireless radio operator during World War II, whom Bishop presented with a Wireless Sparks Badge in 1944. Arthur Bishop, his son used to joke that between him and his father – the legendary World War I Flying Ace Billy Bishop – they shot down 73 planes. That would be 72 planes shot down over the skies of Europe by Billy Bishop, and one by Arthur, himself a fighter pilot during World War II.

Billy Bishop with son Arthur Bishop (in Cockpit).
Image Source: thestar.com

Very Highly Decorated and High Tributes

Bishop's decorations include the Victoria Cross (August 1917), Distinguished Service Order (June 2017) & Bar (September 1917), Military Cross (May 1917), Distinguished Flying Cross (July 1918), légion d'honneur and the Croix de Guerre with palm. He was made a Companion of the Order of the Bath in the King's Birthday Honours List of 1 June 1944.

Billy Bishop's decorations (now part of Canadian War Museum collection) include (left to right) Victoria Cross, Distinguished Service Order with Bar, Military Cross, Distinguished Flying Cross, 1914–1915 Star, British War Medal 1914–1920.
Image Source: Wikipedia

Victoria Cross Citation

The citation for his VC, published in *The London Gazette* on 11 August 1917, read: For most conspicuous bravery, determination, and skill. Captain Bishop, who had been sent out to work independently, flew first of all to an enemy aerodrome; finding no machines about, he flew on to another aerodrome about three miles southeast, which was at least 12 miles the other side of the line. Seven machines, some with their engines running, were on the ground. He attacked these from about fifty feet, and a mechanic, who was starting one of the engines, was seen to fall. One of the machines got off the ground, but at a height of 60 feet, Captain Bishop fired 15 rounds into it at very close range, and it crashed to the ground. A second machine got off the ground, into which he fired 30 rounds at 150 yards range, and it fell into a tree. Two more machines then rose from the aerodrome. One of these he engaged at a height of 1,000 feet, emptying the rest of his drum of ammunition. This machine crashed 300 yards from the aerodrome, after which Captain Bishop emptied a whole drum into the fourth hostile machine, and then flew back to his station. Four hostile scouts were about 1,250 feet above him for about a mile of his return journey, but they would not attack. His machine was very badly shot about by machine-gun fire from the ground.

The Royal Canadian Mint's silver collector coin celebrating the 125th anniversary of the birth of Billy Bishop.
Image Source: prnewswire.com

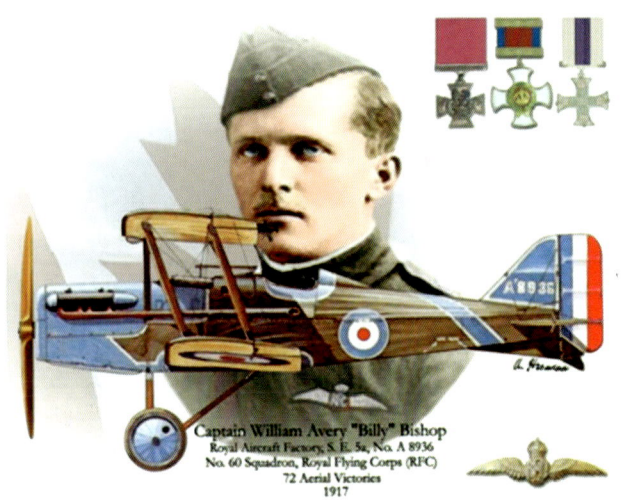

Image Source: Pinterest (small scale art)

Distinguished Flying Cross Citation

His citation for the Distinguished Flying Cross read: A most successful and fearless fighter in the air, whose acts of outstanding bravery have already been recognised by the awards of the Victoria Cross, Distinguished Service Order, Bar to the Distinguished Service Order, and Military Cross. For the award of the Distinguished Flying Cross now conferred upon him he has rendered signally valuable services in personally destroying twenty-five enemy machines in twelve days—five of which he destroyed on the last day of his service at the front. The total number of machines destroyed by this distinguished officer is seventy-two, and his value as a moral factor to the Royal Air Force cannot be over-estimated.

Other Tributes

Bishop also holds a number of non-military awards. In 1967, Bishop was inducted into the International Air & Space Hall of Fame. An award is also named in honour of Bishop. The Air Force Association of Canada approved the establishment of a trophy to commemorate the late Air Marshal W.A. Bishop, VC, in recognition of his "outstanding contribution to the legacy of excellence in Canadian aviation". Billy Bishop's childhood home was re-purposed into the Billy Bishop Home and Museum in 1987. The museum is located in Owen Sound, Ontario. The museum has exhibits on the family, Bishop himself, and veterans. There is a permanent exhibit with information on Bishop at the Grey Roots Museum and Archives, just south of Owen Sound.

The home Bishop grew up in was later re-purposed into the Billy Bishop Home and Museum.
Image Source: Wikipedia

Replica of Bishop's Nieuport 17 fighter at Billy Bishop Toronto City Airport, one of two Canadian airports that bears his name.
Image Source: Wikipedia

Media Coverage

Bishop's life has also been the subject of a number of works in media. *Billy Bishop Goes to War* feature film and Canadian musical, written by John MacLachlan Gray in collaboration with the actor Eric Peterson in 1978. *A Hero to Me: The Billy Bishop Story – WW1 Canadian Flying Ace*, a documentary depicting the story of "Billy" Bishop from the perspective of his granddaughter Diana, was produced for Global Television and TVO in 2003. In addition to television and film, Bishop has also been featured on Canadian stamps. On 12 August 1994, Canada Post issued "Billy Bishop, Air Ace" as part of the Great Canadians series.

Two Airports on His Name

Several places also have honoured Bishop by bearing his name. Two airports in Ontario are named after Bishop. The airport in Owen Sound is officially named "Owen Sound Billy Bishop Regional Airport." Toronto's island airport was renamed Billy Bishop Toronto City Airport in 2009. Although Owen Sound's mayor questioned the change, the proposal was approved by the Toronto Port Authority on 10 November 2009. Having two airports in the province with similar names was a concern. Toronto's Pearson International Airport was originally named *Bishop Field Toronto Airport Malton.*

Bishop's name is featured on the Wall of Honour, at the Royal Military College of Canada.
Image Source: Wikipedia

Other Ways of Bishop Memorializing

"Billy Bishop Private" is a roadway on private land at Ottawa Airport, Ottawa, Ontario, where the "Billy Bishop Room" for visiting dignitaries also exists. "Billy Bishop Way" is a street near the Downsview airport in Toronto, Ontario. "Billy Bishop Park" is a public park in Ottawa, created with the help of the Royal Canadian Legion. "Mount Bishop", a 2,850-metre-high (9,350 ft.) mountain on the Alberta – British Columbia border. "Bishop Building", the 1st Canadian Air Division and the Canadian NORAD Region Headquarters in Winnipeg, Manitoba. "Billy Bishop Legion Branch 176" in Vancouver, British Columbia. "Billy" Bishop was added to the wall of honour at the Royal Military College of Canada in Kingston, Ontario in 2009. Air Force Association of Canada's Air Marshal W. A. Bishop Memorial Trophy is one of the highest awards for aviation in Canada. Bishop's former home in Ottawa, Ontario, constructed in 1905 in the Queen Anne Revival style, has been opened to the public in the annual Doors Open Ottawa showcase of buildings.

Billy Bishop Park – Kanata, Ontario.
Image Source: waymarking.com

Billy Bishop was a genuine hero.
Image Source: amazon.in/Billy-Bishop-Peter-Kilduff

Controversy: Was Bishop a Fake?

It wasn't until long after Bishop's death that any questions were raised about his wartime record. He had always been a self-publicist and his ego was well known, but nobody questioned his integrity or wartime record. After all, the victories were on record. In 1982 his record was challenged. A documentary was released in Canada called *The Kid Who Couldn't Miss.* The documentary made the sensational claim that the national hero wasn't

quite what everybody believed. It questioned the actions that led to his being awarded the Victoria Cross. It was such a sensational and important claim that the Canadian Senate established a Committee to investigate Bishop's wartime record. That committee didn't actually reach any firm conclusions. It couldn't prove that Bishop was or wasn't genuine in the claimed victories or his VC winning combat. An anomaly had been seized upon as evidence of calling the story fake. In truth, the records used to call him a fake, themselves are often error-strewn. Of course, it was in the national interests to retain him as a recognised national hero.

Given the many gaps in British and German records (including the destruction of documents during bombing campaigns in the Second World War), historians have not been able to confirm all of Bishop's combat claims – Kilduff, for example, could only confirm 21 of 72 victories. As the evidence is inconclusive, it is unlikely that the debate will ever be settled. Without a doubt, Bishop was both brave and skilled. Whether or not his combat claims were exaggerated, his daring and his success were an inspiration during the First World War. For many, he was – and is – a Canadian hero.

REFERENCES

1. Veteran Affairs Canada. https://www.veterans.gc.ca/eng/remembrance/people-and-stories/billy-bishop
2. Encyclopedia Britannica https://www.britannica.com/biography/William-Avery-Bishop
3. Billy Bishop, greatest ace or a fake? https://schoolshistory.org.uk/topics/world-history/first-world-war/billy-bishop-greatest-ace-or-a-fake/
4. Library and Archives Canada https://www.bac-lac.gc.ca/eng/discover/military-heritage/first-world-war/100-stories/Pages/bishop.aspx
5. Katie Dahl, Canada's Great War Album https://greatwaralbum.ca/Great-War-Album/About-the-Great-War/Air-Force/Billy-Bishop
6. The Canadian Encyclopedia https://www.thecanadianencyclopedia.ca/en/article/william-avery-bishop
7. Billy Bishop. The Canadians in World War I. https://www.theworldwar.org/explore/exhibitions/past-exhibitions/billy-bishop
8. List of World War I aces from Canada https://en.wikipedia.org/wiki/List_of_World_War_I_aces_from_Canada
9. Billy Bishop, Wikipedia, https://en.wikipedia.org/wiki/Billy_Bishop
10. Billy Bishop – Let's have a proper introduction… https://puttingdownuprising.com/2019/06/01/supplemental-reading-winged-warfare/
11. Five Things I Learned About My Grandfather, Billy Bishop https://www.dundurn.com/news/Five-Things-I-Learned-About-My-Grandfather-Billy-Bishop
12. Lieutenant-Colonel David Bashow, "The Incomparable Billy Bishop: The Man and the Myths", http://www.billybishop.net/incomparable.html
13. Billy Bishop Bio, YouTube, https://www.youtube.com/watch?v=y7ZI0M2QLwg
14. Artist S. J. Payne, For Royal Air Force Museum https://artuk.org/discover/artists/payne-s-j-
15. 3 Squadron – Australian Flying Corps/Royal Australian Air Force. Australian Top Guns. http://www.3squadron.org.au/
16. Captain Steven Dieter, Billy bishop, "Canada's first air force Victoria Cross winner Espirite de corps", *Canadian Military Magazine*, June 5, 2020 http://espritdecorps.ca/army-articles/billy-bishop-canadas-first-air-force-victoria-cross-winner
17. Connor Wilkie Ingenium, Honouring the brave: William Avery "Billy" Bishop was a top Canadian aviator, Ingenium Channel, October 31, 2019, https://ingeniumcanada.org/channel/articles/honouring-the-brave-william-avery-billy-bishop-was-a-top-canadian-aviator

24

RICHARD IRA BONG

The Most Decorated American Fighter Pilot and Top Air Ace

Image Source: Wikipedia

Richard Ira Bong was a United States Army Air Forces Major and Medal of Honour recipient in World War II. He was one of the most decorated American fighter pilots and the country's top flying Ace in the war, credited with shooting down 40 Japanese aircraft, all with the Lockheed P-38 Lightning fighter. He died in California while testing a Lockheed P-80 jet fighter shortly before the war ended.

Early Years: Interest in Flying

Bong was born September 24, 1920, in Superior, Wisconsin, the first of nine children born to Carl Bong, an immigrant from Sweden, and Dora Bryce, who was an American of Scots-English descent. Dick Bong's upbringing epitomized the values and expectations of that era – loyalty to his family and a deep sense of patriotism. Known by the common nickname "Dick", he grew up on a farm in Poplar, Wisconsin, and like all farm children, he had chores to perform and was expected to drive farm machinery at an early age. He hunted and fished in the surrounding woods and streams, played on his school athletic

Richard Ira Bong.
Image Source: thisdayinaviation.com

Richard Bong was born on September 24, 1920, in Superior, Wisconsin.
Image Source: facebook.com/BongVetsCenter

teams, and sang in his church choir. As his 4H project, he planted the extensive evergreen windbreak on the family farm, still in the family. At that time he was like a model all-American boy. He became interested in aircraft at an early age while watching planes fly over the farm carrying mail for President Calvin Coolidge's summer White House in Superior. A skilled hunter, he also built and flew model airplanes. Bong recalled, "I knew then I wanted to be a pilot." Bong entered Poplar High School in 1934, where he played the clarinet in the marching band and participated in baseball, basketball, and hockey. Because Poplar was a three-year school at the time, Bong transferred to Central High School in Superior for his senior year, graduating in 1938. Dick was a good student and finished 18th in his high school class of 428 and commuting, a 44-mile round-trip.

Begins Flying

He began studying at Superior State Teachers College (the current-day University of Wisconsin–Superior) in 1938. While there, Bong enrolled in the Civilian Pilot Training Program and also took private flying lessons, doing his first solo on his 20th birthday and earning a private pilot's license in a Piper Cub. After completing two years of college, Bong enlisted in the U.S. Army Air Forces Aviation Cadet Program at Wausau, Wisconsin on May 29th, 1941, and received orders to the Rankin Aeronautical Academy, a primary flight school near Tulare, California, where he flew solo on Boeing-Stearman PT-13 biplane trainer on June 25th, 1941. He went on to fly Vultee BT-13s at Gardner

Bong in His Lockheed P-38 Lightning.
Image Source: journaltimes.com

Field, California, and North American Texan AT-6s at Luke Field, Arizona. One of Bong's instructors at Luke, Captain Barry Goldwater (later a U.S. Senator from Arizona), who later said of him: "He was a very bright gunnery student. But the most important thing came from a P-38 check pilot who said Bong was the finest natural pilot he ever met. There was no way he could keep Bong from getting on his tail, even though he was flying an AT-6, a very slow airplane."

Joins the United States Army Air Forces

Bong's ability as a fighter pilot was recognized while he was training in northern California. Bong received his fighter pilot Wings and was commissioned a Second Lieutenant in the Army Air Forces Reserves on January 9th, 1942, a month after the attack on Pearl Harbour had plunged America into World War II. But Bong excelled at gunnery so much that his commanding officer kept him at Luke Field, Arizona as an instructor for several months. His first operational assignment was on May 6 to the 49th Fighter Squadron (FS), 14th Fighter Group at Hamilton Field, California, where he learned to fly the twin-engine Lockheed P-38 Lightning.

Loop Around Golden Gate Bridge – Grounded – Reprimanded

Jon Guttman, Research Director at History Net, wrote about "The Spectacular Combat Career of America's Ace of Aces". He said Bong, was quiet, shy, and introverted on the ground; aggressive, hostile and fearless in the air. Major General George C. Kenney, commanding officer of the Fourth Air Force, had had enough. Ever since a certain pilot arrived at Hamilton Field for combat training on May 6, 1942, he had been using nearby San Francisco as his private playground, looping his Lock-heed P-38 Lightning around the Golden Gate Bridge and waving at secretaries as he zoomed past their office windows. On June 12, 1942, Bong flew

Golden Gate Bridge.
Image Source: history.com

very low ("buzzed") over a house in nearby San Anselmo, the home of a pilot who had just been married. But when the young hotshot's prop wash blew a housewife's wet clothes into the dirt and she reported it to his airbase, Kenney called him on the carpet for disciplinary action. "Lieutenant Bong," the general ordered, "Monday morning you check this address out in Oakland, and if the woman has any washing to be hung out on the line...you do it for her. Then, when the clothes are dry, take them off the line and bring them into the house. And don't drop any of them on the ground or you will have to wash them all over again. I want this woman to think we are good for something else besides annoying people. Now get out of here, before I change my mind. That's all!" While 2nd Lt. Richard I. Bong carried out the order, Kenney made a mental note to have that headstrong but undeniably skillful fighter pilot with him at whichever overseas assignment he got. Bong was cited and temporarily grounded for breaking flying rules, along with three other P-38 pilots who had looped around the Golden Gate Bridge on the same day. Within the coming year, Bong would indeed prove himself good for something besides annoying people – except, of course, for the enemy.

Bong Sent to the Southwest Pacific Area

Selected by General Douglas MacArthur to lead the Fifth Air Force in the South Pacific, Kenney wanted 50 of the best P-38 pilots he knew to join him when he took command at Brisbane, Australia, on September 3. Bong was one of them. Kenney later wrote, "We needed kids like this lad." In all subsequent accounts, Bong denied flying under the Golden Gate Bridge. Nevertheless, Bong was still grounded when the rest of his group was sent without him to England in July 1942. Bong then transferred to another Hamilton Field unit, 84th Fighter Squadron of the 78th Fighter Group. From there, Bong was sent to the Southwest Pacific Area.

Senior Allied commanders in New Guinea in October 1942. Left to right: Mr Frank Forde; General Douglas MacArthur; General Sir Thomas Blamey; Lieutenant General George Kenney; Lieutenant General Edmund Herring; Brigadier General Kenneth Walker.
Image Source: Wikipedia

A flight of two camouflaged Lockheed P-38J Lightnings, circa 1943. Dick Bong is flying the closer airplane, P-38J-5-LO 42-67183. Picture from Lockheed Martin.
Image Source: thisdayinaviation.com

War in Pacific: Honing his Combat Skills

On September 10, 1942, Lt. Bong was assigned to the 9th Fighter Squadron, of the 49th Fighter Group, but that unit was still flying Curtiss P-40 Warhawks, based at Darwin, Australia. In November, while the squadron still awaited delivery of the scarce P-38s. In December 1942 Lt. Gen. Kenney attached him temporarily to the 39th Squadron, 35th Fighter Group, based at Laloki airfield near Port Moresby, New Guinea. There, Bong made the acquaintance of Captain Thomas J. Lynch, who had scored three victories the previous May while flying Bell P-39 Airacobras. Hailing from Catasaugua, Pennsylvania, Tommy Lynch was a good pilot and a cool-headed, technically-minded tactician whose aerial audacity never clashed with his sense of responsibility for the men he led. Honing his fighting skills under Lynch's tutelage, Bong came to regard him as both a mentor and a friend.

First Aerial Victories

Dick Bong impressed his squadron mates as someone who was introverted and unobtrusive on the ground but stunningly aggressive in the air, writes Jon Guttman. Japanese army and navy had launched their first major joint air operation in the southwest Pacific on December 27, involving about 40 Mitsubishi A6M2 Zero carrier fighters, Nakajima Ki.43 (Oscar) army fighters, and Aichi D3A1 (Val) navy dive

Japanese Fighter Nakajima Ki-43 Type 1 Army Fighter (AvionsLegendaires.net).
Image Source: thisdayinaviation.com

bombers. As the D3As attacked Allied installations at the newly seized Buna, 12 P-38s of the 39th Fighter Squadron met them. Lynch was leading 2nd Lts. Dick Bong, Kenneth Sparks, and John Magnus down on the Vals when their escorts crossed the Americans' paths. Lynch's gunfire disintegrated one fighter, and then a Zero threatened him. Bong side-slipped, fired at Lynch's assailant and saw it spin away, then sped earthward as three other Zeros moved in on him, finally pulling out, as he described it, "2 inches above the shortest tree in Buna." At that moment he caught a Val just pulling out of its dive and quickly turned it into a fireball. Too low to accomplish anything more, Bong headed back to Port Moresby to report his first two victories—the first credited to a P-38 pilot of the 49th Group. The 39th Squadron claimed a total of 12 victories, including an additional Oscar for Lynch, making him an Ace. Bong thus claimed his initial aerial victories, a Mitsubishi A6M "Zero", and a Nakajima Ki-43 "Oscar". For this action, Bong was awarded the Silver Star.

Major Bong surrounded by newsmen in New Guinea.
Image Source: acesofww2.com

Becomes a Lightning Ace

On January 7, 1943, 36 Curtiss P-40Ks of the 49th Group's 7th and 8th squadrons took off to attack a Japanese convoy. Meanwhile, Lynch led eight P-38s, including Bong and Planck, across the Owen Stanley Mountains to rendezvous with the

P-40s. They ran into the convoy's 11th *Sentai* (army air regiment) air umbrella at 1315 hours. They claimed six Oscars in the fight, including one by Bong after a five-minute duel. Returning to Dobodura to refuel, the Lightnings then took off for Lae, where they encountered another 16 of the 11th *Sentai*'s Ki.43 fighters at 1530. Bong and Planck damaged two Oscars on their first pass, and Bong destroyed one on his second. During a January 8 escort mission, Bong made a frontal attack from above, and the Oscar explode and fell 18,000 feet into Huon Gulf. In only four aerial engagements Bong had become the Fifth Air Force's first Lightning Ace, and General Kenney rewarded him with a trip to Australia for R&R.

Rejoins 9th Fighter Squadron

On February 3, Bong re-joined the 9th Squadron, now equipped with P-38s. The 49th FG was based at Schwimmer Field near Port Moresby. USAAF B-17s were aggressively attacking Japanese ships. On March 3, while escorting B-17s and North American B-25s to the target, Bong saw seven 11th *Sentai* Oscars pass below him, going for American bombers. Dropping behind one, he shot with one burst and watched it crash five miles offshore in Huon Gulf.

Japanese Attack 9th Squadron Airbase

The Japanese struck back on March 11, when a force of Mitsubishi G4M1 "Betty" bombers attacked the 9th Squadron's airstrip at Horanda. The Americans scrambled, and Bong took off just before enemy bombs landed on the strip. Pursuing the bombers, he fired into one without result and twice had to dive away from attacking Zeros. Bong engaged one that was still on his tail, shot him, and then found another Zero coming at him. He fired a short burst, and found seven more attacking him. He later reported, "First two Zeros were burning all around the cockpit and the third was trailing a long column of smoke." Before he escaped the rest in a dive, one Zero shot up his left wing and engine, causing a coolant leak. He reportedly feathered the left engine and landed back safely. He received credit for two confirmed and one probable.

Major Richard Bong and Major Thomas McGuire 15 November 1944 in the Philippines.
Image Source: commons.wikimedia.org

Becomes Leading American Ace in New Guinea

On March 29, 2nd Lt. Clay Barnes led Bong after a suspicious lone airplane. After a long chase at 400 mph, they caught up with the Mitsubishi Ki.46 twin-engine army reconnaissance plane, over the Bismarck Sea. Bong hit the Ki.46's fuel tank, and the plane disintegrated in flames. His ninth victory tied him with Lynch as the leading American Ace in New Guinea. Soon afterward, Kenney promoted Bong to First Lieutenant.

Letter to His Mother: Advice for Brother

Bong was now a well-established and accomplished fighter pilot. Jon Guttman explains, how Bong wrote to his mother on April 10, 1944, that included advice for his younger brother, who was planning to join the Army Air Forces: "He must not get contemptuous of any airplane, no matter how simple and easy it may be to fly. Don't just get in and fly it, but know what makes it tick.... If he forgets, why, any airplane in the world can kill him if he isn't it's complete master."

Image Source: Britannica

Explains Complexity of Air Combat

Bong regarded aerial combat as a game whose risks made life interesting, but he was not above quitting a fight if he judged the odds were too heavy against him. He claimed to be a poor shot, yet his squadron mates stated that he hit whatever he fired at 90 percent of the time. Bong said one secret of his success was a policy of getting close enough to "put the gun muzzles in the Jap's cockpit." Another was his penchant for engaging his opponents head-on, which gave the P-38, a stable gun platform with firepower superior to the Zero and Oscar, a distinct advantage. At least 16 of his victories were attained in head-on gun duels.

Becomes a Double Ace

Early April, the Japanese launched Operation I, a massive air offensive. During a Japanese attack on U.S. shipping in Milne Bay on April 14, Bong became a double Ace, with 10 victories, when he shot down a G4M1 off Cape Frere, it and earned him the Air Medal. Bong went through a "no kill" spell until June 12, when in a series of duels with the Oscars, Bong managed to shoot one. Bong himself returned with a flat right tire and the right tail boom riddled with 7.7mm hits.

Quadruple Success Earns the DSO

On 26 July, ten Lightnings of the 9th Squadron were flying a sweep over the Markham Valley when they encountered 10 Ki.43s and 10 new Kawasaki Ki.61 fighters. Bong shot two of them in two separate high-speed, head-on passes. He shot another two in a turning fight. Bong's quadruple success in that fight was matched by 1st Lt. Jim

The P-38 "Lightning," and Major Richard Ira Bong.
Image Source: archive.jsonline.com

"Duckbutt" Watkins. This action earned him the Distinguished Service Cross. Two days later, the 9th Squadron took on Rabaul-based Ki.43s and claimed seven of them. Bong took five 7.7mm hits in his left wing, but shot one Ki.43. He was now the top-scoring American in the Pacific with 16, and on August 24, he was promoted to Captain. He was sent on a short furlough.

Becomes a Flight Commander

During combat engagement on 06th September, he was credited with two probable, but his right engine got hit. Bong was fortunate to reach Marilinan airstrip before crash landing his P-38H, which was subsequently written off. On October 2, Bong was made a flight commander. On the 29th he shot two Zeros over Rabaul, and two more on 1st November.

Meets His Future Wife: Paints her name on his Aircraft

Based on orders for General Henry H. "Hap" Arnold in Washington, D.C. Bong was sent on two months leave home. Bong got a chance to see his family, enjoy his mother's cooking and hang out with hometown friends.

Richard Ira Bong pointing at the Area of Action.
Image Source: flickr.com

He also met Marjorie Vattendahl, a local girl who had been recently elected homecoming queen at State Teachers College in Superior, Wisconsin. Bong was promptly made king, and much of his leave, the two spent together. He was made to participate in parades, speeches and awards ceremonies to boost public morale. Bong finally returned to the southwest in January 1944. He was put in charge of replacement aircraft for V Fighter Command. He now allotted a brand-new P-38J, one of the first in the area. On the aircraft nacelle, he put a portrait of Marjorie, and called it "Marge." His first victory in *Marge* came on February 15, 1944. Bong was rewarded with another R&R, during which he met with General MacArthur.

Puts Marjorie Vattendahl "Marge" sticker on his P-38.
Image Source: journaltimes.com

Surpasses Eddie Rickenbacker's WW I Record

On April 12, Captain Bong shot down his 26th and 27th Japanese aircraft, surpassing Eddie Rickenbacker's American record of 26 credited victories in World War I. Soon afterwards, he was promoted to Major by General Kenney and dispatched to the United States to see General "Hap" Arnold, who gave him leave. After visiting training bases and going on a 15-state bond promotion tour, Bong returned to New Guinea in September. He was assigned to the V Fighter Command staff as an advanced gunnery instructor with permission to go on missions but not to seek combat. Bong continued flying from Tacloban, Leyte, during the Philippines campaign; by December 17, he had increased his air-to-air victory claims to 40. In addition, he had seven probable victories and 11 enemy planes damaged in two years and 500 combat hours. Kenney told him, "Like it or not, the American Ace of Aces is now going home for good".

General of the Army Henry H. ("Hap") Arnold and Major Richard I. Bong, circa 1945.
Image Source: thisdayinaviation.com

Attack Tactics

Bong considered his gunnery accuracy to be poor, so he compensated by getting as close to his targets as possible to make sure he hit them. In some cases he flew through the debris of exploding enemy aircraft, and on one occasion collided with his target, which he claimed as a "probable" victory.

Bong received the Medal of Honour personally from General Douglas MacArthur in a special ceremony in December 1944.
Image Source: historynet.com

Medal of Honour from MacArthur

On the recommendation of General Kenney, the Far East Air Force Commander, Bong received the Medal of Honour personally from General Douglas MacArthur in a special ceremony in December 1944. His rank of Major would have qualified him for a squadron command, but he always flew as a flight (four-plane) or element (two-plane) leader. Throwing away his prepared speech, the smiling General said, "Major Richard Ira Bong, who has ruled the air from New Guinea to the Philippines, I now induct you into the society of the bravest of the brave, the wearers of the Medal of Honour of the United States of America."

Ace of Aces is Back in the USA: Marries

Bong arrived in the States on Christmas Eve to a hero's welcome. It was time for Bong to bask in the glory of his achievement. He was sent on a propaganda tour which he described as "worse than having a Zero on your tail". On February 10, 1945, Dick Bong married Marge in a ceremony attended by 1,200 guests. Their honeymoon was in California.

Dick Bong married Marge.
Image Source: airpowerasia.com

Early Death in Flight Testing

Bong then became a test pilot assigned to Lockheed's plant in Burbank, California, where he flew P-80 Shooting Star jet fighters at the Lockheed Air Terminal. On August 6, 1945, he took off to perform the acceptance flight of P-80A 44-85048. It was his 12th flight in the P-80; he had a total of four hours and fifteen minutes of flight time in the jet. The plane's primary fuel pump malfunctioned during take-off. Bong either forgot to switch to the auxiliary fuel pump, or for some reason was unable to do so. Eyewitnesses saw puffs of black smoke exit the tailpipe as he climbed to 300 or 400 feet, then the plane rolled right, the canopy flew off and the jet pitched nose-first into the ground. Two minutes after the take-off, Bong's body was found about 100 feet from the engine, partially wrapped in the shrouds of his parachute. Bong cleared away from the aircraft, but was too low for his parachute to deploy. The plane crashed into a narrow field at Oxnard St & Satsuma Ave, North Hollywood. He was just 24 years. His death was front-page news across the country, in national newspapers, even though it occurred on the same day as the atomic bombing of Hiroshima, which in itself was a historic moment. The I-16 fuel pump had been added to P-80s after an earlier fatal crash. Captain Ray Crawford, a fellow P-80 test/acceptance flight pilot who flew on August 6, later said Bong had told him that he had forgotten to turn on the I-16 pump on an earlier flight.

News of Bong's death threw a heavy pall over the U.S. Army Air Force during the waning days of the war. General Kenny was very saddened and said, "We not only loved him, and we boasted about him, we were proud of him. That's why each of us got a lump in our throats when we read that telegram about his death." Bong had survived many air battles only to die in a routine test flight accident. In his autobiography, Chuck Yeager writes that part of the culture of test flying at the time, due to its fearsome mortality rates, was anger toward pilots who died in test flights, to avoid being overcome by sorrow for lost comrades. Bong's brother Carl (who wrote his biography) questions whether Bong repeated the mistake so soon after mentioning it to another pilot. Carl's book—*Dear Mom, So We Have a War* (1991)—contains numerous reports and findings from the crash investigations.

Bong was killed in 1945 while testing a P-80A similar to this one.
Image Source: Wikipedia

His death featured prominently in national newspapers, even though it occurred on the same day as the atomic bombing of Hiroshima.
Image Source: Wikipedia

Test pilot finale.
Image Source: journaltimes.com

Bong was just 24. He was survived by his wife, Marjorie Vattendahl (1923-2003). Bong's funeral in Superior was attended by thousands, and scores more lined the 20-mile route of the funeral cortege from Superior to Poplar, where Bong was buried in the family plot at Poplar Cemetery, Wisconsin. Few names in Wisconsin are more revered than that of Poplar native U.S. Army Air Force Major Richard Ira Bong, one of America's most decorated World War II fighter pilots, and the country's "Ace of Aces". And America's all-time highest-scoring fighter pilot. "Quiet, shy and introverted on the ground," and "aggressive, hostile and fearless in the air."

Bong's aerial victories are well chronicled. All victories were in Lockheed P-38 Lightning aircraft. The variants were P-38F/G/H/L. His aerial victories started on December 27, 1942, when he shot down a Mitsubishi A6M "Zero", and a Nakajima Ki-43 "Oscar" over Buna. He achieved his "Ace" status on January 8, 1943. On one occasion he shot down four aircraft in a day on July 26, 1943. Once he shot down three aircraft in a day on April 12, 1944. On ten occasions he shot two aircraft a day. His last air victory was on December 17, 1944, when he shot a Japanese Nakajima Ki-43 "Oscar". Most of his victims were flying Zeros or Oscars. After two years of combat including over 200 missions, Bong had 40-recorded victories and seven probable victories.

Aerial Victory Credits

August 8, 1945, funeral service for Richard Bong at Concordia Lutheran, the same church where Dick and Marge were married just a few short months earlier.
Image Source: p38assn.org

Military Awards

Medal of Honour

Bong's military decorations and awards included the United States Army Air Force pilot badge (in the USA the flying badge is considered a military award): Medal of Honour, Distinguished Service Cross, Distinguished Flying Cross, and Silver Star, among other campaign medals.

Medal of Honour citation dated December 8, 1944, read:

For conspicuous gallantry and intrepidity in action above and beyond the call of duty in the Southwest Pacific area from October 10, to November 15, 1944. Though assigned to duty as gunnery instructor and neither required nor expected to perform combat duty, Maj. Bong voluntarily and at his own urgent request engaged in repeated combat missions, including unusually hazardous sorties over Balikpapan, Borneo, and in the Leyte area of the Philippines. His aggressiveness and daring resulted in his shooting down 8 enemy airplanes during this period.

Other Honours and Commemorations

Richard Bong was honoured with many commemorations, only a few are listed below

- Richard I. Bong Memorial Bridge along US Route 2 in the Twin Ports of Duluth, Minnesota and Superior, Wisconsin.
- Richard I. Bong Airport in Superior, Wisconsin
- Richard I. Bong Bridge in Townsville, Australia
- Major Richard Ira Bong Squadron of the Arnold Air Society at the University of Wisconsin
- Richard Bong Theatre in Misawa, Japan, and the 613th Air and Space Operations Center, Thirteenth Air Force, Hickam Air Force Base, Hawaii.
- Bong Avenues in many airbases of US Air Force
- Bong Street, Dayton, Ohio, leading to the National Museum of the United States Air Force.
- National Aviation Hall of Fame (1986)
- Wisconsin Aviation Hall of Fame (1987).
- Richard I. Bong Veterans Historical Center in Superior, Wisconsin. Housed in a structure intended to resemble an aircraft hangar, it contains a museum, a film screening room, and a P-38 Lightning restored to resemble Bong's plane.
- Bong was named as the class exemplar at the United States Air Force Academy for the Class of 2003.
- International Air and Space Hall of Fame (2018).

Marge Bong and General Kenney at Bong Memorial.
Image Source: wisconsinhistory.org

Richard Bong Veterans Historical Center.
Image Source: Wikipedia

Still Remembered

Eric Johnson wrote on May 25, 2020, in The Journal Times, Richard I. Bong: Remembering Wisconsin's 'Ace of Aces'. On this Memorial Day, 75 years after Bong's death and the end of World War II, around 300,000 of America's 16 million World War II veterans are still alive according to the U.S. Department of Veterans Affairs. Despite all the ensuing years, Bong is still widely remembered by a number of memorials across Wisconsin. The Bong Memorial Room at his alma mater Superior High School includes in its collection Bong's uniform, all 26 of his military decorations, newspaper clippings, photographs and a fragment of the P-80 in which he was killed.

REFERENCES

1. Jon Guttman, Research Director, History Net, The Spectacular Combat Career of America's Ace of Aces, originally published in the March 2007 issue of *Aviation History* magazine. historynet.com/richard-ira-bong-american-world-war-ii-ace-of-aces.htm
2. Richard I. Bong Veterans Historical Centre https://bongcenter.org/bong-bio/
3. Richard Bong, Wikipedia, https://en.wikipedia.org/wiki/Richard_Bong
4. The National Aviation Hall of Fame, Honouring Aerospace Legends to Inspire Future Leaders, Bong, Richard Ira, Military Combat, Enshrined 1986 (1920-1945) https://www.nationalaviation.org/our-enshrinees/bong-richard/
5. Eric Johnson, The Journal Times, Richard I. Bong: Remembering Wisconsin's 'Ace of Aces', May 25, 2020. http://www.kenoshanews.com/news/richard-i-bong-remembering-wisconsin-s-aces-of-aces
6. Aces of WW 2 https://acesofww2.com/USA/aces/bong/
7. The P-38 National Association and Museum http://p38assn.org/
8. Richard I. Bong https://www.thisdayinaviation.com/6-august-1945/
9. Richard I. Bong Veterans Historical Center on Facebook https://www.facebook.com/BongVetsCenter

25

Ernst Udet

The Highest Scoring Surviving German Air Ace of WW

Ernst Udet was a German pilot during World War I and a Luftwaffe Colonel-General (Generaloberst) during World War II. Udet joined the Imperial German Air Service at the age of 19, and eventually became a flying Ace of World War I, scoring 62 confirmed victories. He was the highest-scoring German fighter pilot to survive that war, and the second-highest scoring after Manfred von Richthofen, his commander in the Flying Circus. Udet rose to become a squadron commander under Richthofen, and later under Hermann Göring. He spent the 1920s and early 1930s as a stunt pilot, international barnstormer, light aircraft manufacturer, and playboy.

In 1933, Udet joined the Nazi Party and became involved in the early development of the Luftwaffe, where he was appointed director of research and development. Influential in the adoption of dive-bombing techniques as well as the Stuka dive bomber, by 1939 Udet had risen to the post of Director-General of Equipment for the Luftwaffe. The stress of the position and his distaste for administrative duties led to Udet developing alcoholism.

The launch of Operation Barbarossa, combined with issues with the Luftwaffe's needs for equipment outstripping Germany's production capacity and increasingly poor relations with the Nazi Party, caused

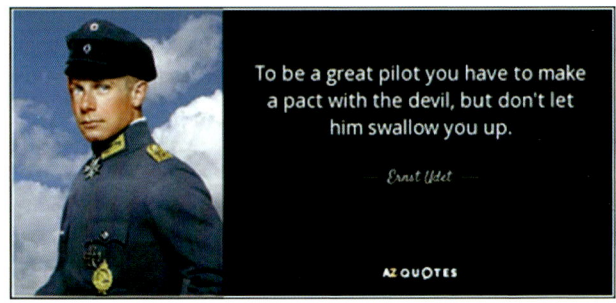

Image Source: azquotes.com

Udet to commit suicide on 17 November 1941 by shooting himself in the head. "Our defeat was caused by Udet," Hitler would claim. "That man concocted the most nonsensical state of affairs ever seen in the history of the Luftwaffe."

The colourful and boisterous Ernst Udet had one of the most remarkable flying careers of the first half of the 20th century.

Early Years: Connect With Aviation

Ernst Udet was born on 26 April 1896 (Sunday), in Frankfurt am Main, German Empire. He was what the Germans call a *Sonntagskind* ('Sunday's Child')—

With 62 victories, Ernst Udet was the second-highest-scoring German Ace of WW1.
Image Source: lapauitgewers.net

Gustav Otto. Image Source: ww2gravestone.com

lucky, happy, and carefree. Udet grew up in Munich, and was known from his early childhood for his sunny temperament and fascination with aviation. In his youth, he hung out at a nearby airplane factory and an army airship detachment. In 1909, he helped found the Munich Aero-Club. After crashing a glider he and a friend constructed, he finally flew in 1913 with a test pilot in the nearby Otto Works owned by Gustav Otto, which he often visited.

World War I: Struggles to Enlist

Shortly after the beginning of World War I, Udet attempted to enlist in the Imperial German Army on 2 August 1914, but at only 160 cm (5 ft 3.0 in) tall he did not then qualify for enlistment. Later that month, when the *Allgemeiner Deutscher Automobil-Club* appealed for volunteers with motorcycles, Udet applied and was accepted. Udet's father had given him a motorcycle when he had passed his first year examination, and along with four friends, Udet was posted to the 26. *Württembergischen* Reserve Division as a "messenger rider." After injuring his shoulder when his motorcycle hit a crater from an artillery shell

13 year old Udet experimented with his own home-made glider in 1909, the results were less than stellar.
Image Credit: Ullstein Bild, Source: historynet.com

explosion, he was sent to a military hospital, and his motorcycle was sent for repairs. When Udet tried to track down the 26th Division, he was unable to find it and decided to serve in the vehicle depot in Namur. During this time, he met officers from the Chauny flying sector, who advised him to transfer as an aerial observer. However, before he received his orders, the army dispensed with the volunteer motorcyclists, and Udet was sent back to the recruiting officials.

Private Flying Training: Joins Air Service

Udet tried to return to the fighting, but he was unable to get into either the pilot or aircraft mechanic training the army offered. However, he learned that if he were a trained pilot, he would be immediately accepted into army aviation. Through a family friend, Gustav Otto, owner of the aircraft factory he had hung out around in his youth, Udet received private flight training. This cost him 2,000 Deutsche Marks (about $400 in 1915 U.S. dollars) and new bathroom equipment from his father's firm. Udet received his civilian pilot's license at the end of April 1915 and was immediately accepted by the Imperial German Air Service.

Aviatik B Artillery Reconnaissance Aircraft.
Image Source: Wikipedia

Starts Flying for Artillery Ranging

Udet at first flew in *Feld Flieger-Abteilung* 206 (FFA 206) – a two-seater artillery observation unit, as an *Unteroffizier* (non-commissioned) pilot with observer *Leutnant* Justinius. He and his observer won the Iron Cross (2nd class for Udet and 1st class for his Lieutenant) for nursing their damaged Aviatik B.I two-seater back to German lines after a shackle on a wing-cable snapped. Justinius had climbed out to hold the wing and balance it rather than landing behind the enemy lines and being captured. Later, after yet another similar incident, the Aviatik B aircraft was retired from active service.

Court-Martialled for Losing an Aircraft

Later, Udet was court-martialled for losing an aircraft in an incident the flying corps considered a result of bad judgment. Overloaded with fuel and bombs, the aircraft stalled after a sharp bank and plunged to the ground. Miraculously, both Udet and Justinius survived with only minor injuries. Udet was placed under arrest in the guardhouse for seven days.

After an Incident: Transferred for Fighter Flying

On his way out of the guardhouse, he was asked to fly *Leutnant* Hartmann to observe a bombing raid on Belfort. A bomb thrown by hand by the *Leutnant* became stuck in the landing gear, but Udet performed aerobatics and managed to shake it loose. As soon as the Air Staff Officer heard about Udet's performance during the incident, he ordered Udet transferred to the fighter command.

Fighter Pilot Udet: Initial Disasters

His aggressive style and eagerness for battle resulted in his quickly being promoted to Staff Sergeant. Udet was assigned a new Fokker to fly to his new fighter unit – FFA 68 – at Habsheim. Mechanically defective, the plane crashed into a hangar when he took off, and was then given an older Fokker to fly. In this aircraft, he experienced his first aerial combat, which almost ended in disaster. While

In December 1915, a young Udet experienced his first one-on-one combat while flying this Fokker Eindecker E.I monoplane.
Picture Source: historynet.com

lining up on a French Caudron, Udet found he could not bring himself to fire on another person and was subsequently fired on by the Frenchman. A bullet grazed his cheek and smashed his flying goggles. Udet survived the encounter but from then on learned to attack aggressively and began scoring victories.

O'Brien Browne, describes this most aptly in "Ernst Udet: The Rise and Fall of a German World War I Ace" written for History Net. He writes, "On a pale December morning in 1915 a single Fokker Eindecker monoplane sailed high above the clouds, hunting for prey over the Vosges sector of the Western Front. Its young, inexperienced German pilot, his face greased for protection from the cold, felt snug in his thick flight suit and sheepskin-lined boots. Eyes alert, he carefully scanned the vast expanse of seemingly empty blue sky. Suddenly, a glint of silver caught the pilot's eye, moving towards him from the west. It was the enemy. Instead of manoeuvring above and behind his opponent, the novice pilot forgot all his combat training and simply flew head-on at the oncoming aircraft. As the enemy neared, the German recognized it as a French Caudron G.IV, a queer-looking machine with a twin-boom lattice tail section and a truncated tub between the plane's two engines carrying the pilot and observer. As the German pilot reached for the firing button on the joystick, his mouth became dry at the prospect of his first aerial battle. The Frenchmen flew directly at him, looming so close that the observer's head was clearly visible. The German pilot poised his thumb over the firing button, muscles tense. The moment of truth: kill or be killed. But as the two planes came within point-blank range of each other, paralyzing fear gripped the young German and he froze. He stared at his opponent, helpless. A second later, he heard popping noises and felt his Fokker shudder. Something slapped hard against his cheek and his goggles flew off. His face was sprayed with broken glass, and blood trickled down his cheek. With the French observer still firing, the German dived into a nearby cloud and limped back to his airfield. Once his wounds had been dressed, he secluded himself in his room and spent a sleepless night berating himself for cowardice and stupidity. Such was the inauspicious beginning of one of the most remarkable flying careers of the first half of the 20th century. The young pilot's name was Ernst Udet".

French Caudron G.IV. Image Source: commons.wikimedia.org

First Aerial Victory

After a period of intense soul-searching, Udet determined that he would succeed as a fighter pilot. He had his squadron's mechanics construct a model of a French plane against which Udet could fly practice attacks, honing both his shooting and combat flying skills. The additional training soon paid off. He downed his first French opponent on 18 March 1916. On that occasion, he had scrambled to attack two French aircraft, instead he found himself facing a formation of 22 enemy aircraft of various types. He dived from above and behind, giving his Fokker D.III biplane full throttle, and opened fire on a Farman F.40 from very close range. Udet pulled away, leaving the flaming bomber trailing smoke, only to see, to his horror, the observer fall

Udet with his Fokker D.III biplane.
Image Source: in.pinterest.com/bcfarrant/ww1-aircraft

from the rear seat of the stricken craft. The victory won Udet the Iron Cross First Class, later describing it: "The fuselage of the Farman dives down past me like a giant torch... A man, his arms and legs spread out like a frog's, falls past–the observer. At the moment, I don't think of them as human beings. I feel only one thing – victory, triumph, victory."

Becomes an Air Ace

That year, FFA 68 was renamed *Kampfeinsitzer Kommando Habsheim* and later *Jasta 15* on 28 September 1916. Udet's second victory was a Bréguet-Michelin bomber, brought down during a massive bombing raid on Oberndorf by French and British units, escorted by four Nieuports of the American volunteer *Escadrille N.124*, on October 12. Udet forced a French Breguet to land safely in German territory, then landed nearby to prevent its destruction by its crew. The bullet-punctured tires on Udet's Fokker flipped the plane forward onto its top wings and fuselage. Udet and the French pilot eventually shook hands next to the Frenchman's aircraft. He finished his score for 1916 with a Caudron G.IV on December 24. In January 1917, Udet was commissioned as a *Leutnant der Reserve* (Lieutenant of Reserves). The same month, *Jasta* 15 re-equipped with the Albatros D.III, a new fighter with twin synchronized Maschinengewehr 08 machine guns. On February 20, he forced down a Nieuport 17 into the French lines. Its pilot, Sergeant Pierre Cazenove de Pradines of N.81, survived to eventually become a seven-victory Ace. On April 24, Udet shot down a Nieuport fighter, which burst into flames after a short dogfight, and he destroyed one of the new Spad VII fighters on May 5. Udet claimed five more victories, before transferring to *Jasta* 37 in June 1917.

French Nieuport 17 C.1 Fighter.
Image Source: Wikipedia

Encounter with French Hero and Ace Georges Guynemar

During his service with *Jasta* 15, Udet later wrote, he had encountered Georges Guynemer, a notable French Ace, in combat at 5,000 m (16,000 ft). Guynemer, who preferred to hunt enemy planes alone, by this time was the leading French Ace with more than 30 victories. Udet saw Guynemer and they circled each other, looking for an opening and testing each other's turning abilities. They were close enough for Udet to read the "*Vieux*" of "*Vieux Charles*" written on Guynemer's Spad S.VII. The opponents tried every aerobatic trick they knew and Guynemer fired a burst through Udet's upper wing. However, Udet maneuvered for advantage. Once Udet had Guynemer in his sights, his machine guns jammed and while pretending to dogfight he pounded on them with his fists, desperate to unjam them. Guynemer realized his predicament and instead of taking advantage of it, simply waved a farewell and flew away. Udet wrote of the fight, "For seconds, I forgot that the man across from me was Guynemer, my enemy. It seems as though I were sparring with an older comrade over our own airfield." Udet

Ace vs. Ace: Guynemer against Udet.
Image Source: Real photos Photoshop composition flickr.com/photos/nicolas_grignon/39690270275

felt that Guynemer had spared him because he wanted a fair fight, while others have suggested that the French Ace was impressed with Udet's skills and hoped they might meet again on equal terms. To the end of his life, Udet never forgot that act of chivalry.

Transferred to Jasta 37: Elevated to Squadron Commander

Eventually, every pilot in *Jasta* 15 was killed except Udet and his commander, Heinrich Gontermann, who said to Udet: "The bullets fall from the hand of God … Sooner or later they will hit us." Udet applied for a transfer to *Jasta* 37, and Gontermann was killed three months later when the upper wing of his new Fokker Dr.1 tore off as he was flying it for the first time. Gontermann lingered for twenty-four hours without awakening and Udet later remarked, "It was a good death." The new location did him good, and he brought his score up to nine by the end of August. By late November, Udet was a triple Ace. He was already modelling his attacks after those of Guynemer, coming in high out of the sun to pick off the rear aircraft in a squadron before the others knew what was happening. Having witnessed one of these attacks, his commander in *Jasta* 37 Kurt Grasshoff, on being transferred, selected Udet for command over more senior men. Udet's ascension to command on 7 November 1917, was followed six days later by the award of the Knight's Cross of the Order of Hohenzollern. Despite his seemingly frivolous nature, drinking late into the night, and womanizing lifestyle, Udet proved an excellent Squadron Commander. He spent many hours coaching new fighter pilots, and like many successful Aces, emphasized good marksmanship over flashy stunt flying. He enjoyed the star status that came with being a pilot and often dressed in a dapper style, a cigarette usually poised carefully in one hand. He still displayed the disdain for authority and routine that had characterized him as a child. And he enjoyed being curt and cheeky to pompous officers, his ranking position and success as a fighter pilot usually saving him from reprimand. By year's end, he was a 16-victory Ace and a highly decorated pilot.

House Order of Hohenzollern.
Image Source: military.wikia.org

Selected For the Flying Circus

Udet's success attracted attention for his skill, earning him an invitation to join the "Flying Circus", *Jagdgeschwader* 1 (JG 1), an elite unit of German fighter Aces under the command of Manfred von Richthofen, popularly known as the Red Baron. Richthofen drove up to Udet one day as he was trying to pitch a tent in Flanders in the rain. Pointing out that Udet had 20 kills, Richthofen said, "Then you would actually seem ripe for us. Would you like to?" Udet accepted. After watching him shoot down an artillery spotter by frontal attack, Richthofen gave Udet command of *Jasta* 11, von Richthofen's former squadron command. The group commanded by Richthofen also contained *Jastas* 4, 6 and 10. Udet's enthusiasm for Richthofen was unbounded, who demanded total loyalty and dedication from his pilots, immediately cashiering anyone who fell out of line. At the same time, Richthofen treated them with every

Manfred von Richthofen's "Flying Circus".
Image Source: Wikipedia

consideration, and when it came time to requisition supplies he traded favours for autographed photos of himself that read: "Dedicated to my esteemed fighting companion." Udet remarked that because of the signed photographs, ".... sausage and ham never ran out." One night, the squadron invited a captured English flyer for dinner, treating him as a guest. When he excused himself for the bathroom, the Germans secretly watched to see if he would try to escape. On his return the Englishman said, "I would never forgive myself for disappointing such hosts"; the English flyer did escape later from another unit.

Richthofen Killed in Action

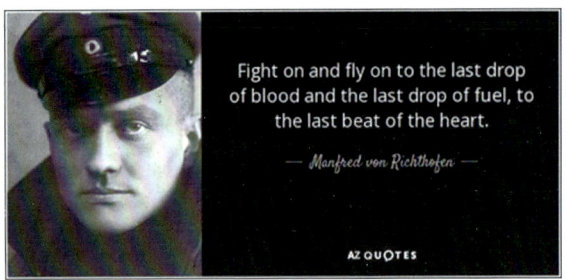

Image Source: azquotes.com

After joining *Jasta* 11, Udet began flying multiple patrols daily, although he was increasingly troubled by intense pain in his ears. Nevertheless, he pushed his victory score up to 23 before the pain became so intolerable that Richthofen ordered him to take sick leave. This time off was vital for Udet's war-shattered nerves. Despite a doctor's warning that he would never fly again, Udet's ears began to improve. In addition, he received news that he had been awarded one of Germany's highest military awards, the *Ordre Pour le Mérite*, generally referred to by its nickname, the 'Blue Max.' Richthofen was killed on April 21, 1918, in France. Shaken by the death of the man whom he later described as 'the greatest of soldiers' – a man many had believed was indestructible, Udet said about Richthofen: "He was the least complicated man I ever knew. Entirely Prussian and the greatest of soldiers."

He returned to JG 1 against the doctor's advice and remained there until the end of the war, commanding *Jasta* 4. The conflict was entering its last, dreadful months, which would see some of the most intense fighting of the entire war. His unit was now equipped with the formidable Fokker D.VII, the plane generally considered the finest fighter of WWI.

Ernst Udet with his childhood sweetheart, Eleanor "Lo" Zink.
Image Source: combatace.com

Meets Childhood Sweetheart

While at home, Udet had reacquainted himself with his childhood sweetheart, Eleanor "Lo" Zink. Notified that he had received the *Pour le Mérite*, he had one made up in advance so that he could impress her, and painted her name on the side of his Albatros fighters and Fokker D VII. Also on the tail of his Fokker D VII was the message "*Du doch nicht*" – "Definitely not you" – a taunt and challenge to Allied pilots.

Oberleutnant Ernst Udet in his Siemens Schuckert D.III. The LO! Inscription on the sides of his aircraft was dedicated to his then fiancée and later wife, Eleanore Zink.
Image Source: flickr.com/photos/drakegoodman

Early Flier to Parachute to Safety

During the spring and early summer, Udet's score rose to 35. On 29 June 1918, Udet was one of the early fliers to be saved by parachuting from a disabled aircraft, when he jumped after a clash with a French Breguet. His harness caught on the rudder and he had to break off the rudder tip to escape. His parachute did not open until he was 250 ft (76 m) from the ground, causing him to sprain his ankle on landing.

First Encounter with U.S. Army Air Service

On July 2, JG.I had its first encounter with the U.S. Army Air Service and shot down two Nieuport 28s of the 27th Aero Squadron. One of the pilots, 2nd Lt. Walter B. Wanamaker, was brought down injured by Udet, who gave him a cigarette and chatted with him until the medics arrived. On a whim, Udet cut the serial number, N6347, from the rudder of Wanamaker's plane. When the two met again at the Cleveland Air Races on September 6, 1931, Udet returned the trophy to his former opponent. It can still be seen at the U.S. Air Force Museum in Dayton, Ohio.

On Udet's aircraft tail of his Fokker D VII was the message "*Du doch nicht*" – "Definitely not you" – a taunt and challenge to Allied pilots.
Image Source: albiondesigncentre.com

Udet returning the fabric trophy to Wanamaker.
Image Source: National Museum of the U.S. Air Force

Becomes National Hero

War was now going badly for the Germans. The German air force was hampered by a lack of fuel, equipment, and new recruits. "The war gets tougher by the day," Udet wrote. "When one of our aircraft rises, five go up on the other side." Udet meanwhile was reaching new heights of achievement. Udet scored 20 victories in August 1918 alone, mainly against British aircraft. Between July 1 and September 26, he downed 26 Allied aircraft, bringing his total to 62 and he became a national hero. During his last air battle on 28 September 1918, in which he brought down two Airco DH.9 bombers, Udet was wounded in the thigh, for which he was still recovering on Armistice Day, 11 November 1918, when the war ended in Germany's defeat.

Inter-War Years: Lots of Action

The adventures of Udet's life continued without pause after the war. On his way home from the military hospital, he had to defend himself against a Communist who wished to rip the medals off his chest. Udet and Robert Ritter von Greim performed mock dogfights at weekends for the POW Relief Organization, using surplus aircraft in Bavaria. After the war, he was initially active in the Richthofen Veterans' Association. He was invited to start the first International Air Service between Germany and Austria, but after the first flight the Entente Commission confiscated his aircraft.

Marries and Divorces Eleanor "Lo" Zink

Udet married Eleanor "Lo" Zink on 25 February 1920, however, the marriage lasted less than three years and they were divorced on 16

Udet, a national hero.
Image Source: welkinlions.tumblr.com

Original autograph Real Photo Postcard (RPPC) of German WWI fighter pilot and later Luftwaffe General Ernst Udet.
Image Source: germanpostalhistory.com

February 1923. His independent nature, his many affairs, and disdain for routine led to the breakup of his marriage.

Multi-Talented and Multi-Faceted

His talents were numerous – among these were juggling, drawing cartoons, and party entertainment. Udet was known primarily for his work as a stunt pilot and for playboy-like behaviour. Udet flew in air shows and races, performing throughout Latin America and Europe. Udet's flamboyant lifestyle flourished. He became a well-known womanizer and a hard drinker, a party boy who loved to dine and share a laugh with an international group of friends. He spent money as quickly as it came in. He enjoyed the company of movie stars, film producers, and other public figures. Flying always remained his greatest passion. He flew for movies and for airshows and flying stunts such as picking a cloth from the ground with his wingtip, flying under low bridges, and completing loops only several meters from the ground. One stunt only Udet performed was successive loops with the last complete after turning off the engine mid-air and landing the aircraft in a sideways glide. He appeared with Leni Riefenstahl in three films: *The White Hell of Pitz Palu* (1929), *Stürme über dem Mont Blanc* (1930), and *S.O.S. Eisberg* (1933).

Biplane Dead Stick Landing Ernst Udet – Chicago International Air Races 1933.
Image Source: youtube.com/watch?v=hz4J8f6pkIk

Udet's stunt pilot work in films took him to California. In the October 1933 issue of *New Movie Magazine*, there is a photo of Carl Laemmle, Jr.'s party for Udet in Hollywood. Laemmle was head of Universal Studios which made *SOS Eisberg*, a US-German co-production. Udet was invited to attend the National Air Races at Cleveland, Ohio. In 1935 he appeared in *Wunder des Fliegens: Der Film eines deutschen Fliegers* (1935) directed by Heinz Paul. His co-star Jürgen Ohlsen, who had previously starred (uncredited) in the extremely popular Nazi propaganda film *Hitlerjunge Quex: Ein Film vom Opfergeist der deutschen Jugend*, played a youth who lost his pilot father in World War I and was befriended and encouraged by Udet, his idol.

Ernest Udet in the cockpit of his aptly named Udet U 12 "Flamingo," an airplane designed specifically for his spectacular airshow performances.
Image Source: commons.wikimedia.org

This was probably the happiest time of Udet's life. He was reeling in money. His autobiography, *Mein Fliegerleben* (English title: *Ace of the Iron Cross*), was a hit, selling more than 600,000 copies by the end of 1935. He was arguably the most famous stunt pilot of his day.

Attempts at Aircraft Manufacturing

American films were good publicity for Udet. An American, William Pohl of Milwaukee, telephoned him with an offer to back an aircraft manufacturing company. Udet Flugzeugbau was born in a shed in Milbertshofen. Its intent was to build small aircraft that the general public could fly. It soon ran into trouble with the Entente Commission and transferred its operations to a beehive and chicken coop factory. The first airplane that Udet's company produced was the *U2*. Udet took the second model, the *U4*, to the Wilbur Cup race in Buenos Aires at the expense of Aero Club Aleman. Finally up to *U12* model were made. The club wanted him to do cigarette commercials to reimburse them for the expense, but he refused. He was rescued by the Chief of the Argentinian Railways, a man of Swedish descent named Tornquist, who settled the debt. In 1924, Udet left Udet Flugzeugbau when they decided to build a four-engine aircraft, which was larger and not for the general population.

His autobiography, Mein Fliegerleben (English title: Ace of the Iron Cross)

More Aviation-Linked Activities

He and another friend from the war, Angermund, started an exhibition flying enterprise in Germany, which was also successful, but Udet remarked, "In time this too begins to get tiresome. We stand in the present, fighting for a living. It isn't always easy. But the thoughts wander back to the times when it was worthwhile to fight for your life." Udet and another wartime comrade, Suchocky, became pilots to an African filming expedition. The cameraman was another veteran, Schneeberger, whom Udet called "Flea," and the guide was Siedentopf, a former East African estate owner. Udet described one incident in Africa in which lions jumped up to claw at the low-flying aircraft, one of them removing a strip of Suchocky's wing surface. Udet engaged in hunting while in Africa.

Schneeberger and Suchocky inspect the damage done to the plane by the lioness.
Image Source: archivaria.com/Udet/Udet9.html

Building the *Luftwaffe*

Udet joined the Nazi party in 1933 when Hermann Göring promised to buy him two new U.S.-built Curtiss Hawk II biplanes (export designation of the F11C-2 Goshawk Helldiver). Though not interested in politics, in 1934 Udet made the difficult decision to join the new *Luftwaffe*. Whatever his misgivings about the Nazis, he realized that they had an iron grip on power in his country. The planes were used for evaluation purposes and thus indirectly influenced the German idea of dive-bombing aeroplanes, such as the Junkers Ju 87 (*Stuka*) dive bombers. They were also used for aerobatic shows held during the 1936 Summer Olympics. Udet piloted one of them, which

Reichsmarschall Herman Göring (left) and Udet, head of the technical office of the air ministry, observe aerial manoeuvres by the new Luftwaffe on June 16, 1938.
Picture Credit: Ullstein, historynet.com

Udet in Developmental Aircraft Cockpit.
Image Source: Pinterest Uploaded by Allister

survived the war and is now on display in the Polish Aviation Museum. Earlier in 1934, Udet taught Aviation Minister Erhard Milch to fly. And as the top pilot in the country, Udet's opinion was considered quite significant when matters of aviation policy were discussed. It was flattering to be listened to by those in positions of authority. Patriotism, the challenge of rebuilding the air force he had so loved, plus a sense of stability and security offered by the prospect of a normal job, all played a part in helping him make up his mind.

Aircraft Development Tasks

He was promoted rapidly from *Oberstleutnant* (Lieutenant Colonel) to *Oberst* (Colonel) and then inspector of fighter and dive-bomber pilots. In the summer of 1936, Udet was pressured by Göring into becoming the head of the technical office of the Reich's air ministry, a position of weighty organizational responsibilities. Despite his new duties, Udet, who had always shunned paper-pushing, seemed able to find the time to test-fly the industry's newest designs, such as the Messerschmitt Bf-109, as well as the latest from Focke Wulf and Heinkel.

After the trials of the Ju 87, a confidential directive issued on 9 June 1936 by *Generalfeldmarschall* Wolfram von Richthofen called for the cessation of all further Ju 87 development, although the Ju 87 had been awarded top marks and was about to be accepted. However, Udet immediately rejected von Richthofen's instructions and Ju 87 development continued. In this post, Udet was finally responsible for the introduction of the Junkers Stuka and the Messerschmitt Bf 109. Udet became a major proponent of the dive bomber, taking credit for having introduced it to the *Luftwaffe*. Udet also shunned the bureaucracy in his new job, and the pressure led to him developing an addiction to alcohol, drinking large amounts of brandy and cognac.

Udets's board bar from his Siebel Fh 104 A-0 on display in the *Deutsches Technikmuseum* Berlin.
Image Source: Wikipedia

Director-General of Equipment

In January 1939, Udet visited Italian North Africa, accompanying Marshal of the Italian Air Force, Italo Balbo on a flight, because at the time there were distinct signs of the German military and diplomatic co-operation with the Italians. In February 1939 Udet became *Generalluftzeugmeister,* Director-General of Equipment of the *Luftwaffe.* Udet was not adept at the political intrigue that characterizes all bureaucracies. Increasingly, he was outmanoeuvred by his onetime friend Erhard Milch. Ambitious and scheming, Milch resented Udet's special relationship with Göring and craved the power and prestige attendant on Udet's job. Nevertheless, Udet continued to reap honours from Hitler, who was most likely unaware of the interdepartmental in-fighting. On June 21, 1940, Udet was one of the few people who witnessed the French surrender to the Germans. A month later, he was awarded the Knight's Cross of the Iron Cross and promoted to *Generaloberst* (Colonel General).

WW II: High Aircraft Requirements: Despair

When World War II began, his internal conflicts grew more intense as aircraft production requirements were much more than the German industry could supply, given limited access to raw materials such as aluminium. Hermann Göring responded to this problem by simply lying about it to Adolf Hitler, and after the *Luftwaffe*'s defeat in the Battle of Britain, Göring tried to deflect Hitler's ire by blaming Udet. On 22 June 1941, the launch of Operation Barbarossa, the German invasion of the Soviet Union, drove Udet further into despair. In April and May 1941, Udet had led a German delegation inspecting Soviet aviation industry in accordance with the Molotov-Ribbentrop Pact. Udet informed Göring that the Soviet air force and aviation industry were very strong and technically advanced. Göring decided not to report this to Hitler, hoping that a surprise attack would quickly

Ernst Udet with Bomber Pilots Galland and Mölders / Photo 1940.
Image Source: akg-images.com

destroy Russia. Udet realized that the upcoming war on Russia might destroy Germany. He tried to explain this to Hitler but, torn between truth and loyalty, suffered a psychological breakdown. At the end of August, Udet had a long, private talk with Göring in which he tried to resign. Göring refused, knowing that such resignation from a top Luftwaffe official would create bad publicity. Göring kept Udet under control by giving him drugs at drinking parties and hunting trips. Udet's drinking and psychological condition became a problem, and Göring used Udet's dependency to manipulate him.

In his article titled "The Nazi Blame Game", Dwight Jon Zimmerman wrote in July 2017, "The problem with the *Luftwaffe* was that it was a tactical air force increasingly tasked with a strategic mission. And blamed for that failing was its director of aviation armaments, *Generaloberst* Ernst Udet, who turned out to be the wrong man in the wrong place at the wrong time for the *Luftwaffe*". As historian Leonard Mosely wrote, "Udet was bold, brave, and a first-class flier, he was, unfortunately, no planner." Udet filled his staff with wartime friends unqualified for their design and production roles. Accounts of Udet's managerial incompetence began reaching Göring. As the man who hired Udet and who was himself a poor (and corrupt) administrator, Göring was in an awkward position and at first, chose to ignore the problem. Albert Speer, who would later take over all war production, noted that "Göring was not actually blind to reality. I would occasionally hear him make perceptive comments on the situation. Rather, he acted like a bankrupt who up to the last moment wants to deceive himself along with his creditors."

Göring and Udet in 1938, before the cracks started to show.
Bundesarchive photo. Image Source: defensemedianetwork.com

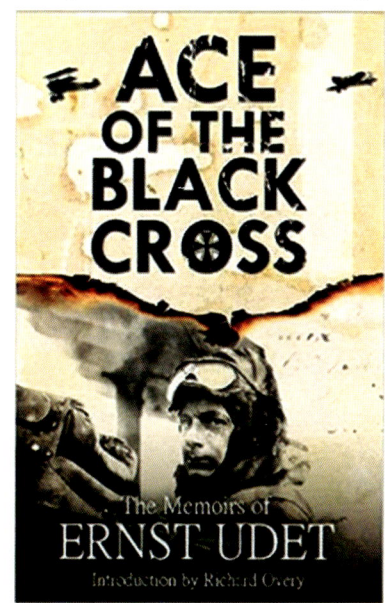

Memoirs of Ernst Udet, the German 'Ace of Aces'

Commits Suicide

On 17 November 1941, at age 45, Ernst Udet committed suicide in Berlin, by shooting himself in the head while on the phone with his girlfriend, Inge Bleyle. Udet's suicide was concealed from the public, and at his funeral, he was lauded as a hero who had died in flight while testing a new weapon. The circumstances of Udet's death were kept secret, and he was given a state funeral attended by Adolf Hitler, Göring, and other top officials. On their way to attend Udet's funeral, the World War II fighter Ace Werner Mölders died in a plane crash in Breslau, and high *Luftwaffe* executive *General der Flieger* (General of Aviators) Helmuth Wilberg died in another plane crash near Dresden. Udet was buried next to Manfred von Richthofen in the *Invalidenfriedhof* Cemetery in Berlin. Mölders was buried next to Udet.

Luftwaffe honour guard including Ace Major Adolf Galland at left—escorted his remains to their final resting place.
Image Source: ww2gravestone.com

Suicide Note

According to Udet's biography, *The Fall of an Eagle*, he wrote a suicide note in red pencil which among other things said, "Ingelein, why have you left me?" and "Iron One, you are responsible for my death." "Ingelein" referred to his girlfriend, Inge Bleyle, and "Iron One" to Hermann Göring. The book *The Luftwaffe War Diaries* similarly states that Udet wrote "*Reichsmarschall*, why have you deserted me?" in red on the headboard of his bed. It is possible that an affair Udet had with Martha Dodd, daughter of the U.S. ambassador to Germany and Soviet sympathizer during the 1930s might have had some importance in these events. Records made public in the 1990s confirm Soviet security involvement with Dodd's activities. Evidence indicates that Udet's unhappy relationship with Göring, Erhard Milch, and the Nazi Party, in general, was the cause of a mental breakdown.

Adolf Hitler and Hermann Göring at the funeral of General Ernst Udet.
Image Source: hitler-archive.com

Portrayals

Handsome, dashing, and a skilled raconteur, Udet was Germany's greatest living fighter Ace from World War I. Carl Zuckmayer's 1946 play *Des Teufels General* ("The Devil's General") was a fictional treatment of Udet's final days. *Des Teufels General* was a 1955 film version of the Zuckmeyer play, with Curd Jürgens in the title role. In the film *Von Richthofen and Brown* (1971), Udet was portrayed by Robert La Tourneaux. The character of "Ernst Kessler" in the 1975 film *The Great Waldo Pepper* is clearly based upon Ernst Udet. Kessler was portrayed by actor Bo Brundin. It also contains dogfighting scenes between a Fokker Dr.I and a Sopwith Camel. In the movie *The Red Baron*, Udet is portrayed by Jiøí Laštovka. Udet is also featured in the Knights of the Sky video game as an enemy German pilot.

Ernst Udet's grave in Invalidenfriedhof Cemetery, Berlin.
Image Source: Wikipedia

REFERENCES

1. O'Brien Browne, Ernst Udet: The Rise and Fall of a German World War I Ace. History Net, This feature was originally published in the November 1999 issue of Aviation History magazine. https://www.historynet.com/ernst-udet-the-rise-and-fall-of-a-german-world-war-i-ace.htm
2. Spartacus Educational https://spartacus-educational.com/FWWudet.htm
3. Who's Who – Ernst Udet, Firstworldwar.com https://www.firstworldwar.com/bio/udet.htm
4. Dwight Jon Zimmerman, The Fall of Ernst Udet and Gerd von Rundstedt – The Nazi Blame Game, July 06, 2017 https://www.defensemedianetwork.com/stories/the-nazi-blame-game/
5. Eddie Elbert, Udet, Ernst, WW2 Grave Stone https://ww2gravestone.com/people/udet-ernst/
6. Willie Bodenstein, Ernst Udet-WWI Pilot-Film Actor and Luftwaffe Officer. http://lapauitgewers.net/arn0001181
7. My Life as a Pilot – Four Men in Africa http://www.archivaria.com/Udet/Udet9.html
8. Ernst Udet, Wikipedia, https://en.wikipedia.org/wiki/Ernst_Udet

INDEX

Adenauer, Konrad, German Chancellor, 9
AEW&C, 107
Air Force Reserve Officer Training Corps (AFROTC), 106, 111
Air Insurgency, 152
Air Vice Marshal James Edgar "Johnnie" Johnson, 33-43
   Air Victories, 42
   Bar to DFC, 39
   Becomes Flying Ace, 38
   Becomes Highest Scoring Ace, 39
   Becomes Wing Commander, 39
   Broken Shoulder
     Tough Days on Spitfire, 35
   Encounter with Adolf Galland, 38
   Evolving New Combat Tactics and Formations, 36
   Fight to Germans, 36
   First Air Victory, 37
   First Combat Embarrassment, 36
   German Focke-Wulf Fw 190 Introduced, 39
   Initial Tactical Flying, 35
   Injury Resurfaces
     Shoulder Operation, 35
   Introduces New Tactics, 39
   Korean War, 40
   Last Victory War, 40
   Mission when Bader Bailed Out, 38
   Passes On, 43
   Personal Life, 41
   Persuasion and War Clouds Supports Joining RAF, 34
   Post War
     Permanent Commission, 40
   Post-Retirement Life, 41
   RAF Service, 41
   Struggle to Join RAF, 34
   Time for First Action, 36
   Victories Follow, 37
   War Flying, 42
   Youth, 34
Alexei Solomatin, 68
Arado Ar 68 aircraft, 47
Area Bombing Directive, 182
Auxiliary Air Force (AAF), 34
Avro 504, 13

Barkhorn, Gerhard, 220-27
   Ace a Day, 222
   Car Accident Death, 225
   Critical Milestones, 226
   Crosses Century Mark, 222
   Defence of Reich
     Eased From Command Due Health, 223
   Early Life and Career, 221
   First Combat, 221
   First German to Cross 1,000 Missions, 222
   Flight Test Assignment, 224
   Higher Appointments, 225
   Honours/Awards, 226
   Hospitalised, 223
   Initial Civil Life and Re-joins Service, 224
   Last Missions, 224
   Moved to Eastern Front: Opens Air Victory Account, 221
   Moved to New Me-262 Jet Fighter, 224
   Operational Career Summary, 225
   Second German to Cross 250 Victories, 223
   Selected For Fighter Stream, 221
   Shoots Hero of Soviet Union, 222
   The Book, 227
Battle of Britain, 220
Battle of France, 220
Big Blow, 58
Bishop, William Avery, 255-66
   Aerial Victory, 258
   Aviation Love, 256
   Becomes Air Ace, 258
   Becomes Icon
     Asked to Return to Canada, 260
   Books, Movies and Civil Aviation, 262
   Controversy
     Was Bishop a Fake?, 265
   Distinguished Flying Cross Citation, 264
   Early Life, 255
   Family of Aviators, 262
   Final Score, 261
   First Combat Mission, 257
   High Tributes, 263
   Highest Scoring RFC Ace
     Third Highest War, 260
   Initial Combat Mess-ups, 258
   Injured
     Misses Somme Battle, 257
   Joins Horse Cavalry: Excelled Firing Range, 256
   Last Days, 262
   Leading Front: Bounty on His Head, 259
   Media Coverage, 264
   Post-WW I Career, 261
   Promoted Air Marshal, 262
   Promoted Lieutenant-Colonel, 261
   Return to Europe
     Squadron Command – More Victories, 260
   Survives Encounter with Red Baron, 259
   Switches to RFC
     Air Observer, 257
   Tributes, 264

Two Airports on His Name, 265
Very Highly Decorated, 263
Victoria Cross and Controversy, 259
Victoria Cross Citation, 263
Ways of Bishop Memorializing, 265
Wedding: Assignment in Washington, 260
Black Thursday, 122
Blenheim, Bristol, 93
Bloody April, 27, 244
Blue Bandits, 108
Blue legs squadron, 170
Bong, Richard Ira, 267-78
   Ace of Aces is Back USA: Marries, 274
   Aerial Victory Credits, 270, 276
   Attack Tactics, 274
   Becomes a Double Ace, 272
   Bong Sent Southwest Pacific Area, 269
   Early Death, Flight Testing, 274
   Explains Complexity Air Combat, 272
   Flight Commander, 273
   Flying Interest, 267-68
   Honours and Commemorations, 277
   Japanese Attack 9th Squadron Airbase, 271
   Joins United States Army Air Forces, 268
   Leading American Ace in New Guinea, 271
   Letter to His Mother: Advice for Brother, 271
   Lightning Ace, 270
   Loop Around Golden Gate Bridge – Grounded – Reprimanded, 269
   MacArthur Medal Honour, 274
   Medal Honour, 276
   Meets His Future Wife, 273
   Military Awards, 276
   Pacific War: Honing Combat Skills, 270
   Quadruple Success Earns DSO, 272
   Rejoins 9th Fighter Squadron, 271
   Still Remembered, 277
   Surpasses Eddie Rickenbacker's WW I Record, 273
Botha, Andre, 97
Brewster Buffalo, 215
Brig. Gen. Giora "Hawkeye" Epstein, 194-203
   Air Flying, 201
   Begins Fighter Flying, 196
   Early Years, 195
   Feelings Fighter Pilot, 201
   First two Su-7 Kills, 198
   Honour at 80: Brigadier General Rank, 202
   Israeli Flying Aces, 202
   Post Air Force Career, 201
   Post War Career, 201
   Professional Modesty, 201
   Pushes Case for Fighter Flying, 195
   Recalls Beginning of Yom Kippur War, 196
   Seeks Permission to Leave HQs and Join War, 197
   Six-Day War, 196
   Struggles Become Pilot Due Medical State, 195
   Wall of Fire, 198
   What about Us? We want to join the Battle, 200
   When Fought off 11 Enemy Jets Alone, 198
   Yom Kippur War, 197, 200
Brig. Gen. Jalil Zandi, 204-10
   AIM-54A Targets, 208
   Depleting F-14s, 208
   Early Aerial Victories, 205
   Eight Years Action, 205
   F-14 Spares, Modifications, 208
   Final Engagement, 206
   Iran, Tomcat F-14, 210
   Iranian Tomcat Units Combat, 207
   Iran-Iraq War: Aerial Victories, 206
   Ouster Shah: Jalil Put Behind Bars, 205
   Personal Impressions, 206
   Post-War Days, 210
   Service Career, 204
   Spares Shortage, 208
   Tomcat Shoots Three MiG 23s with One Missile, 209
British *Boelcke*, 25
Bubi, 3

Capt. (later Group Capt.) Arthur Peck, 117
Capt. (later Squadron Leader) Edward Dawson Atkinson, 117
Capt. DeBellevue, 109
Capt. Douglas Carbery, 117
Capt. George M. Cox, 117
Capt. John A. Madden, Jr., 109
Capt. Lawrence Percival Coombes, 117
Capt. Maurice Douglas Guest Scott, 117
Capt. Robert W. "Smitty" Smith, 129
Cecil Golding, 97
Cherniy Chort (Black Devil), 5
China, 173
Circus Attacks, 36
Circus Offensive, 36
Coc, Nguyen Van, 145-56
   Aerial Trap, Early Ejection, 147
   Aerial Victories, 147
   Air Victory, 147
   Becomes an Ace, 149
   Col. Olds and New Tactics, 153
   Col. Robin Olds vs. Nguyen Van Coc, 153
   Combat Flying Moved out, 156
   Daunting Odds, 146
   Double Attack, 148
   Early Family Tragedies, 145
   Hit and Run VPAF, 152
   Honours and Awards, 156
   Joins Air Force
   MiG 17 and MiG 21 Conversion, 146
   Operation Bolo, 151
   Operation Rolling Thunder, 151
   Post War Years, 156
   U.S. Armed Forces Overwhelming Superiority Assets, 152
   UAVs Last Victories, 151
   Victories, 149-50
   Vietnamese Aggression, 152
   VPAF Increased Losses Again, 148
Col. Hagerstrom, 130
Col. Harrison Thyng, 130
Col. René Paul Fonck, 243-54
   Air Challenges Interest, 246
   Avenges Guynemer Death, 247
   Becomes Flying Ace, 246
   Becomes Leading Allied Ace, 249
   Book on Fonck, 253
   Citations, 253
   Civil Life, 250
   Clinical Professionalism, 247
   Early Life, 243
   First Victory, 245
   Finally Joins Military Aviation, 244
   Ascetic and Withdrawn, 250
   Great Way , 252
   Last Days, 252
   Legion D'Honneur, 247
   Mes Combats, 245
   More Actions, 245

Most Decorated French War Heroes, 254
Nazi Links Controversy, 251
Post-World War II
  Cleared For Loyalty, 251
  Returns to Military Aviation, 251
SPAD XII, 248
Spectacular 2018: Ace-in-a-Day, 248
Surpasses Legendary Guynemer, 249
Transatlantic Air Race, 251
Transferred Ex Bomber Unit, 244
Unit Switches to G.4s, 245
Winning Streak, 246
Col. V. Garrison, 130
Collier's Weekly, Lineman of the Year, 76
Combat Tree, 106
Cooper, Tom
  Iranian F-14 Tomcat Units in Combat, 207
Corkscrew Loop, 171

Dassault *Ouragons*, 196
De Havilland Tiger Moth biplane, 35
DeBellevue, Charles B., 105-112
  Air Combat First Major Day, 107
  Air Force Cross Citation, 112
  Competition to Become USAF's first Vietnam "Ace", 108
  Decorations, 111
  Early Years: Joins USAF, 106
  Explains Two Engagements, 110
  Honour, 110-11
  MiG Credits:
    Pilots, Aircraft and Weapons, 110
  MiG Kills, 106-9
  Post-Vietnam War, 111
Der Adler (The Eagle), 98
Der Stahlhelm, Bund der Frontsoldaten (The Steel helmet, League of front-line Soldiers), 229
Der Stern von Afrika (The Star of Africa), 102
Deutsche Luftstreitkräfte, 23
Deutsches Jungvolk, 179
Distinguished Flying Cross, 38, 131
Distinguished Service Cross, 131
Distinguished Service Order, 15
Double Fighter Knight, 215, 218

EFTS (Elementary Flying Training School), 34
European Theatre of Operations (ETO), 123

F-16 *Fighting Falcons*, 196
Fähnrich, 91
Feld Flieger-Abteilung 206 (FFA 206), 280
Fighter Interceptor Squadron, 126
Fighter Interceptor Wing (FIW), 126-29
  1st Lt. Anthony Kulengosky, 129
  1st Lt. Harry Shumate, 129
  1st Lt. Joe L. Cannon, 129
Fliegerausbildungs Regiment, 179
Flugzeugführerschule, 179
Flying Counting Machine, 97
Fokker Tri Plane, 24
Freedom of Information Act (FOIA), 206
Fritz Oblesser, 8
Full Circle, 41

Gabreski, Gabby, 118-33
  Aerial Victories Summary, 130
  Aggressive Commander
    Overflies China, 128
  All Set for Marriage, Becomes a Prisoner of War, 124
  Becomes Jet Ace, 127
  Becomes an Ace
    Has a Close Shave, 122
  Becomes Leading American Ace in European Theatre, 123
  CO 61st Fighter Squadron, 121
  Combat Grace and Magnanimity, 129
  Early Years, 119
  First Fighter Squadron
    Meets His Future Wife, 119
  First Victory, 121
  Flying, 125
    Interest, 119
  Humane Side, 130
  Joins USAAF, 119
  Korea, Mission, 130
  Korean War, Initial Days, 126
  Legacy, 132
  MiG Alley, 126
  Military Honours and Awards, 131
  Offers to Serve as Liaison Officer to Polish Squadrons, 120
  Opinion His Wingmen, 128
  Personal Life and Death, 132
  Post-Korea, 130
  Post-Retirement Assignments, 131
  RAF Duty Polish Squadron, 120
  Repatriation, Fighter Command, 125
  Responsibilities 56th FG, 122
  Stops Logging Sorties to Avoid Return to USA, 130
Galland, Adolf Josef Ferdinand, 44-63
  Aerial Victory Claims, 63
  Air Combat Champagne on board, 53
  Air Ops Full Command, 55
  Aircraft Chased and Shot, 49
  Bails out and is Injured, 54
  Becomes Wing Commander, 51
  Begins Flying Career, 45
  Britain Battle, 50
  Britain Last of Battle, 52
  Channel Front, 53
  Conflict with Göring, 56
  Crash and Injury, 46-47
  Disagreements, Me 262 Aircraft, 57
  Dismissal and Relieved Command, 58
  Early Life, 45
  Engagements, 62
  Family of Aces, 45
  Fighter Pilots Revolt, 59
  Galland's New Tactics, 52
  German Pilots Fatigue, Low Morale, 52
  German Surrender
    Offer to Join Americans against Soviet Union, 59
  Germany Collapsed: Command of a New Unit, 59
  Göring Summoned, 50
  Hitler Audience, 51
  in Argentina, 61
  Initial Flying Interest, 45
  Iron Cross, 48
  Joins Luftwaffe, 46
  Knight's Cross, Iron Cross, 50
  Looks after Douglas Bader, 54
  Mölders' Tactics, 49
  Own Aircraft Consultancy, 62
  Parachute Shooting Down Pilot, 51
  Poland Invasion, 48
  Posted Flight Testing, 48
  Posting High Command, 54
  Posting Mediterranean, 55
  RAF Offensive, 53
  Resign Offers, 56
  Return to Germany
    Denied Job, West German Air Force, 61
  Rheumatism Excuse Ground Attack, 48

Self-appraisal and Introspection, 60
Spanish Civil War, 47
Spanish Cross, 47
Spitfire Encounter, 49
Suggested Operational Aircraft Modifications, 47
Survives Again, 54
Three Marriages, Last Days, 62
Western Europe Invasion, 49
Gathering of Eagles, 62
Gen. Günther Rall, 228-42
  100th Victory
    Oak Leaves Presented by Hitler, 234
  Aerial Victories, 241
  American 'Zemke Fan' Tactic, 237
  Appointed Head of Fighter Wing, 238
  Awarded Knight's Cross, Iron Cross, 234
  Awards, 242
  Back to Action, 233
  Back to Air Victories after Short Leave - 250th Mark, 236
  Battle of Britain, Faulty Tactics, 231
  Becomes an Ace, 232
  Claim Disputes, 235
  Crimean Campaign, 233
  Crosses 200 Victories, Third to Reach, 236
  Decides Military Career, Later Luftwaffe, 229
  Defence of Reich
    Heavy German Losses, 237
  Early Life, 229
  Eastern Front
    Air Victories Begin, 232
  First Air Victory
    Lots of Bullets but Renewed Self-Confidence, 231
  Force Landing, Mid Air Collision, 235
  Forced Retirement, 240
  Group Commander, 235
  Hospitalisation, 233
  India Visit, 240
  Initial Action, 230
  Kharkov First Battle, 232
  Kuban Bridgehead
    First Real Test for German Fighters, 234
  Meets Future Wife, 233
  Memoir Flight Log Book, 241
  Moves as Instructor, 238
  Nazi Schooling, 229
  Prisoner of War, 238
  Rall Gets Hit
    Bails Out – Injured, 237
  Rebuilding German Air Force, 240
  Re-joins German Air Force, 239
  Senior Command, 240
  Service and Death, 241
  Shot Down and Crashes
    Survives with Broken Back, 233
  Trains with USAF, 239
  Transferred to Greece and Romania, 231
  Unit High Losses, 238
  Victories Start Slowing, 236
  Wife' Jewish Connection, 239
German Navy's Operation Cerberus, 55
German, 178
  POW, 12
German Reich Party (DRP), 143
Gneisenau, 181
Göring, Hermann, 52
Gp. Capt. Sir Douglas Robert Bader, 11-21
  Air Crash, Amputation of Legs, 13
  Bader's Hawker Hurricanes, 16
  Bar to DSO, 17
  Battle of Britain, 15
  Big Wing Tactic, 15
  Combat Credo, 18
  Distinguished Flying Cross, 16
  Early Years, 12
  Escape from Hospital, 18
  France Battle, 14
  Getting into Action, 14
  Honours and Awards, 20
  Invalided From Service, 13
  Joining RAF, 12
  Last Combat
    Who Shot Bader Controversy, 17
  Last Flight, 21
  Personal Charm and Traits, 20
  Personal Life, 20
  Post RAF Career, 19
  Prisoner of War, 18
  RAF
    Last Years, 19
    Return, 13
  Wing Leader, 16
Great Purge, 64
Great Terror, 65
Ground-Controlled Intercepts GCI, 153
Gruppenkommandeur, 139

Guttman, Jon
  Military History, 213
Guynemer, Georges, French Air Ace, 246

Hartmann, Erich Alfred, 1-9
  Aerial Combat, 3
  Aerial Victories Authenticated, 8
  Assessment Enemy Air, 8
  Counter Check His Claims, 4
  Diamonds Knight's Cross, 6
  Drunk Awards Ceremony, 5
  Fighting Techniques, 4
  Hitler Impressions, 9
  Initial Tactics, Grooming, 3
  Knight's Cross Iron Cross, 4
  Last Combat Missions, 6
  Life/Career, 2
  Method of Attack, 8
  Post War Years, 7
  Prisoner of War, 7
  Roll of Honour, 9
  Securing Release Soviet Prison, 9
  Top Scoring Ace, 5
  War Crimes Charges, 7
Himmelbett, 180
Hiroyoshi Nishizawa, 176

Hitler, Adolf, 6, 54, 213, 234, 290
Honourable Artillery Company (HAC), 114

IAI *Neshers* and *Kfirs*, 196
IJNAS, 167
Ilmavoimat (Finnish Air Force), 211
Ilyushin DB-3 2M-87, 214
Imperial Defence College (IDC), 41
Imperial German Army, 279
Imperial Iranian Air Force (IIAF), 204
Imperial Japanese Navy Air Service (IJNAS), 166
Iran-Iraq War, 204
Islamic Republic of Iran Air Force (IRIAF), 204, 207-8
Israel, 195
Israeli Aircraft Industries (IAI), 196
Israeli Defence Force (IDF), 195
Iwamoto, Tetsuzo, 166-77
  Aerial Victories Claimed in His Diary, 172
  Air Kills Debate
    Yet Greatest Japanese Fighter Ace of All Time, 173
  Awards, 174

Becomes Top IJNAS Ace in China, 170
Caroline and Philippine Islands, 171
China Front, 168
Combat Mission: Four Victories, 169
Death, Medical Complications, 175
Famous Japanese Units, 176
Flying Groups Merge, 170
Gets Enrolled for Flight Training, 168
Japanese Imperial Navy Air Aces, Mitsubishi A6M Zero, 175-76
Joins Naval Air Group, 168
Military Career, 168
Pacific War: Pearl Harbour & Carol Sea, 170
Post-war Life, 174
Promotions, 174
Rabaul, New Britain, 171
Tactics, 171
Young Days, 167
Zero Fighting Tiger, 173

Jagdfliegerschule Werneuchen (fighter pilot school), 230
Japan, 167, 169, 171
Junichi Sasai, 176
Juutilainen, Eino Ilmari "Illu", 211-19
    1/6 Shared Victory, Meaning, 214
    Ace Day Flying Bf 109G-2, 217
    Awards, 218-19
    Becomes Ace on Brewster Buffalo, 216
    Between Wars, 215
    Continuation War, 216
    Continues to Fly Till Old, 219
    First Encounter with I-16, 214
    Flying Interest, 212
    Flying Qualities Summary, 218
    Fokker D.XXI, 214
    His Brother: Terror of Morocco, 212
    Immola: Completes 22 Victories, 217
    Initial Reaction to Outbreak of War, 213
    Initial Tactics, 213
    Initial Training, 212
    Love of Flying and Aerial Victories, 218
    Major Encounter: Three Victories, 216
    Other Types of Missions, 215
    Quiet Death: Great Legacy, 219
    Service, 212
    Soviet–Finnish War, 213
    Tactics, 215
    Victories, Brewster B-239, 217
    Winter War: Initial Air Success, 214

Kameradenwerk, 134
Kammhuber Line, 180
Kanonenvogel (Cannon bird), 137
Kazuo Sugino, 176
Khomeini, Ayatollah, 205
King Henry VIII, 114
Kozhedub, Ivan Nikitovich, 157-65
    Aerial Victories, 164
    Alleged Shooting down of two USAAF P-51 fighters, 164
    Be Kind to the Aircraft He Will Be Kind to You, 160
    Early Life, 157
    Fly Begins, 158
    Greatest Soviet Pilot, 162
    Higher Command, 163
    Honours and Awards, 164
    Inspired Great Soviet Aviators, 158
    Joins Red Army, 158
    Korean War, 163
    Legacy, 165
    Lone-Wolf Operations, 160
    New Plane Gets Gift from Farmer, 160
    Promoted Junior Lieutenant, 160
    Recalled His Younger Days Later, 158
    Recalls, Importance Battle for Kursk, 163
    Recalls, Tactics and Air Power, 162
    Shoots Down a Me-262, 161
    Stamp Release, 165
    Transferred Operational Unit, 159
    War Action Begins, 159
    World War II Summary, 162
Kuban tactics, 107

Lappin, Yaakov, 201
Lavochkin La-5, 159
Litvyak, Lydia Vladimirovna, 64-72
    Assigned Men's Regiment, 66
    Change of Unit to Stay on Yak-1s, 67
    Death Controversy, 70
    Dramatization and Books, 72
    Early Years, Harsh Reality, 64
    Femininity and Fashion, 71
    Fiancé and Love, 72
    First Women to Score Aerial Victory, 66
    Flight Commander, 69
    Free Hunter Concept, 68
    Last Mission, 69
    Lev Lvovich Shestakov, 67
    Love with Her Leader, 68
    Marina Mikhaylovna Raskova, 65
    Night Witches, 65-66
    Number of Kills and Awards, 71
    Russian Blondes, 70
    Search and Recognition, 70
    Shooting a Manned Observation Balloon, 68
    Shoots a Knight's Cross Holder, 67
    Stage Play, 72
    Strong Character Yet believed in Luck, 71
    White Lily and Red Rose, 72
    Who Shot Her, 69
    Women's Regiment, 65
    Wounded First Time, 68
Lt. Col. George A. Davis, 130
Lt. Col. John F. Bolt, 130
Lt. Indra Lal Roy, 113-17
    Family, 114
    Final Resting Place, 115
    Flying Aces from British India, 117
    Honours and Awards, 116
    Initial Years, 114
    Killed in Action,. 115
    Aerial Victories, 116
    Roy Joins RFC, 114
    Stamp Release, 116
Lt. Thomas Cecil Silwood Tuffield, 117
Luftflotte (Air Fleet), 55
Luftwaffe Structure, 230

Mackay Trophy, 110
Maj. Erich Hartmann, 1
Marseille, Hans-Joachim, 4
    Star of Africa, 221
Marseille, Hans-Joachim, 88-104
    100th Victory: Oak Leaves and Swords, 97
    Air Combat, Style and Idea, 96
    Amazing and Ingenious Combat Pilot, 97
    Apolitical Core, 103
    Autopsy Report, 101
    Bailout Fatal Fall, 100
    Better and More Western Aircraft Combat Strain, 99
    Bristol Blenheim Downs, 93
    Britain Battle, First Engagement, 90
    Claims Four Hurricanes a Day, 94

Combat: Another Rebuke, 91
Controversies, 102
Dismissed for Squadron, 90
Famous Through Propaganda, 98
Fighter Wing Hits Low Morale
 Marseille's Leadership Style, 101
Fractures Arm
 Continues Fly and Score, 98
Funeral and Inquiry, 101
High Proportion Victories, 97
Highest Italian Military Award for Bravery, 97
in Media, 102
Joins Luftwaffe, 90
Low Kill Rate and Four Crashes, 93
Lufbery Circles, 94
Marseille Introspects: Creates Unique Self-Training Program, 93
Memorials: Grave, Undefeated, 104
Mercy Mission and Letter of Regret, 96
North Africa Arrival
 Force Landing in Desert, Hitch Hiked to Airbase, 92
North Africa, Initial Victories, 92
Own Special Tactics, 95
Personal Fitness Training, 94
Plays American Jazz in Presence of Hitler, 103
Promotions and Added Responsibility, 96
Punishment for Insubordination Move to New Wing, 92
Refuses Accompany Erwin Rommel to Berlin, 100
Regular Victories Now On
 Flies Pick Downed Pilots-Penance, 93
Reluctant Use New Aircraft, 99
Returns to Combat
 17 Victories Day, 98
Shot Down
 Narrow Escape, 96
Toughest Adversary, 99
Troubled Childhood
 Parents Separate, 89
Unorthodox Combat
 One man Show, 95
Victory Claims, 102
Where I Go, Mathias Goes, 103
Wrote-Off Four Aircraft, 91
Youth, Family, 89
Matthews and Foreman, *Luftwaffe Aces: Biographies and Victory Claims*, 241
Mickey Mouse emblem, 63
MiGCAP, 107, 108, 109
Mirage IIIs, 196

NATO, 241
Nazi Crimes, 239
Nazi Germany, 140

Olds, Robin, 74-87
 Aerospace Safety Director, 84
 Air Force Academy 1967–71, 84
 Air Force Cross Citation, 85
 Awards and Decorations, 85
 Command in Thailand
  Vietnam War, Blackman and Robin, 81
 Command Squadron: Ace on Two Types, 79
 Commander 81st Tactical Fighter Wing: Near Court Martial, 80
 Curriculum Cut Short Due War, 76
 Dead Stick Shoot, 77
 Demonstration Flights, 79
 Deputy Chief, Air Defence Division, 80
 Dies of Cancer: Made a Class Exemplar, 86
 Dog Fight Advocate: But No Gun Pods, 83
 Enters West Point, 75
 Europe Second Tour, 78
 Final Mission, 82
 Final WW II Kill, 78
 First Flight, 75
 First Jet Aerobatics Team, 79
 Flying P-38 Lightning: First Kill, 77
 Flying Summary, 82
 Football, Hall of Fame, 76
 Inspector General Tour, 1971, 84
 Korean War, Missed Service, 80
 Leaves Air Force when Refused another Tour to Vietnam, 84
 Marriage Hollywood Actress, 85
 MiG Killer: Operation Bolo, 81, 83-84
 More Kills for Olds: Triple Ace, 82
 Mustang Pilot, 78
 National Aviation Hall of Fame, 86
 Olds' Moustache: Showing the Middle Finger, 83
 Post WW II Assignments
  Professional vs. Career Struggle, 79
 Promoted Colonel
  Unenthusiastic Staff Appointments, 80
 Retired Life, Alcohol, 86
 Strafing Credits, 78
 TV Episodes, 84
 USAF/RAF Exchange Programme, 80
 West Point Days, 76
 Young Days, 75
Operation Barbarossa, 135, 221, 278
Operation Big Week, 123
Operation Bolo, 153
Operation Gisela, 190
Operation Gomorrah, 185-86
Operation Kutuzov, 235
Operation Linebacker, 84
Operation Millennium, 182
Operation Overlord
 Normandy Landing, 188
Operation Polkovodets Rumyantsev, 235
Operation Rolling Thunder, 151
Operation Typhoon, 135
Operation Vengeance, 130
OTU (Operational Training Unit), 34

Pahlavi, Mohammad Reza, Shah of Iran, 205
Panzerknacker (Tank cracker), 137, 139
Polish Flight, 123
Prinz Eugen, 181

Quality before Quantity, 191

Reach for the Sky, 20
Red Air Force, 232
Red Army, 233, 235
Red Baron, 115
Reich Labour Service (RAD), 135
Richthofen, Manfred Albrecht Freiherr von, 23-32
 Aircraft His Choice, 26
 Aircraft Painted Red, 26
 Bloody April, 27
 Brilliant Tactician, 27
 Ceremonial Burial, 30
 Early Years, 24
 Fatal Wound, 28
 Final Cemetery, 30
 Final Combat, 28
 First Confirmed Aerial Victory, 25
 Flying Circus, 27

Flying Pilot Career, 24
Great Air Ace, 31
Honours/Tributes, 31
Initial Tactics, 25
Initial War Service, 24
Last Combat Theories, 29
Legend/Hero, 28
Richthofen's Victories Authenticated, 30
Shooting British Ace Major Hawker, 25
The Blue Max, 26
Who Shot Red Baron?, 29
Why He Remained a Captain Only, 28
Richthofen's Circus, 23
Robin Pare, 97
Royal Air Force (RAF), 11-12, 30, 33, 49, 80, 90, 166, 181
  Bomber Command, 180, 187
Royal Air Force Volunteer Reserve (RAFVR), 34
Royal Australian Air Force (RAAF), 93, 96
Royal Canadian Military Institute (RCMI), 31
Royal Engineers, 12
Royal Flying Corps (RFC), 113-14, 257
Royal Navy, 166
Rudel, Hans-Ulrich, 134-44
  Aerial Victory, 139
  Appointed Leader of SG 2, 140
  Battle of Kursk, 137
  Becomes Arms Dealer and Military Adviser, 142
  Clandestine Move to Argentina, 141
  Daring Rescue of a Downed Pilot behind Enemy Lines, 139
  Death and Funeral, 143
  Forced Landing Highway, 138
  Gets a Troop Doctor as His New Gunner, 140
  Gunner a Knight's Cross, 138
  Honours and Awards, 144
  Joins Luftwaffe, 135
  Ju 87G: More Success – More Awards, 137
  Leg is Amputated: Hits Many More Targets, 141
  Military Career, 143
  Multiple Marriages, 143
  Offers to Fly Out Hitler to Safety, 141
  Philosophy, 144
  Refuses Hitler to Be Ground for Injury, 140
  Repeat Injuries but Flies on Regard Less, 140
  Returns to Germany, Joins Politics, 143
  Rudel's Gunner – Erwin Hentschel, 136
  Saves His Life, 136
  South America: Reignites Nazi Past, 141
  Stalingrad Battle, 136
  Stuka Loyalty, 138
  Surrenders to Americans, 141
  World War II Begins: Becomes Stuka Pilot, 135
  Writes Wartime Memoirs: Stuka Pilot, 142
Russia, 7, 9, 24, 167

Saburo Sakai, 176
Scharnhorst, 181
Schnaufer, Heinz-Wolfgang, 178-93
  Advanced Flying Training, 180
  Aerial Victories Summary, 192
  Appendicitis Operation: Break from Flying Operations, 188
  Awarded Knight's Cross Iron Cross with Oak Leaves and Swords, 189
  Awards, 193
  Back in Action: More Aerial Victories, 183
  Battle of Ruhr: Action Begins, 184
  Becomes Air Ace, 184
  Bomber Allied Raid Operation Millennium, 182
  British Honorary Title, The Spook of St. Trond, 190
  British Plan Area Bombing, 182
  Channel Dash Operation, 181
  Dies Car Accident, 192
  Early Years, Nazi Schooling, 179
  Failed Attempt to Go to South America, 191
  Family of Wine Merchants and Winery, 178
  First Aerial Victory: Ground Controlled Intercept, 183
  Flying Interest, 179
  Gets Hit: Lands Back Safely – Hospitalised, 183
  Group Commander IV/NJG 1: Ace in a Day, 188
  Initial Night Intercept Concept, 180
  Initial Y Control Missions, 187
  Joins Night Fighter Squadron, 180
  Knight's Cross, Iron Cross, 42 Victories, 187, 191
  Lean Period, 184
  Luftwaffe Joins, 179
  Night Fighter Volunteers, 180
  Night Fighter, Becomes Leading, 190
  Offered Post Inspector Night Fighter Force, Declines, 190
  Operation Donnerkeil, 181
  Post-War Recognition, 193
  Prisoner of War, 191
  Promoted First Lieutenant, 185
  Repeat Unit Relocations, 181
  Rumpelhardt Temporary Grounded, 184
  Second Time Ace in a Day – Night 9 Victories, 190
  Squadron Leader, 12.Staffel/NJG 1, 186
  Swords and Diamonds, 189
  Unit Moves to Belgium, 181
  Unit Relocated Due Heavy Allied Bombing, 189
  Upward-firing Auto-cannon Schräge Musik, 187
  Winery Business, 191
  Wing Commander of Nachtjagdgeschwader 4, 189
  Y-Control Fighter Guidance System, 186
Schröer, Werner, 97
Sector Offensive Sweeps, 36
See–Decide–Attack–Reverse, 4
Shoichi Sugita, 176
Shturmoviks., 139
Sir Alan Smith, 38
Sito Origami, 176
South African Air Force (SAAF), 94
Soviet Red Air Force, 220
Soviet Union, 160, 216
Spanish Civil War, 67
Split S, 50
Staffel (flying unit), 135
Staffel, Marseille, 104
Stern, German weekly magazine, 241
Stilwell, Blake, *Mighty History*, 209
Stuka Pilot, 142
Super Mysteres, 196
Supermarine Spitfire, 50

Tactical Fighter Wing (TFW), 80
Takeo Okumura, 176
Tetsuzo Iwamoto, 176
The Flying Circus, 23
The Hardest Day, 50
The Philippines, 176
The Red Baron, 26
The Red Battle Flyer, 24
The Red Fighter Pilot, 24
The Truth Is This, 175
TOPGUN Programme, 84
Toshio Ohta, 176
Toshiyuki Sueda, 176
Triple Nickel Squadron, 106

U-Boat War, 9
Udet, Ernst, 278-91
    Aircraft Development Tasks, 288
    Attempts at Aircraft Manufacturing, 287
    Aviation-Linked Activities, 287
    Becomes an Air Ace, 282
    Becomes National Hero, 285
    Building Luftwaffe, 287
    Commits Suicide, 290
    Connect with Aviation, 279
    Court-Martialled for Losing an Aircraft, 280
    Director-General of Equipment, 288
    Early Flier to Parachute to Safety, 284
    Encounter, French Hero and Ace Georges Guynemar, 282
    Fighter Pilot Udet: Initial Disasters, 280
    First Aerial Victory, 281
    First Encounter with U.S. Army Air Service, 285
    High Aircraft Requirements, 289
    Inter-War Years: Lots of Action, 285
    Marries and Divorces Eleanor "Lo" Zink, 285
    Meets Childhood Sweetheart, 284
    Multi-Talented/Multi-Faceted, 286
    Portrayals, 291
    Private Flying Training: Joins Air Service, 280
    Richthofen Killed in Action, 284
    Selected Flying Circus, 283
    Starts Flying for Artillery Ranging, 280
    Struggles Enlist, 279
    Suicide Note, 290
    Transferred Fighter Flying, 280
    Transferred Jasta 37, 283
Ukraine, 157
Undergraduate Navigator Training (UNT), 106
Undergraduate Pilot Training (UPT), 106
US, 120, 134, 148-49, 151-52, 155, 205-6, 208, 225
US 90th Infantry Division, 6
US Air Force (USAF), 31, 40, 62, 74-75, 80, 81, 84, 106, 107, 111, 118, 147, 151, 239
US Air Medal and Legion of Merit, 40
US Army, 11, 19, 59
US Army Air Force (USAAF), 102, 118, 123, 164, 171, 185, 237
US Army Air Forces (USAAF), 75, 105
US Central Intelligence Agency (CIA), 134
US Navy, 107, 171, 173, 205
USS Enterprise, 150, 210
USSR, 65, 216

Vesa, Emil, 216
Vietnam War, 148, 156
Vietnamese People's Air Force's (VPAF), 145, 147, 150, 152, 155, 156
Voyenno Vozdushny Sily (Red Army air force, or VVS), 213

Weapon Systems Officer (WSO), 106
Wehrmacht (Nazi German Armed Forces), 49, 51, 235
Wehrmachtbericht (Nazi propaganda communiqué), 236
West Force, 153
Why so many Vietnamese Aces?, 155
Wikiwand, 69
Wilde Sau (Wild Boar), 185
Witwenmacher (widow maker), 240
Wolf Pack, 147, 151
World War I, 8-9, 12, 23-25, 29, 31, 52, 89-90, 94, 105, 113, 115-17, 123, 166, 179, 221, 229, 243-45, 249-52, 255-56, 263, 266, 273, 278-79, 281, 286, 291
World War II, 1-2, 6, 11, 31, 33-34, 36, 43, 63-66, 67, 70, 74-75, 77-78, 81-82, 88, 90, 94, 105, 107, 114, 118, 125, 128-30, 132, 134-35, 143, 157, 162, 164, 166-67, 172-73, 178, 189, 191-92, 201, 211-12, 214, 216, 219, 221, 227, 230, 238, 241, 251, 255, 262-63, 266-68, 276-77, 289-90

Yellow 14, 100

Zahme Sau (Tame Boar), 185
Zerstörerschule, 180